For other titles published in this series, go to
www.springer.com/series/8623

# Protocols for Neural Cell Culture

Edited by

## Laurie C. Doering, PhD

*McMaster University, Hamilton, ON, Canada*

Fourth Edition

*Editor*
Laurie C. Doering
Department of Pathology &
    Molecular Medicine
McMaster University
1200 Main Street West
Health Sciences Centre
Hamilton ON L8N 3Z5
Canada
doering@mcmaster.ca

ISBN 978-1-60761-291-9        e-ISBN 978-1-60761-292-6
DOI 10.1007/978-1-60761-292-6

Library of Congress Control Number: 2009932368

# Preface

Ross G. Harrison's founding experiments with nerve cell cultures between 1907 and 1910 were motivated by the same concerns of Santiago Ramón y Cajal centered on the "Neuron Doctrine." Cajal used impressive silver impregnations in chick spinal cord to illustrate axonal growth in vivo. Harrison, using pieces of frog neural tube nourished in lymph within a depression slide, solved the problems of his predecessors and devised a reproducible method of tissue culture experimentation. For the first time, the outgrowth rates of individual fibers and their growth cones were observed in real time under the microscope. In essence, Harrison established a direct in vitro test to prove Cajal's axon outgrowth theory. Harrison overcame basic tissue culture problems and created a culture technique others could follow. Since the beginning of nervous system tissue culture with Ross Harrison's vision, now just over 100 years ago, the scientific community has established numerous protocols to generate the current wide variety of cell and tissue culture technology.

The rapid growth of neuroscience during the 1980s and a highly acclaimed, intensive tissue culture course held in Saskatoon, Saskatchewan, for several years formed the concept for this protocol series in neural tissue culture. In 1992, Dr. Sergey Fedoroff (University of Saskatchewan) founded this series with the publication of the first edition. The textbook continued to grow in popularity with subsequent editions, and the book is now established as a popular neuroscience protocol reference in numerous laboratories throughout the world.

The fourth edition of *Protocols for Neural Cell Culture* represents a turning point in this series from editorial and content perspectives. Since the publication of the third edition in 2001, neuroscience has continued to expand with discoveries that utilize tissue culture methodologies. The refinement of existing protocols and the emergence of new techniques and culture media formulations are linked with advances in neuroscience. This edition includes three chapters from leading companies who specialize in neural tissue culture biotechnology and contribute significantly to the products used by many scientists.

Updates on the experimental procedures for many of the classical tissue culture preparations in neuroscience are highlighted. In view of the implications for regenerative medicine, methods to grow and expand embryonic and adult neural stem cells are included in this edition. This volume covers the isolation, expansion, and cryopreservation of neural tissue from mouse, rat, and human sources. Immunocytochemistry is a subcomponent of many chapters as it continues to be a solid method to identify cells at developmental points within specific lineages. The basic techniques for establishing specified neuronal and glial preps are complimented by many new chapters including methods to assess aspects of cell function (calcium imaging) and cell death.

I am privileged to have the support from Dr. Fedoroff (a great neuroscience mentor) to continue the Editorship of this collection. I am very grateful to all the authors for their commitment and time. Their dedication has produced a text with exceptional quality and coverage in tissue culture protocols for today's neuroscience. A special thanks to Sharon Ralph for countless hours of editorial assistance, management, and organization in the production of this book.

*Hamilton, ON*                                                                                          *Laurie C. Doering*

# Contents

# Contributors


JACK P. ANTEL • *Neuroimmunology Unit, Montreal Neurological Institute, Montreal, QC, Canada*

BEHNAM A. BAGHBADERANI • *Pharmaceutical Production Research Facility (PPRF), Schulich School of Engineering, University of Calgary, Calgary, AB, Canada*

DOMINIQUE BAGNARD • *Physiopathologie du Système Nerveux, INSERM U575, Strasbourg, France*

SABINE BAVAMIAN • *Department of Cell Physiology and Metabolism, University of Geneva Medical School, Genève, Switzerland*

LEO A. BEHIE • *Pharmaceutical Production Research Facility (PPRF), Schulich School of Engineering, University of Calgary, Calgary, AB, Canada*

BENEDIKT BERNINGER • *Department of Physiological Genomics, Institute of Physiology, Ludwig-Maximilians University Munich, Schillerstrasse 46, D-80336 Munich, Germany*

MANON BLAIN • *Neuroimmunology Unit, Montreal Neurological Institute, Montreal, QC, Canada*

ALEXANDRE BONNIN • *Zilkha Neurogenetic Institute, Keck School of Medicine of University of Southern California, 1501 San Pablo Street, Room 105, Los Angeles, CA 90089-2821*

MICHEL CAYOUETTE • *Cellular Neurobiology Laboratory, Institut de Recherches Cliniques de Montréal, Montréal, QC, Canada; Département de Médecine, Université de Montréal, Montréal, QC, Canada; Division of Experimental Medicine and Anatomy and Cell Biology Department, McGill University, Montréal, QC, Canada*

RUTH COLE • *Mental Retardation Research Center, Semel Institute for Neuroscience and Human Behavior, David Geffen School of Medicine at UCLA, Neuroscience Research Building, Los Angeles, CA, USA*

MARCOS R. COSTA • *Edmond and Lily Safra International Institute of Neuroscience of Natal, Natal, RN 59066-060; Universidade Federal do Rio Grande do Norte, Brazil*

SEAN P. CREGAN • *Robarts Research Institute, London, ON, Canada*

QIAO-LING CUI • *Neuroimmunology Unit, Montreal Neurological Institute, Montreal, QC, Canada*

PETER J. DARLINGTON • *Neuroimmunology Unit, Montreal Neurological Institute, Montreal, QC, Canada*

LAURIE C. DOERING • *Department of Pathology and Molecular Medicine, McMaster University, Hamilton, ON, Canada*

MIRELLA DOTTORI • *Centre for Neuroscience and Department of Pharmacology, The University of Melbourne, Parkville, VIC, Australia*

SERGEY FEDOROFF • *University of Saskatchewan, Saskatoon, SK, Canada*

EVA L. FELDMAN • *Department of Neurology, University of Michigan, Ann Arbor, MI, USA*

SILVIA FERNANDEZ-ILLESCAS • *Stanford University, Stanford, CA, USA*

CRISTINA A. GHIANI • *Mental Retardation Research Center, Semel Institute for Neuroscience and Human Behavior, David Geffen School of Medicine at UCLA, Neuroscience Research Building, Los Angeles, CA, USA*

FRANCISCO L.A.F. GOMES • *Cellular Neurobiology Laboratory, Institut de Recherches Cliniques de Montréal, Montréal, QC, Canada*

JESSE E. HANSON • *Stanford University, Stanford, CA, USA*

SHELLEY JACOBS • *Department of Pathology and Molecular Medicine, McMaster University, Hamilton, ON, Canada*

RAVI JAGASIA • *Junior Research Group Adult Neurogenesis and Neural Stem Cells, Institute of Developmental Genetics, Helmholtz Zentrum München, German Research Center for Environmental Health, Munich-Neuherberg, Germany*

DAVID JUDD • *Invitrogen Corporation, Grand Island, NY, USA*

MICHAEL S. KALLOS • *Pharmaceutical Production Research Facility (PPRF), Schulich School of Engineering, University of Calgary, Calgary, AB, Canada*

JOSEF P. KAPFHAMMER • *Anatomisches Institut, Universität Basel, Basel, Switzerland*

NAVJOT KAUR • *Invitrogen Corporation, Frederick, MD, USA*

HAESUN A. KIM • *Department of Biological Sciences, Rutgers University, Newark, NJ, USA*

PAUL DE KONINCK • *Cellular Neurobiology Division, Centre de Recherche, Université Laval Robert Giffard, Québec, QC, Canada*

CAROLINE LAMBERT • *Neuroimmunology Unit, Montreal Neurological Institute, Montreal, QC, Canada*

SHARON A. LOUIS • *STEMCELL Technologies Inc., Vancouver, BC, Canada*

DANIEL V. MADISON • *Stanford University, Stanford, CA, USA*

RICHARD G. DEL MASTRO • *Invitrogen Corporation, Grand Island, NY, USA*

PATRICE MAUREL • *Department of Cell Biology, Smilow Neuroscience Program, New York University School of Medicine, New York, NY, USA*

PAOLO MEDA • *Department of Cell Physiology and Metabolism, University of Geneva Medical School, Genève, Switzerland*

VERONIQUE E. MIRON • *Neuroimmunology Unit, Montreal Neurological Institute, Montreal, QC, Canada*

ARNAUD MULLER • *Physiopathologie du Système Nerveux, INSERM U575, Strasbourg, France*

FRANCINE NAULT • *Cellular Neurobiology Division, Centre de Recherche, Université Laval Robert Giffard, Québec, QC, Canada*

AKIKO NISHIYAMA • *Department of Physiology and Neurobiology, University of Connecticut, Storrs, CT, USA; University of Connecticut Stem Cell Institute, University of Connecticut, Storrs, CT, USA*

ADRIENNE L. ORR • *Stanford University, Stanford, CA, USA*

ALICE PÉBAY • *Centre for Neuroscience and Department of Pharmacology, The University of Melbourne, Parkville, VIC, Australia*

MARTIN F. PERA • *Center for Stem Cell and Regenerative Medicine, University of Southern California, Los Angeles, CA, USA*

DAVID R. PIPER • *Invitrogen Corporation, Madison, WI, USA*

BRENT A. REYNOLDS • *Department of Neurosurgery, McKnight Brain Institute, University of Florida, Gainesville, FL, USA*

ARLEEN RICHARDSON • *University of Saskatchewan, Saskatoon, SK, Canada*

MORGANE ROTH • *Physiopathologie du Système Nerveux, INSERM U575, Strasbourg, France*

PHILIPPE SAIKALI • *Neuroimmunology Unit, Montreal Neurological Institute, Montreal, QC, Canada*

SAULIUS SATKAUSKAS • *Physiopathologie du Système Nerveux, INSERM U575, Strasbourg, France; Department of Biology, Vytautas Magnus University, Kaunas, Lithuania*

MAKOTO SAWADA • *Department of Brain Function, Research Institute of Environmental Medicine, Nagoya University, Nagoya, Japan*

ELIANA SCEMES • *The Dominick P. Purpura Department of Neuroscience, Albert Einstein College of Medicine, Bronx, NY, USA*

ARINDOM SEN • *Pharmaceutical Production Research Facility (PPRF), Schulich School of Engineering, University of Calgary, Calgary, AB, Canada*

SOOJUNG SHIN • *Department of Stem Cell and Regenerative Medicine, Invitrogen Inc., Frederick, MD, USA*

HIROMI SUZUKI • *Department of Brain Function, Research Institute of Environmental Medicine, Nagoya University, Nagoya, Japan*

RYUSUKE SUZUKI • *Department of Physiology and Neurobiology, University of Connecticut, Storrs, CT, USA*

DAVID V. THOMPSON • *Invitrogen Corporation, Madison, WI, USA*

RICARDO A. VALENZUELA • *Stanford University, Stanford, CA, USA*

JEAN DE VELLIS • *Mental Retardation Research Center, Semel Institute for Neuroscience and Human Behavior, David Geffen School of Medicine at UCLA, Neuroscience Research Building, Los Angeles, CA, USA*

MOHAN VEMURI • *Department of Stem Cell and Regenerative Medicine, Invitrogen Inc., Frederick, MD, USA*

ANDREA M. VINCENT • *Department of Neurology, University of Michigan, Ann Arbor, MI, USA*

INA B. WANNER • *Mental Retardation Research Center, Semel Institute for Neuroscience and Human Behavior, David Geffen School of Medicine at UCLA, Neuroscience Research Building, Los Angeles, CA, USA*

V. WEE YONG • *Departments of Clinical Neurosciences and Oncology, University of Calgary, Calgary, AB, Canada*

XIAOQIN ZHU • *Department of Physiology and Neurobiology, University of Connecticut, Storrs, CT, USA*

HAO ZUO • *Department of Physiology and Neurobiology, University of Connecticut, Storrs, CT, USA*

# Chapter 1

# Neurosphere and Neural Colony-Forming Cell Assays

Sharon A. Louis and Brent A. Reynolds

## Abstract

Our understanding of the biology of stem cells and their therapeutic potential relies heavily on robust functional assays that can identify and measure stem cell activity in vivo and in vitro. In the mammalian central nervous system (CNS), neural stem cells (NSC) are often isolated and studied using a culture system referred to as the neurosphere assay. Since its introduction in 1992, the neurosphere system has been used to isolate, expand, and identify the presence of an NSC population, but also to enumerate NSC frequency. Furthermore, the neurosphere system has been used to investigate the effects of exogenous signaling molecules, genetic alterations, and as an assay to evaluate cell purification strategies. Recently, we challenged the central premise of the neurosphere assay – that all neurospheres are derived from an NSC – is not true, thereby precluding the use of the neurosphere assay to accurately measure NSC numbers. These results implied that while the neurosphere culture system provides a simple means to isolate and expand NSC harvested from the embryonic and adult mammalian CNS, its application as a quantitative in vitro assay for measuring NSC frequency is limited. In order to address the need for an assay that can reliably detect alterations in NSC frequency, we developed a new single-step semisolid-based assay, the neural colony-forming cell (NCFC) assay, which allows discrimination between NSC and progenitors by the size of colonies they produce (i.e., their proliferative potential). We anticipate the NCFC assay will provide additional clarity in discerning the regulation of NSC, thereby facilitating further advances in the promising application of NSC for therapeutic use.

**Key words:** Neural stem cells, clonal analyses, neurosphere assay, neural colony-forming cell assay, neural stem cell frequency, rat, mouse.

---

## 1. Introduction

In the simplest definition, multipotent stem cells are characterized as undifferentiated cells with the capacity for extensive proliferation that gives rise to more stem cells (exhibit self-renewal) as well as progeny that will terminally differentiate into cell types of the tissue from which they are obtained (1, 2). These features make

L.C. Doering (ed.), *Protocols for Neural Cell Culture*, Springer Protocols Handbooks,
DOI 10.1007/978-1-60761-292-6_1, © Humana Press, a part of Springer Science+Business Media, LLC 1992, 1997, 2001, 2010

stem cells important elements in embryonic development and in adult tissue for maintaining cell number following injury and disease or natural cell turnover. Due to the general lack of unique cell surface markers and the absence of a distinct and discernable morphological phenotype, stem cells are typically defined and studied based on functional criteria. Implicit to identifying a stem cell in vitro is the discovery of culture conditions that permit competent cells to exhibit the cardinal stem cell properties of (i) self-renewal over an extended period of time, (ii) generation of a large number of progeny, and (iii) multilineage differentiation potential.

In the CNS embryogenesis, development occurs primarily in two waves: a prenatal wave where most of the neurons are generated and an early postnatal wave where most of the astrocytes and oligodendrocytes are generated, after which the developed adult CNS was thought to remain relatively quiescent. Over the past decades, it has become convincingly clear that the adult brain of both rodents and primates contains areas of active neurogenesis within the subventricular zone (SVZ) and the subgranular zone (SGZ) of the dentate gyrus of the hippocampus (3–5). In these restricted regions of the adult brain, it is believed that neural stem cells (NSC) continue to generate new neurons and glia to areas such as the olfactory bulb, the corpus callosum, and the hippocampus. While the putative NSC within the adult brain has been extensively confirmed in vivo, their functional characterization often relies on in vitro culture systems which provide a retrospective read-out of their function and frequencies. It is therefore crucial that the in vitro methodologies designed to study NCS be robust, meaningful, and the user clearly understands what the methodologies are purported to measure.

The discovery of culture conditions for the expansion of neural stem and progenitor cells from the mammalian CNS (4) provided, for the first time, a relatively simple and robust means to investigate the activity and regulation of these precursor cells. Although other methods are available to culture NSC (5, 6), the neurosphere culture system remains the most frequently adopted method to enrich, expand, and even calculate the frequency of NSC (7). In addition, the serum-free growth conditions of the neurosphere culture are also being employed to promote sphere formation by stem cells and measure their frequency from a variety of tumors (e.g., breast (8) and brain (9, 10) and normal tissues (e.g., breast (11), skin (12), cardiac (13), pancreatic (14), and embryonic (6, 15, 16).

To generate neurospheres, tissue from the embryonic or adult CNS is microdissected, dissociated, and plated in a defined serum-free medium in the presence of a mitogen (e.g., EGF ± FGF-b) (**Fig. 1.1A**). The NCS (as well as neural progenitors) begin to proliferate after about 24 h in culture and form small clusters of cells by 2–3 days. The clusters continue to grow in size and by

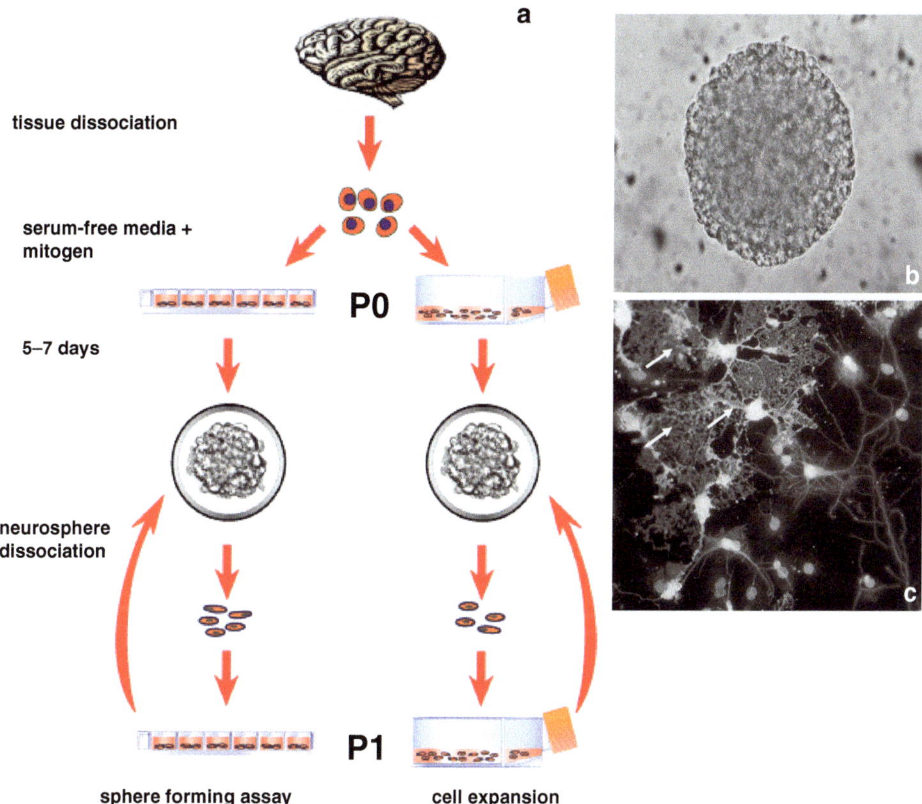



Fig. 1.1. The neurosphere culture procedure (**A**). The diameter of the neurospheres increase over time and neurospheres are detached from the substrate and float in suspension. A floating 7 days in vitro neurosphere. Magnification: ×200 (**B**) Cells within the neurospheres differentiate when the mitogens are removed and serum added to the medium. Phase-contrast micrograph shows cells with long processes (neurons) and lattice and web-like structures (oligodendrocytes). Neurons were identified with a fluorescent-labeled antibody raised against Beta-tubulin (a neuron-specific antigen). Oligodendrocytes (*arrows*) labeled with an antibody against myelin basic protein (MBP) (**C**). Magnification: ×400.

day 3–5 the majority of the clusters detaches from the surface and float in suspension. By day 7 the clusters, called neurospheres, typically measuring 100–200 μm in diameter (**Fig. 1.1B**), are composed of approximately 10,000 cells and are ready to be passaged or subcultured. Neurospheres are individually, or as a population, dissociated into a single cell suspension and re-plated under the same conditions as the primary culture. The neural stem (and even progenitor) cells begin to proliferate within the first 24 h and again form new clusters that are ready to be passaged 5–7 days later. NCS will self-renew and proliferate resulting in a relatively consistent arithmetic increase in cell number. Neurospheres derived from embryonic mouse CNS tissue treated in this manner have been passaged for up to 10 weeks with no loss in their proliferative ability, resulting in a $10^7$-fold increase in total cell number. Cells within the neurospheres can be induced to differentiate by removal of the mitogens and plating either as intact neurospheres or dissociated cells on an adhesive substrate in a low serum-containing medium. After several days, virtually all of the stem cell progeny will differentiate into the three primary cells types found in the CNS: neurons, astrocytes, and oligodendrocytes (**Fig. 1.1C**). Cells isolated from different regions of the rat and human CNS have also been shown to possess the ability to form neurospheres in similar serum-free culture conditions and mitogens; however, the rate of success in continually generating neurospheres for extended periods of time upon subculture has been variable; therefore the need for specific techniques for growing neurospheres from each species has to be strictly adhered to.

While the neurosphere culture system has been the method of choice for the isolation, expansion, and even the enumeration of neural stem and progenitor cells, several recent publications have highlighted some of the limitations of the neurosphere culture system as a read-out for NSC numbers (17, 18). One of the limitations of the neurosphere culture system is the significant degree of neurosphere fusions which occurs in this suspension culture, which questions the accuracy of the common practice of using neurosphere size and numbers as an indicator of stem cells (19). There is also growing evidence that not all neurospheres are derived from neural stem cells, as progenitors are capable of generating new neurospheres sometimes up to 3–4 culture passages, therefore suggesting that measuring neurosphere numbers and/or using secondary or tertiary neurosphere formation data is not a reliable measure of stem cell activity.

We have recently challenged one of the central tenets of the neurosphere assay (that all neurospheres are derived from an NSC) and concluded that an exclusively one-to-one relationship between neurosphere formation and NSC does not exist, thereby suggesting that this tenet is incorrect (17). We have addressed some of the shortcomings of the neurosphere assay by the development of a new assay called the neural colony-forming cell (NCFC) assay which is able to more accurately discriminate between neural stem and progenitor cells compared to the neurosphere assay (20). We have validated the NCFC assay and shown that colony size is an indicator of proliferative potential and cells from colonies >2 mm in diameter meet all the functional criteria for an NSC, which includes the ability to self-renew, generate large numbers of progeny, and maintain multipotency over an extended period of time. Cells from colonies <2 mm lack self-renewal ability and are likely progenitors. The NCFC assay therefore gives a more quantitative read-out for NSC compared to number of neurospheres or secondary neurosphere formation ability measurements that are routinely practiced in the field.

This chapter begins by describing detailed procedures for culturing neurospheres from mouse and rat CNS, which is a very useful culture system for the expansion of large numbers of cells. The original protocols and medium formulations are based on the work of Reynolds and Weiss (4), and the reader is also referred to (21–23). The objective of this section of the chapter is to provide standardized, accurate, and detailed procedures to apply to the neurosphere culture system so that reproducible cultures for each of these species can be obtained consistently. Next we outline the procedure for the new NCFC assay. The development, validation, and applications of the NCFC assay are described in Louis et al. (20).

## 2. Neurosphere Cultures from Primary Embryonic Day 14 Mouse CNS Cells

### 2.1. Reagents and Equipment

#### 2.1.1. Dissection Equipment

1. Large scissors (1) and small fine scissors (1).
2. Extra-fine spring microscissors (1) (cat. no. 15396-01, Fine Science Tools).
3. Small forceps (1) (cat. no. 11050-10, Fine Science Tools).
4. Small fine forceps (1) (cat. no. 11272-30, Fine Science Tools).
5. Ultra-fine curved forceps (1) (cat. no. 11251-35, Fine Science Tools).
6. Bead sterilizer (cat. no. 250, Fine Science Tools).
7. Dissecting microscope (e.g., Stemi 2000, 1:7 zoom, Zeiss).
8. Phosphate-buffered saline (PBS) (e.g., cat. no. 37350, STEMCELL Technologies Inc.) containing 2% glucose, cold, sterile.
9. Plastic Petri dishes, 100 mm, sterile (4–5 plates to hold uteri, embryos, heads, and brains) (cat. no. 27125, STEMCELL Technologies Inc.).
10. Plastic Petri dishes, 35 mm, sterile (4–5 to hold dissected brain regions) (e.g., cat. no. 21700, STEMCELL Technologies Inc.).
11. Tubes, 17 × 100 mm polystyrene test tubes, sterile (e.g., cat. no. 2057, Falcon).

#### 2.1.2. General Equipment

1. Biological safety cabinet (e.g., Canadian Cabinets) certified for Level II.
2. Low-speed centrifuge (e.g., Beckman TJ-6) equipped with biohazard containers.

3. 37°C incubator with humidity and gas control to maintain >95% humidity and an atmosphere of 5% $CO_2$ in air (e.g., Forma 3326).

4. Pipette-aid (e.g., Drummond Scientific).

5. Hemacytometer (e.g., Brightline).

6. Trypan blue (e.g., cat. no. 07050, STEMCELL Technologies Inc.).

7. Light microscope with 5× and 10× objectives for hemacytometer cell counts.

8. Inverted microscope with flatfield objectives and eyepieces to give object magnification of approximately ×20–×30, ×80, and ×125 (e.g., Nikon Diaphot TMD).

9. Pasteur glass pipettes, sterile.

*2.1.3. Tissue Culture Equipment (Neurosphere Cultures)*

1. T-25 $cm^2$ tissue culture flask (Nunc Catalog #156367 or VWR Catalog #15708-130) or T-162 $cm^2$ Flask (cat. no. 3151, Corning).

2. Tubes, 17 × 100 mm polystyrene test tubes, sterile (e.g., cat. no. 2057, Falcon).

3. Tubes, 50 mL, polypropylene, sterile (e.g., cat. no. 2057, Falcon).

4. 24-Well culture dishes (e.g., cat. no. 3526, Corning).

*2.1.4. Media and Supplements (Neurosphere Cultures)*

1. 30% Glucose (cat. no. G-7021, Sigma). Mix 30 g of glucose in 100 mL of distilled water. Filter-sterilize and store at 4°C.

2. 7.5% Sodium bicarbonate ($NaHCO_2$) (cat. no. S-5761, Sigma). Mix 7.5 g of $NaHCO_2$ in 100 mL of distilled water. Filter-sterilize and store at 4°C.

3. 1 *M* HEPES solution (cat. no. H-0887, Sigma).

4. 10 × STOCK solution of DMEM/F12. Mix five 1 L packages each of DMEM powder (high glucose with L-glutamine, minus sodium pyruvate, minus sodium bicarbonate (cat. no. 12100-046, Invitrogen) and F12 powder [contains L-glutamine, no sodium bicarbonate (cat. no. 21700-075, Invitrogen) in 1 L water. Filter-sterilize and store at 4°C.

5. 3 mM Sodium selenite ($Na_2SeO_3$) (cat. no. S-9133, Sigma). Mix 1 mg of $Na_2SeO_3$ with 1.93 mL of distilled water. Filter-sterilize and store at –20°C.

6. 2 mM Progesterone (cat. no. P-6149, Sigma). Inject 1.59 mL of 95% ethanol to a 1 mg stock of progesterone and mix. Aliquot into sterile tubes and store at –20°C.

7. Bovine transferrin iron poor (APO) (cat. no. 820056-1, Serologicals). Dissolve 400 mg of Apo transferrin directly to the 10× hormone mix.

8. Insulin: Dissolve 100 mg insulin in 4 mL of sterile 0.1 $N$ HCl. Mix in 36 mL of distilled water and add entire volume directly to the 10× hormone mix.

9. Putrescine (cat. no. P-7505, Sigma). Dissolve 38.6 mg putrescine in 40 mL of distilled water and add entire volume directly to the 10× hormone mix.

10. 200 mM L-glutamine (e.g., cat. no. 07100, STEMCELL Technologies Inc.).

11. *Basal medium.* To prepare 450 mL of basal medium, the individual components are added in the following order: 375 mL of ultrapure distilled water, 50 mL of 10× DMEM/F12, 10 mL of 30% glucose, 7.5 mL of 7.5% sodium bicarbonate, 2.5 mL of 1 $M$ HEPES, and 5 mL of 200 mM L-glutamine. Mix components well and filter-sterilize.

    **Note:** An optimized basal medium for the culture of neurospheres from embryonic and adult mouse CNS cells is available, NeuroCult® NSC Basal Media (Mouse) (cat. no. 05700, STEMCELL Technologies Inc.).

12. *10× Hormone mix.* To prepare 10× hormone mix, the individual components are added in the following order: 300 mL of ultrapure distilled water, 40 mL of 10× DMEM/F12, 8 mL of 30% glucose, 6 mL of 7.5% sodium bicarbonate, and 2 mL of 1 $M$ HEPES. Mix components well at this point. The following components are then added to above mixture in the order listed: 400 mg of Apo transferrin, 40 mL of 2.5 mg/mL insulin stock, 40 mL of 10 mg/mL putrescine stock, 40 μL of 3 mM sodium selenite, and 40 μL of 2 mM progesterone. Mix all components well and filter-sterilize. Aliquot into 10 or 50 mL volumes in sterile tubes and store at −20°C.

    **Note:** An optimized 10× hormone mix for the culture of neurospheres from embryonic and adult mouse CNS cells is available, NeuroCult® NSC Proliferation Supplements (Mouse) (cat. no. 05701, STEMCELL Technologies Inc.).

13. *Basal-Hormone Mix Media.* This media is prepared as follows: Thaw an aliquot of the 10× hormone mix from **item 12**. Add 50 mL of the 10× hormone mix to 450 mL of basal medium from **item 11** to give a 1:10 dilution. The hormone-supplemented neural culture media should be stored at 4°C and used within 1 week.

14. Human recombinant epidermal growth factor (rhEGF) (cat. no. 02633, STEMCELL Technologies Inc.). A stock solution of 10 μg/mL of rhEGF is made up in 0.1 mL sterile 10 mM acetic acid containing at least 0.1% bovine serum albumin, then adding 19.9 mL of the basal-hormone mix

media from **item 13** and stored as 1 mL aliquots at –20°C until required for use.

15. Human recombinant fibroblast growth factor (rhFGF) (for rat cells; cat. no. 02634, STEMCELL Technologies Inc.). A stock solution of 10 µg/mL of rhFGF is made up in the basal-hormone mix media from **item 13** and stored as 1 mL aliquots at –20°C until required for use.

16. 0.2% Heparin (for rat cells) (cat. no. 07980, STEMCELL Technologies Inco). Mix 100 mg of heparin (cat. no. H-3393, Sigma-Aldrich) in 50 mL of distilled water. Filter-sterilize. Store aliquots of 1 mL at 4°C.

    *Note*: Both EGF and bFGF have been shown to be mitogens for CNS stem cells. In general, the number of neurospheres generated and the rate of expansion are enhanced when the two mitogens are used simultaneously; however, each growth factor can act on different populations of stem cells. EGF is routinely used for embryonic day 14 mouse CNS cultures, while EGF, FGF and heparin are required for culture of embryonic rat, fetal human CNS cells, and adult mouse subventricular zone cells. Heparin facilitates high-efficient binding of FGF to the FGF receptor.

17. *"Complete" NSC Proliferation medium*. Add 2 µL of rhEGF to every 1 mL of the basal-hormone mix media from **item 13** to give a final concentration of 20 ng/mL of EGF.

    *Note*: The performance of media prepared in the laboratory is highly dependent on the quality and purity of the water and raw materials. If media is prepared in the laboratory, use only tissue-culture-grade materials, and if necessary, source various suppliers to determine the best quality reagents as there is significant batch-to-batch variability in some critical reagents. To avoid variability in media performance, STEMCELL provides optimized and standardized media for the proliferation and differentiation of neural stem and progenitor cells from the mouse, rat, and human CNS.

---

## 3. Primary Neurosphere Cultures from Embryonic Day 14 Mouse CNS

The microdissection of the CNS tissue from mouse embryos is extensively referenced (20, 21), and various videos available online (e.g., www.stemcell.com) therefore will not be described in this chapter. All culture procedures including dissections of CNS regions should be performed in Level II biosafety cabinets using aseptic technique and universal safety precautions.

1. Once dissections are complete, collect all the tissues (obtained from 12–30 embryos) with a glass pipette and transfer the tissues and collection media (e.g., PBS + 2% glucose is commonly used) into 14 mL tube and allow tissues to settle. Pipette off supernatant.

2. Resuspend tissues in 2 mL of basal-hormone mix media (**Section 2.1.4., item 13**).

3. Using a Pasteur glass pipette or a plastic disposable tip attached to a P1000 micropipettor set at 1 mL; triturate the tissue approximately ten times until a single-cell suspension is achieved.
   **Note:** To triturate, slightly tilt the tip and press it against the bottom or side of the tube to generate resistance to break up the tissue. The mechanical dissociation of cells by trituration with a fire-polished pipette or disposable plastic tip is known to cause cell death. While earlier published protocols suggested the use of fire-polished glass Pasteur pipettes, non-fire-polished glass Pasteur pipette and disposable plastic tips work well too. However, some precautionary steps can be performed during trituration to diminish the negative effects. For example, avoid forcing air bubbles into the cell suspensions. Also, it is important to wet the pipette with a small amount of media before sucking the cells into the pipette to reduce the number of cells sticking to the glass or plastic surface.

4. If undissociated tissue remains, allow the suspension to settle for 1–2 min and then pipette off the supernatant containing single cells into a fresh tube.
   **Note:** Trituration must be repeated until cell clumps and intact neurospheres are dissociated. Because clumps of cells are heavier than single cells, these will settle to the bottom of the tube when left standing for about 5 min. If all of the collected cells are not required for subculturing, the clumps can be allowed to settle, then the single cell suspension can be removed to a fresh sterile tube and used for subsequent cultures, leaving the undissociated clusters at the bottom of the tube.

5. Add more medium to the undissociated cells for a total volume of 2 mL. Continue to triturate, transfer, and pool the supernatant containing single cells. Repeat trituration if necessary.

6. Centrifuge the cells at 800 rpm ($110g$) for 5 min. Remove supernatant and resuspend the cells with a brief trituration in 2 mL of medium.

7. Measure the precise volume and count cell numbers using a dilution in trypan blue (1/5 or 1/10 dilution) and hemacytometer.

8. For primary cultures, seed cells at a density of $2 \times 10^6$ cells per 10 mL or 80,000 cells/cm$^2$ (T-25 cm$^2$ flask) or $8 \times 10^6$ cells in 40 mL media (T-162 cm$^2$ flask), in complete NSC medium.

***3.1. Subculturing of Mouse Neurospheres***

The procedures outlined below can be performed repeatedly over multiple culture passages for neurospheres derived from embryonic or adult mouse CNS cells to generate long-term neurosphere cultures. If stem cells are maintained during the subculture, they will self-renew and generate large numbers of progeny over time, and the accumulative total cell numbers at each passage plotted against time (or passage number) will give a log expansion in accumulative total cell numbers. Progenitors can also form new neurospheres for up to 4–5 subcultures; however, because of the loss of proliferative potential, these cells eventually do not produce more progeny. Therefore, secondary or tertiary neurospheres formation cannot be reliably used as an indication of self-renewal ability. Self-renewal ability can be determined if single cells are deposited into wells and long-term cultures (>5 passages) generated from these cells (24).

1. Observe the neurosphere cultures under a microscope to determine if the neurospheres are ready for passaging. If neurospheres are attached to the culture substrate, tapping the culture flask against the bench top should detach them.

   *Note*: Initially, single cells should proliferate to form small clusters or aggregates of cells that might lightly adhere to the culture vessel, which will eventually lift off from the substratum as the density of the aggregates increases. Viable neurospheres are characterized by their semitransparent appearance, with many of the cells on the outer surface displaying microspikes (*see* **Fig. 1.1**). Neurospheres should be passaged before they grow too large (>150–200 μm in diameter). The size of the neurospheres can be estimated by sampling ~10 μL from the culture in a hematocytometer (**Fig. 1.2**). In the larger neurospheres, the cells within the core of the neurospheres lack appropriate gas and nutrient/waste exchange and become necrotic and appear dark and are also more difficult to dissociate. It is important that cultures be monitored regularly, and with increased experience, healthy neurospheres cultures will become recognizable.

2. Remove medium with suspended cells and place in an appropriately sized sterile tissue culture tube. If some cells remain attached to the substrate, detach them by shooting a stream of media across the attached cells. Spin at 400 rpm (75*g*) for 5 min.

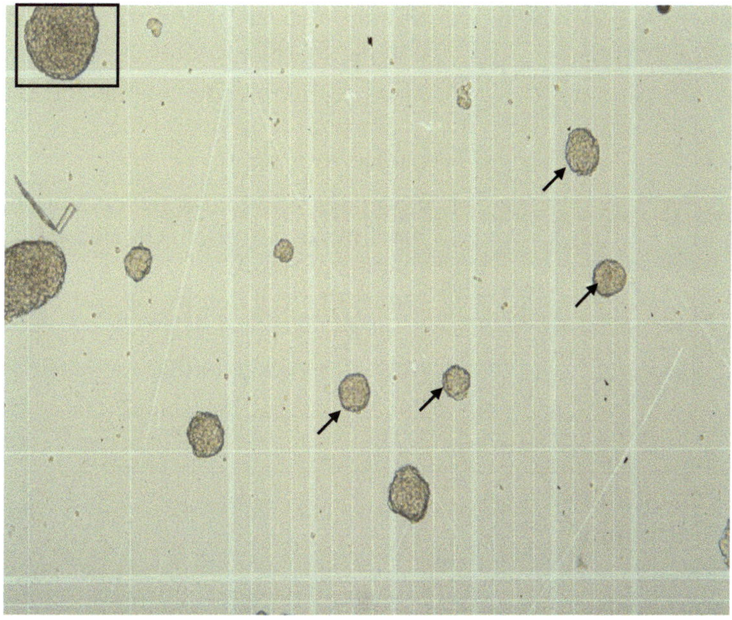

Fig. 1.2. Sizing a neurosphere using a hemacytometer. A neurosphere that is 100 μm in diameter should not be bigger than a quarter of a small square. The neurosphere outlined with a square is approximately 200 μm in diameter. *Arrows* indicate neurospheres that are approximately 100 μm in diameter.

3. Remove the supernatant and resuspend cells in a maximum of 2–3 mL of complete NSC medium.

4. If more than one tube was used to harvest cultures, resuspend each pellet in a small volume (0.5 mL) of complete NSC medium and pool all cell suspensions.

5. With a Pasteur pipette, triturate the neurosphere until single cell suspension is achieved.
   **Note:** Other non-mechanical methods to dissociate neurospheres to produce a homogenous single cell suspension are available. For example, the NeuroCult® Chemical Dissociation Kit Catalog (cat. no. 05707, STEMCELL Technologies Inc.) offers a non-mechanical, non-enzymatic alternate procedure for dissociating neurospheres derived from embryonic and adult mouse CNS cells, which yields greater cell viabilities compared to the trituration method. The use of trypsin is not recommended as it digests certain cell surface proteins and effects stem cell properties. The commercially available product Accutase™ (cat. no. 07920, STEMCELL Technologies Inc.) can be used for dissociating neurospheres derived from embryonic and adult mouse CNS without adversely affecting stem cell function.

6. Centrifuge pooled single-cell suspension at 800 rpm (110$g$) for 5 min. Remove the supernatant and resuspend cells by

tituration in an appropriate (approx 1–2 mL) volume of complete NSC medium.

7. Measure the precise volume and count cell numbers using a dilution in trypan blue (1/5 or 1/10 dilution) and hemacytometer.

8. Set up the cells for the next culture passage in complete NSC medium at $5 \times 10^5$ cells per 10 mL or 20,000 cells/cm$^2$ (T-25 cm$^2$ flask).
   **Note:** The cell density for plating primary striatum, cortex, ventral mesencephalon, or other regions of the E14 mouse brain is higher than that for subculturing conditions.

## 4. Establishment and Subculture of Primary Neurospheres from Embryonic Day 18 Rat Cortex

The basic procedure to culture neurospheres from embryonic and adult mouse CNS cells can be applied to the culture of rat neural stem and progenitor cells; however, NSC from rat CNS tissue have different growth characteristics compared to mouse NSC. The rate of proliferation of E18 rat cortical cells is higher (varies from 2- to >20-fold expansion of total cells every 3–4 days of culture) during early passages (P2–P5). A critical phase in the growth curve occurs between P5 and P9 (20–36 days) in culture, when a significant decrease in terms of fold expansion (average between two- and fivefold) is observed. During this critical period, cells start to attach to the flask and differentiate (**Fig. 1.3**). To reduce attachment of spheres to the flask, it is important to perform partial medium changes every 2 days and subculture every 3–4 days.

### 4.1. Preparation of Medium, Supplements, and Growth Factors for Embryonic Day 18 Rat Cells

1. Prepare complete proliferation medium for rat cells by mixing basal medium and 10× hormone mix in a 1:10 ratio (as in **item 13 Section 2.1.4. above**)

2. To the basal-hormone media mix, add 20 ng/mL rhEGF, 10 ng/mL bFGF, and 0.0002% heparin. This media will henceforth be called "complete proliferation medium (rat)".
   **Note:** Because of the different requirements for the culture of neurospheres from rat CNS, STEMCELL Technologies Inc. has developed optimized media specifically for rat cells (www.stemcell.com).

### 4.2. Primary Neurosphere Cultures from Embryonic Day 18 Rat CNS Cells

1. When dissections of E18 cortices (from 12 embryos) are complete, transfer tissue in PBS with 2% glucose into a 14 mL conical tube, allow tissues to settle and pipette off supernatant.

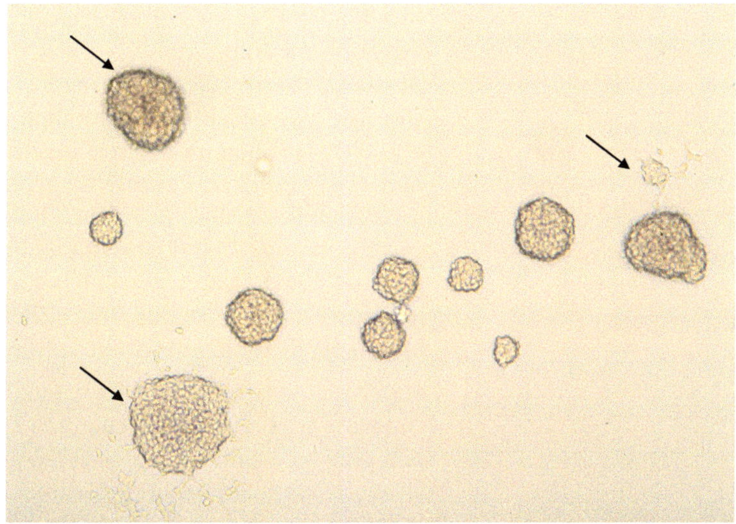

Fig. 1.3. Neurospheres derived from embryonic day 18 rat cortex. The culture conditions were sub-optimal, resulting in a few of the neurospheres (*arrows*) starting to attach to the flask and differentiate. If left unattended, the degree of attachment will increase over time, making it more difficult to generate new neurospheres from rat cells upon sub-culture.

2. Resuspend the tissue in 1 mL complete proliferation medium (rat).

3. Use a 1 mL pipettor with sterile pre-wetted plastic tip to triturate the tissue approximately 5–10 times or until single cell suspension is achieved.
   **Note:** Do not introduce air bubbles into the cell suspension. To maintain high viability, avoid using fire-polished glass pipettes to disaggregate neurospheres derived from rat cells.

4. Add 1 mL complete proliferation medium (rat) to the single cell suspension and mix carefully to avoid creating any bubbles. Leave for about 1 min to allow the undispersed pieces of tissue to settle.

5. Transfer supernatant to a new sterile 14 mL tube. Discard undissociated tissue.

6. Centrifuge supernatant at $110g$ ($\sim$800 rpm) for 5 min. Discard supernatant.

7. Resuspend cells with a brief trituration (two times), with a disposable pre-wet plastic pipette tip attached to a 1 mL micropipettor in 1 mL complete proliferation medium (rat).

8. Measure the exact volume and perform a viable cell count on a hemacytometer using a dilution (1/5 or 1/10, depending on amount of tissue dissected) in trypan blue.

9. Seed cells at a density of $6 \times 10^4$ viable cells/mL (2400 cells/cm$^2$) or $1.2 \times 10^5$ viable cells/mL (4800 cells/cm$^2$)

in a T-25 cm$^2$ tissue culture flask in 10 mL of complete proliferation medium (rat).

**Note:** A lower volume of media (5 mL) can be used in the first day of culture, and then an additional 5 mL of complete proliferation medium (rat) added after 2 days and this would be considered replenishing or performing a half media change (*see* **Note below**). It is important to use the correct brand of tissue cultureware as suggested in **Section 2.1.3**. The use of ultra-low adherent dishes is not necessary if the procedures are adhered to and the appropriate media formulation for rat cultures is used.

10. Incubate cultures in a 5% $CO_2$ incubator at 37°C.

    **Note:** A partial medium change (25–30% of total volume) is highly recommended 2 or 3 days after plating, to prevent the medium from becoming acidic. To change the medium, position the flask in an upright position and let the cells and spheres settle to the bottom (2–3 min). Slowly remove ~3 mL of medium being careful not to remove the cells, and replace the volume with fresh complete proliferation medium (rat). If spheres appear small (~<50 μm), it may not be necessary to perform a half media change.

### 4.3. Subculturing of Rat Neurospheres

One of the common problems encountered when culturing neurospheres from the rat CNS is the loss of self-renewal properties of the stem cells and increased differentiation. If more than 20% of the neurospheres are attached to the bottom of the flasks, it is unlikely that long-term subculturing of these cells will be achieved and obtaining a fresh supply of primary rat cortices would be recommended. As long as strict adherence to the protocols below is followed, it is possible to generate new neurospheres from rat CNS cells for multiple passages (>20 passages) and still maintain self-renewal and multilineage differentiation.

1. Observe the neurosphere cultures under a microscope to determine if the neurospheres are ready for passaging.

   **Note:** Cultures of rat E18 cortical or striatal neurospheres should be passaged every 3–4 days compared to every 5–7 days for their mouse counterparts. Similar to neurospheres derived from mouse CNS cells, rat neurospheres should be passaged before the diameter exceeds 100 μm to avoid hypoxic cells in the centre of the spheres.

2. Remove neurospheres and medium and place in an appropriately sized sterile tissue culture tube (e.g., 14 mL tube).

3. Spin at 75$g$ (~400 rpm) for 5 min. Remove supernatant, leaving behind approximately 150–180 μL medium.

4. Set the volume of a P200 micropipettor attached to a sterile plastic tip to slightly less than the approximate volume of the remaining medium.

**Note:** If the volume of remaining medium is 180 μL, set the volume of the pipettor to 160 μL to avoid creating bubbles.

5. Pre-wet the tip with medium to reduce cells sticking inside the tip.

   *Note*: To prevent cell sticking to the plasticware and hence cell loss during cultures, it is important to pre-wet all pipette tips or disposable pipette with media prior to aspirating cells through the pipette.

6. Gently triturate the cell pellet 10–15 times.

   *Note*: Slightly tilt the tip and press it against the bottom or side of the tube to generate resistance in order to break up the neurospheres. Rinse the side of the tube during trituration to remove the remaining neurospheres that are attached to the side of the tube. If some neurospheres remain undissociated after 15 triturations (this usually occurs at later passages), trituration can be extended to a maximum of 25–35 times. To maintain high cell viability, do not use fire-polished glass pipette to disaggregate the neurospheres.

7. Measure the volume of cells and medium. Count viable cells using trypan blue exclusion assay (1/10 dilution at earlier passages, lower dilution for later passages) on a hemacytometer.

8. Seed cells at a density of $1.25 \times 10^4$ or $2.5 \times 10^4$ viable cells/mL (*see* **Note below**) in a T-25 $cm^2$ tissue culture flask containing 10 mL of complete proliferation medium (rat).

9. Incubate cultures in a 5% $CO_2$ incubator at 37°C.

10. Cultures should be examined under the microscope regularly. A partial medium change should be performed at day 2 after culture setup as described in the above section.

   **Note:** It is suggested that more than one flask is cultured during the critical passages, P5–P9. Two different seeding densities, $1.25 \times 10^4$ cells/mL and $2.5 \times 10^4$ cells/mL, should be used.

---

## 5. NCFC Assay Setup

The NeuroCult® NCFC Assay allows for the identification and discrimination of NSC and progenitor cells from mouse or rat primary CNS tissue and cultured neurospheres based on their proliferative potentials (**Fig. 1.4**). Primary or cultured neural cells are suspended in serum-free medium containing optimized levels of growth supplements and recombinant cytokines.

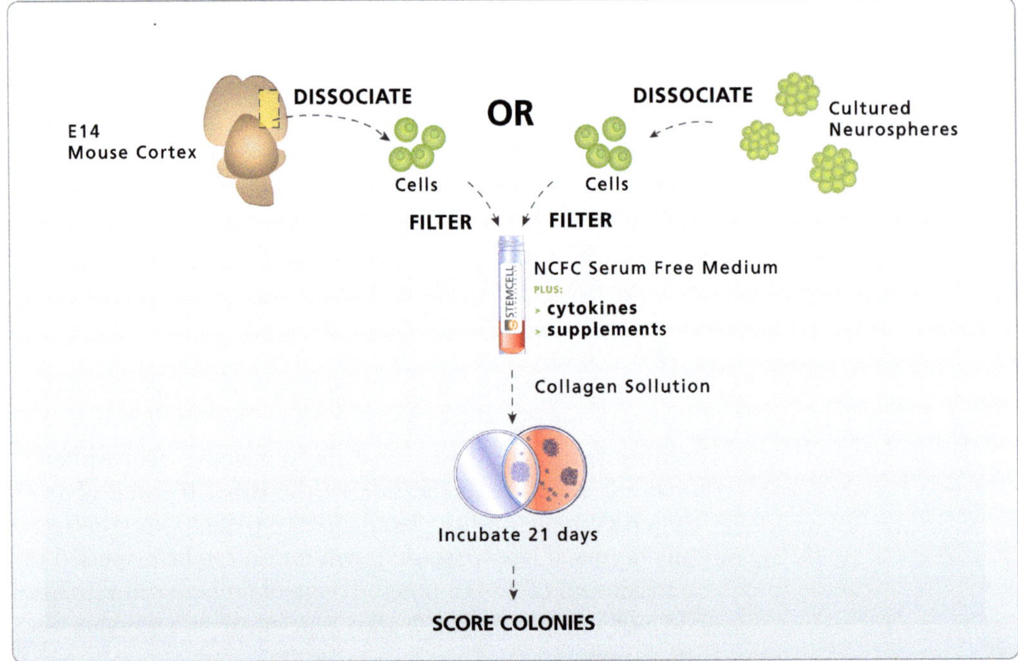



Fig. 1.4. The procedure to set up NCFC cultures from primary CNS tissue and cells derived from neurosphere cultures.

Collagen is then mixed with the cell-medium suspension and dispensed into 35 mm culture dishes. At the end of the 21 day culture period, clonally derived colonies of different sizes are scored.

**5.1. Media, Supplements, and Tissue Culture Equipment for Neural Colony-Forming Cell Assay**

Note that while we originally developed the media formulation and procedures for the NCFC assay, the reagents to perform this assay are now available through a commercial supplier and therefore media formulations remain proprietary. The neural colony-forming assay called The NeuroCult® NCFC assay kit (Mouse) is currently supplied as a ready-to-use kit from STEMCELL Technologies Inc. cat. no. 05740. The kit contains the following components:

- 05720 NeuroCult® NCFC serum-free medium without cytokines supplied as 50 mL/bottle
- 05701 NeuroCult® NSC proliferation supplements (Mouse) supplied as 50 mL/bottle
- 05700 NeuroCult® NSC basal medium (Mouse) supplied as 450 mL/bottle
- 04902 collagen solution supplied as 35 mL/bottle
- 07914 NeuroCult® collagenase solution 5 mL/bottle
- 27100 35 mm culture dishes 7 packs (10 dishes/pack)
- 27500 gridded scoring dishes 5 dishes
- 40 μm cell strainer (cat. no. 27305 STEMCELL Technologies Inc.)

- 24- or 96-well culture dish (24-well cat. no. 3527 or 96-well cat. no. 3596, Corning)
- 6-well culture dish (cat. no. 3506, Corning)
- T-25 cm$^2$ flasks (cat. no. 156367, Nunc)
- 245 mm square bioassay dish (STEMCELL Technologies Inc., cat. no. 27130)

1. Thaw bottles or aliquots of NeuroCult® NCFC serum-free medium without cytokines and NeuroCult® NSC proliferation supplements (Mouse) at room temperature or overnight at 4°C.

2. Place thawed medium and supplements at 37°C and the collagen solution on ice.

   *Note*: It is important to keep the collagen on ice throughout the culture setup to prevent the collagen from gelling.

3. Filter the single cell suspension of neural cells (derived from cultured neurospheres or primary embryonic cells and prepared as described in **Sections 2 and 3 above**) through a 40 μm cell strainer to remove any undissociated cells or clumps of cells.

   **Note:** The NCFC assay is based on the formation of colonies from a single or clonal colony-forming cell; therefore, it is critical that the initial single cell suspension is homogenous for this step. In mixing experiments, the frequency of chimera (non-clonal) colonies in the NCFC assay was estimated to be <1% compared to >20% chimera neurospheres observed in the neurosphere cultures.

1. *Primary embryonic cells*: Dilute primary embryonic cells to 6.5 × 10$^5$ cells per mL in complete proliferation medium, which will give a final cell plating density of 7500 cells per 35 mm culture dish in a 25 μL volume.

2. *Cells derived from neurosphere cultures*: Dilute these cells to 2.2 × 10$^5$ cells per mL, which will give a final cell plating density of 2500 cells per 35 mm culture dish in a 25 μL volume.

   *Note*: The cell plating density was determined by performing titration curves and determining the linearity ranges for primary cells isolated from normal embryos of pregnant CD1 albino mice or cells from neurospheres derived from normal E14 CD1 albino mice cortices and/or striata cultured for two passages. It may be necessary to perform a titration curve within the range of 5000–50,000 cells per dish for primary cells or 1000–5000 cells per dish cultured cells when different species and transgenic animals are used. It is possible that the cloning efficiency is changed for these cells types. Adjust the initial concentration of the cells so that 25 μL volume of cells is always added to the medium mixture described in **step 5** below to maintain accurate media concentrations.

3. Prepare and label the correct number of 35 mm dishes required for the intended experiment. The volumes of reagents listed below are designed for duplicate 35 mm plates.

4. Place a sterile 14 mL tube or the appropriate number of tubes for dispensing the NCFC assay reagents and cells for each test condition in a tube rack.

5. To make the semisolid collagen NCFC medium (allowing duplicate 35 mm culture dishes), add the following components in the given order: 1.7 mL of NeuroCult® NCFC serum-free medium without cytokines, 0.33 µL of NeuroCult® NSC proliferation supplements (Mouse), 6.6 µL of a (rhEGF) stock solution of 10 µg/mL) and then 25 µL of the cell suspension either at a concentration of $6.5 \times 10^5$ cells/mL (for primary cells) or $2.2 \times 10^5$ cells/mL (for cultured cells).
   **Note:** Do not add the collagen solution yet. If multiple tubes are being set up, add cells to a single tube and then add collagen and plate cells. Do not let cells sit in NCFC medium for an extended period of time before plating.

6. Mix the medium containing cells (~2.1 mL total) by pipetting with a disposable 2 mL pipette.

7. Using a separate sterile 2 mL pipette, transfer 1.3 mL of cold collagen solution to the tube and mix again by pipetting. Using the same 2 mL pipette, remove 1.5 mL of the final culture mixture and dispense this volume into a 35 mm culture dish. Dispense another 1.5 mL in the same manner into a second 35 mm dish. Remove any air bubbles by gently touching bubble with the end of the pipette.
   **Note:** The collagen starts to gel within several minutes following the addition to the cell suspension. If more than one tube is being set up, collagen should be added to the first tube only, and the contents dispensed into dishes before proceeding to the next tube.

8. Gently tip each culture dish using a circular motion to allow the mixture in the dishes to spread evenly over the surface.

9. Place the 35 mm culture dishes in a 100 mm Petri dish. This Petri dish must also contain an open 35 mm culture dish filled with 3 mL of sterile water to maintain optimal humidity during the prolonged incubation period. Replace the lid of the 100 mm Petri dish.
   **Note:** If many dishes are used, these dishes can also be placed in a covered 245 mm square bioassay dish with two or three open 35 mm culture dishes containing sterile water.

10. Transfer the plates to an incubator set at 37°C, 5% $CO_2$ and >95% humidity.

**Note:** Gel formation will occur within approximately 1 h. It is important not to disturb the cultures during this time.

11. Culture cells for 21 days (differences in colony size can be clearly discerned after 21 days).

    **Note:** Refer to **Fig. 1.5** showing the formation of a colony over time.

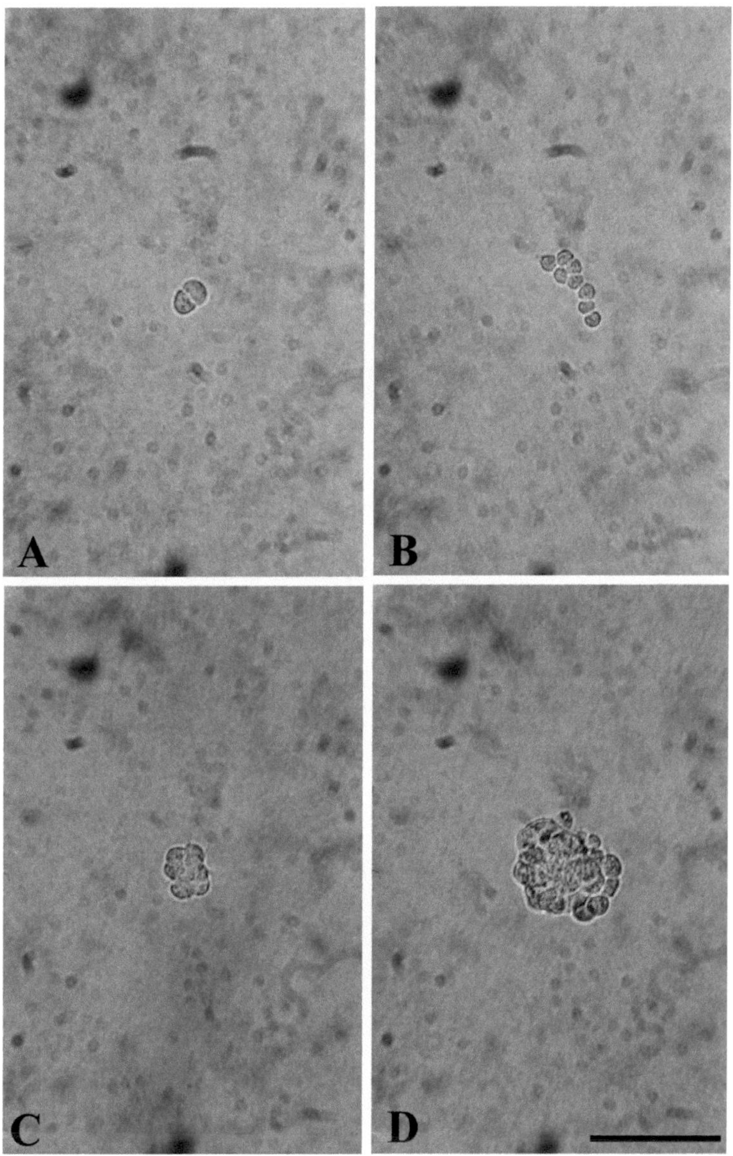

Fig. 1.5. Formation of a colony in the N-CFC assay. (**A**) Two days after plating, small colonies are found in the semisolid collagen-based medium. (**B**) Seven days after plating, many of the colonies have reached 80–100 μm in diameter and contain several hundred cells. (**C**) After 14 days, colonies of different sizes can be observed. Some colonies appear to stop growing beyond 7 and 14 days while others continue to increase in size. Scale bar 10 μm (a), (b) and 400 μm (c).

12. As cultures are incubated for an extended period of time (21–28 days), cultures should be replenished with complete replenishment medium once a week to avoid depletion of culture media.

    **Note:** Make up 10 mL of complete replenishment medium by mixing 9 mL of basal medium and 1 mL of 10× hormone mix. To this, add 500 μL of the 10 μg/mL stock solution of hEGF. The complete replenishment medium contains 1:10 basal and 10× hormone mix and 0.5 μg/mL of hEGF. Store the prepared media at 4°C for up to 3 weeks of replenishing.

13. Gently (so as not to disrupt gel) add 60 μL of complete replenishment medium into the center of each NCFC dish once every 7 days during the entire NCFC culture incubation (21 days).

14. Cultures should be visually assessed regularly for overall colony growth and morphology using an inverted microscope.

    **Note:** Do not leave cultures at room temperature for extended periods of time as the collagen gel will begin to liquify.

*5.1.1. Categorizing NCFC Colonies and Procedure for Scoring NCFC Colonies*

Within 4–7 days of plating in the NCFC Assay, neural stem and progenitor cells begin to proliferate forming small colonies (**Fig. 1.5**). By day 14, these small colonies have grown in size and differences can be discerned between colonies. A number of colonies appear to stop growing after approximately 10–14 days while other colonies continue to expand. By day 21–28, colonies can be classified into four categories: (1) less than 0.5 mm in diameter, (2) 0.5–1 mm in diameter, (3) 1–2 mm in diameter, and (4) 2.0 mm or >2 mm in diameter. Refer to **Fig. 1.6** for representative examples of different colony sizes and **Figs. 1.7** to **1.10** for their morphologies in each of the categories.

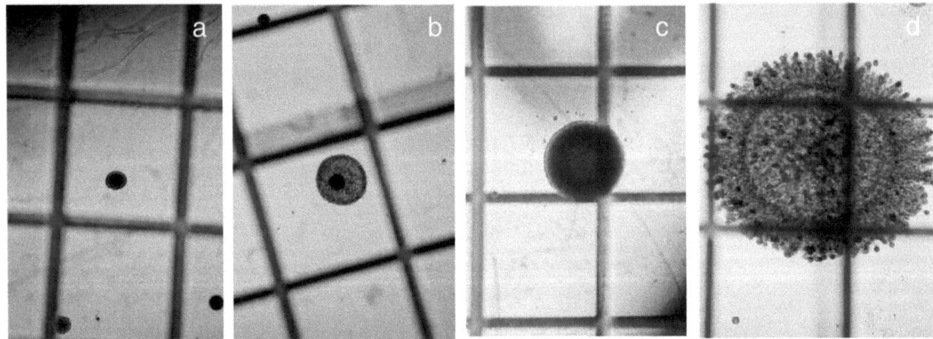

Fig. 1.6. Representative colony morphologies and size categories derived from embryonic day 14 mouse cortical cells. Colonies were categorized into four size groups: **(a)** less than 0.5 mm in diameter, **(b)** 0.5–1 mm in diameter, **(c)** 1–2 mm in diameter, and **(d)** greater than 2 mm in diameter. The grid measures 2 mm × 2 mm.

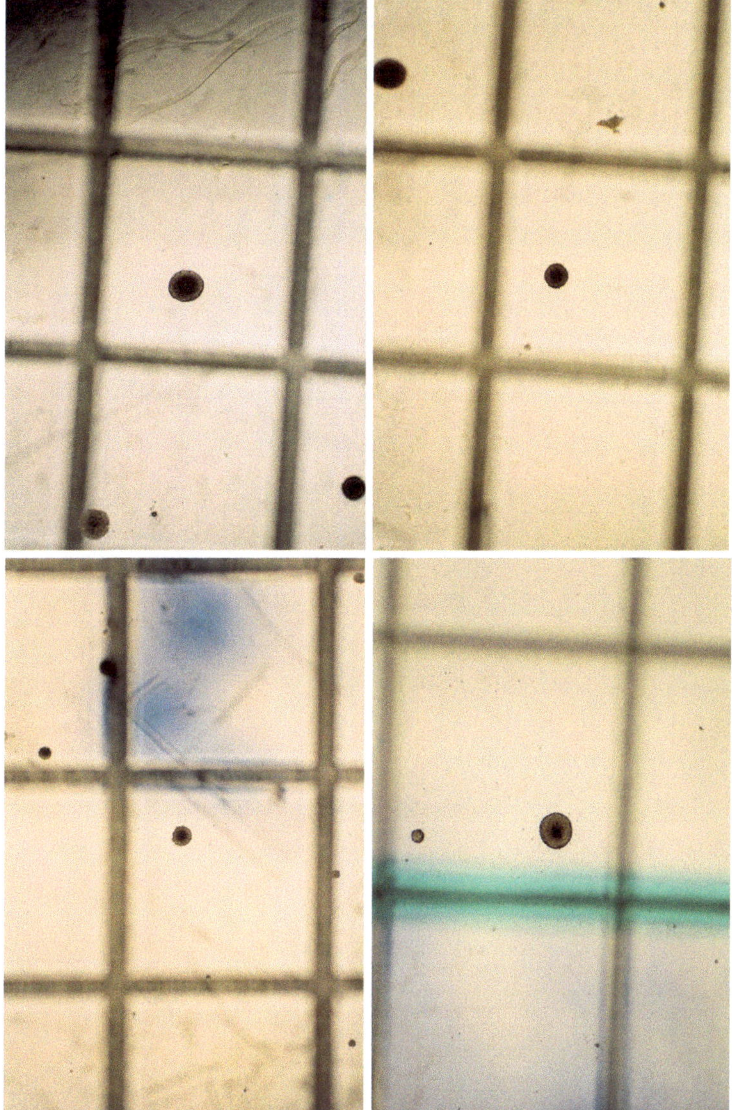

Fig. 1.7. Representative colony morphologies observed in colonies less than 0.5 mm in diameter derived from embryonic day 14 mouse cortical cells.

1. Place an individual 35 mm culture dish on a gridded scoring dish and then place both the culture dish and gridded dish on the dissecting microscope stage.

2. First scan the entire dish using a low-power $(2.5\times–5\times)$ objective lens, noting the relative proximity of the colonies to each other. Scoring can then be performed with the same lens. Use a higher-power $(10\times)$ objective to examine colonies in greater detail.

   Classify colonies into four categories:

Fig. 1.8. Representative colony morphologies observed in colonies 0.5–1 mm in diameter derived from embryonic day 14 mouse cortical cells.

a. Less than 0.5 mm in diameter (**Fig. 1.7** )

b. 0.5–1 mm in diameter (**Fig. 1.8** )

c. 1–2 mm in diameter (**Fig. 1.9** )

d. 2.0 mm or >2 mm in diameter (**Fig. 1.10** )

*5.1.2. Estimation of NSC or Neural Progenitors*

The following criteria are applied for the quantification of NSC and progenitor cells from *primary embryonic cells* or cultured neurospheres derived from *embryonic cells:*

The original cell that forms a colony 2.0 mm or >2 mm in diameter is referred to as a neural colony-forming cell–neural stem

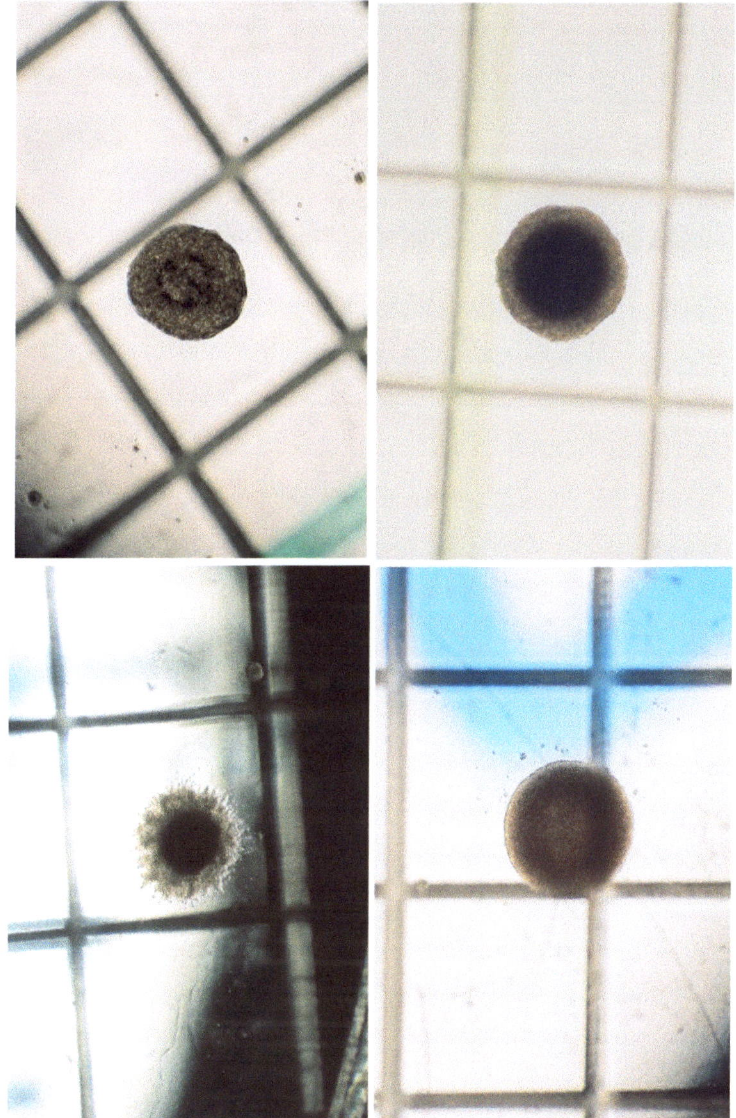

Fig. 1.9. Representative colony morphologies observed in colonies 1–2 mm in diameter derived from embryonic day 14 mouse cortical cells.

cell (NCFC-NSC) as this cell has high proliferative potential and multilineage potential. Colonies <2 mm contain cells that lack self-renewal ability and multipotency and are likely produced by a progenitor.

The NCFC assay can also be used to measure total colony numbers, which is a similar read-out to total sphere numbers read-out in the neurosphere cultures that detects the cells which have colony-forming ability or sphere-forming ability.

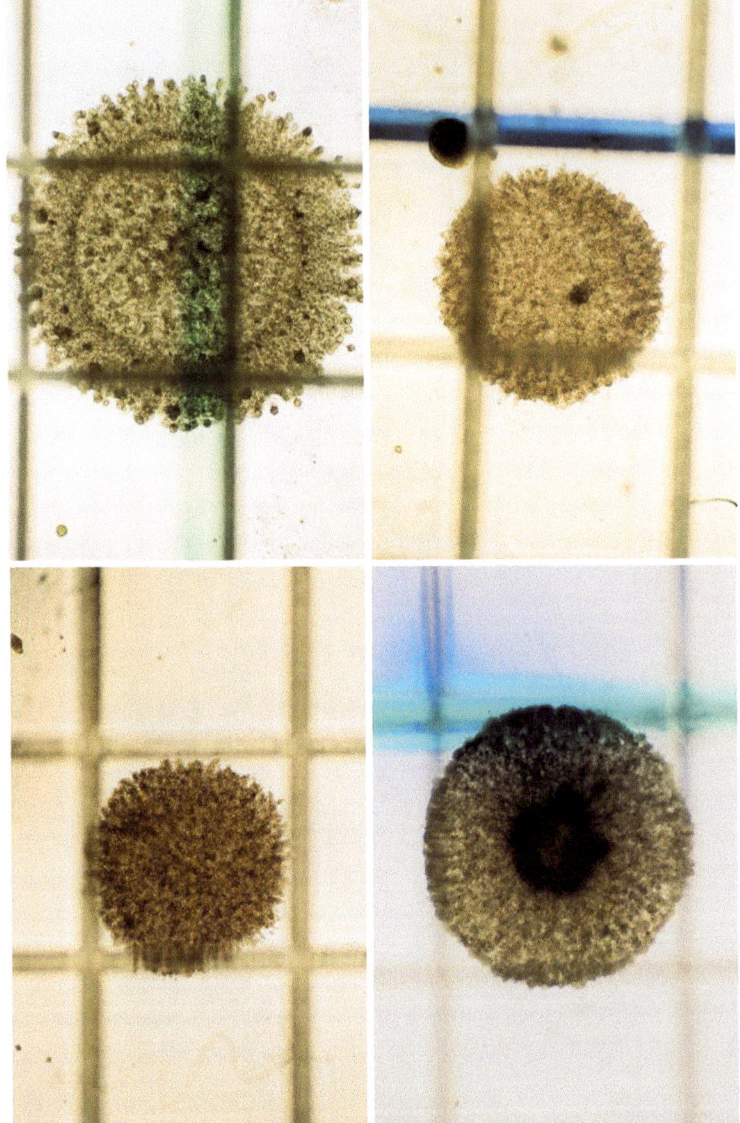

Fig. 1.10. Representative colony morphologies observed in colonies >2 mm in diameter derived from embryonic day 14 mouse cortical cells.

## 6. Isolation of Colonies from the NCFC Assay

The NCFC assay was validated for "normal or unmanipulated" stem cells and progenitors. If the NCFC assay is simply used as a routine read-out for neural stem cells and neural progenitor cells, isolation of the cells within the different colonies does not need to be performed each time the assay is applied. However, in some cases where the cells studied in the NCFC assay were obtained from a transgenic animal or treated animal, it is possible that the regulation of stem or progenitor cells is altered. It is important to investigate if colony size still remains an accurate indicator of proliferative potential, i.e., does NSC still generate large colonies and the progenitors smaller colonies as categorized in the scoring criteria.

Colonies from the different size categories are harvested from the collagen matrix by cutting the colonies out with a pair of sterile extra-fine spring microscissors and placing the embedded colony in collagenase solution to digest the matrix. The cells are dissociated from the collagen matrix, and all the cells from each colony of the different size categories plated in a single well of a 96-well plate in proliferation medium. The cells are cultured until secondary neurospheres are observed (between 7 and 10 days). If secondary neurosphere are observed, these neurospheres are collected, dissociated into a single cell suspension, and the resulting cells re-plated in fresh medium. Tertiary neurospheres is determined and, if neurospheres are detected, these are collected and re-plated under the same culture conditions again using an increasing sized culture vessel at each subsequent passage to generate long-term neurosphere cultures. If long-term cultures were established from the cells from the original colony, it would indicate that the initial neural colony-forming cell (NCFC) had high proliferative potential. On the other hand, if cells from the colonies only generated neurospheres for limited passages in neurosphere cultures, it would suggest that the original NCFC was a progenitor or differentiated cell.

***6.1. Procedure to Isolate NCFC Colonies from Collagen Gels***

1. Mark the colony that is to be isolated with a marker pen on the bottom of the dish. To isolate colonies from the four size categories, set up the dissecting microscope and sterilize a pair of extra-fine spring microscissors in a bead sterilizer or 70% ethanol.

2. Aliquot 100 μL of the 0.25% solution of collagenase solution into individual sterile 1.5 mL eppendorf tubes with one tube for each colony.

3. Look through the dissecting microscope and cut single colonies from the collagen matrix with the sterilized microscissors. Transfer the individual colonies into the tubes containing the collagenase with the microscissors.

4. Gently shake the microscissors in the collagenase solution to dislodge the harvested colony.

5. Incubate the collagen-embedded colonies for 10 min at 37°C.

   **Note:** Do not exceed 15 min for the digestion as the collagen may digest cell surface proteins.

6. After the incubation, mechanically disrupt the excised colony with a plastic disposable pipette tip attached to a P200 micropipettor, breaking up the matrix and producing a single cell suspension.

   **Note:** Sometimes, it is difficult to obtain a single cell suspension from the colonies. Excessive mechanical treatment will result in cell death; therefore, the presence of small clumps of cells is sometimes unavoidable in the initial cultures of cells from the harvested colonies. Viable cells will proliferate and form neurospheres, which can then be mechanically dissociated at end of the culture period.

7. Centrifuge cells in the eppendorf tubes at 800 rpm ($150g$) for 5 min in a microcentrifuge, and remove supernatant.

8. Add 1 mL of fresh complete proliferation medium (prepared as in **Section 2**) and resuspend the cells by gently triturating with a plastic disposable pipette tip attached to a P200 micropipettor.

9. Repeat the centrifugation at 800 rpm ($110g$) for 5 min, remove medium and resuspend the cells in 100 μL of fresh medium.

10. Depending on the number of colonies harvested, dispense an appropriate volume of complete proliferation medium into each well of a 24- or 96-well culture dish.

    **Note:** Generally, more cells are obtained from colonies >2 mm in diameter; therefore, they are dispensed into 24-well plates while cells from colonies <2 mm in diameter are dispensed into 96-well plates. Because of the low cell number obtained from the primary NCFC colonies, cell counts may not be possible. Cell counts can be performed at the end of the initial culture and in subsequent cultures.

11. Plate all the cells from a single colony into an individual well of a 24- or 96-well culture dish containing complete proliferation medium.

12. Monitor cultures regularly in the next few days to determine which of the wells contain neurospheres.

13. After 7–14 days, new neurospheres may form from the original cells from the different sized colonies.

    **Note:** Cells from colonies >2 mm in diameter may generate neurospheres within 7 days while the cells from the colonies <1 mm in diameter may form fewer neurospheres after a longer incubation time. As soon as neurospheres (~100–150 μm in diameter) are formed, these should be collected and dissociated into a single cell suspension for further applications.

14. Collect the culture media and neurospheres from the 24- or 96-well culture dishes using either a disposable plastic

tip attached to a P1000 micropipetor set at ~900 mL (24-well) or a disposable plastic tip attached to a P200 micropipetor set at ~0.180 mL.

15. Centrifuge the neurospheres at 800 rpm ($110g$) for 5 min and carefully remove the supernatant.

16. Add approximately 200 µL of complete proliferation medium and mechanically dissociate the neurospheres with a disposable plastic tip attached to a P200 micropipetor set at ~0.180 mL.

    **Note:** A cell count based on trypan blue exclusion can be performed at this stage; however, the dilution factor (1:2 or 1:3) will need to be worked out for the relatively lower number of cells.

17. This subculture procedure is repeated every 7–14 days, passaging cells into an appropriate sized culture vessel depending on numbers of cells generated (e.g., 6-well dishes or T-25 cm$^2$ flasks).


## References

1. Weissman IL. Stem cells: units of development, units of regeneration, and units in evolution. Cell 2000;**100**:157–68.
2. Potten CS and Loeffler M. Stem cells: attributes, cycles, spirals, pitfalls and uncertainties. Lessons for and from the crypt. Development 1990;**110**:1001–20.
3. Gross CG. Neurogenesis in the adult brain: death of a dogma. Nat Rev 2000;**1**:67–73.
4. Reynolds BA and Weiss S. Generation of neurons and astrocytes from isolated cells of the adult mammalian central nervous system. Science 1992;**255**:1707–10.
5. Ray J, Peterson DA, Schinstine M and Gage FH. Proliferation, differentiation, and long-term culture of primary hippocampal neurons. Proc Natl Acad Sci USA 1993;**90**:3602–6.
6. Conti L, Pollard SM, Gorba T et al. Niche-independent symmetrical self-renewal of a mammalian tissue stem cell. PLoS Biol 2005;**3**:e283
7. Pevny L and Rao MS. The stem-cell menagerie. Trends Neurosci 2003;**26**:351–9.
8. Ponti D, Costa A, Zaffaroni N et al. Isolation and in vitro propagation of tumorigenic breast cancer cells with stem/progenitor cell properties. Cancer Res 2005;**65**:5506–11.
9. Galli R, Binda E, Orfanelli U et al. Isolation and characterization of tumorigenic, stem-like neural precursors from human glioblastoma. Cancer Res 2004;**64**:7011–21
10. Singh SK, Hawkins C, Clarke ID et al. Identification of human brain tumour initiating cells. Nature 2004;**432**: 396–401.
11. Dontu G, Abdallah WM, Foley JM et al. In vitro propagation and transcriptional profiling of human mammary stem/progenitor cells. Genes Dev 2003;**17**:1253–70.
12. Toma JG, Akhavan M, Fernandes KJ et al. Isolation of multipotent adult stem cells from the dermis of mammalian skin. Nat Cell Biol 2001;**3**:778–84.
13. Beltrami AP, Barlucchi L, Torella D et al. Adult cardiac stem cells are multipotent and support myocardial regeneration. Cell 2003;**114**:763–76.
14. Seaberg RM, Smukler SR, Kieffer TJ et al. Clonal identification of multipotent precursors from adult mouse pancreas that generate neural and pancreatic lineages. Nat Biotechnol 2004;**22**:1115–24.
15. Benzing C, Segschneider M, Leinhaas A et al. Neural conversion of human embryonic stem cell colonies in the presence of fibroblast growth factor-2. Neuroreport 2006;**17**:1675–81.
16. Smukler SR, Runciman SB, Xu S and van der Kooy D. Embryonic stem cells assume a primitive neural stem cell fate in the absence of extrinsic influences. J Cell Biol 2006;**172**:79–90.
17. Reynolds BA and Rietze RL. Neural stem cells and neurospheres–re-evaluating the relationship. Nat Methods 2005;**2**:333–6.

18. Rietze RL and Reynolds BA. Neural stem cell isolation and characterization. Methods Enzymol 2006;**419**:3–23.

19. Singec I, Knoth R, Meyer RP et al. Defining the actual sensitivity and specificity of the neurosphere assay in stem cell biology. Nat Methods 2006;**3**:801–6.

20. Louis SA, Rietze RL, Deleyrolle L et al. Enumeration of neural stem and progenitor cells in the neural colony-forming cell assay. Stem Cells. 2008;26(4):988–96. epub 2008 Jan 24

21. O'Connor TJ, Vescovi AL and Reynolds BA. Isolation and propagation of stem cells from various regions of the embryonic mammalian central nervous system.In: Celis, JE (eds.) Cell Biology: A Laboratory Handbook. 2nd edition, Academic Press, London, 1998;Vol. 1:149–53.

22. Gritti A, Galli R, and Vescovi AL.Cultures of stem cells of the central nervous system. In: Federoff, S and Richardson, A (eds.) Protocols for Neural Cell Culture. Marcel Dekker, New York, 2001:173–97.

23. Louis SA and Reynolds BA. In: Helgason, CD and Miller, CL (eds.) Methods Mol Biol. Generation and differentiation of neurospheres from murine embryonic day 14 central nervous system tissue. Humana Press, 2005;**290**:265–80.

24. Gritti A, Galli R and Vescovi AL. In: Weiner, LP (eds.) Methods Mol Biol. Clonal analyses and cryopreservation of neural stem cell cultures. Humana Press 2008;**438**:173–83.

# Chapter 2

# Directed Neuronal Differentiation of Embryonic and Adult-Derived Neurosphere Cells

Marcos R. Costa, Ravi Jagasia, and Benedikt Berninger

## Abstract

Neurosphere-forming cells can be isolated from virtually all neural tissues during embryonic development as well as some neural tissues in the adult such as the adult subependymal zone (SEZ) and spinal cord. While these cells share in common the ability of generating clonal aggregates, the so-called neurospheres, that display self-renewal capacity and cellular multipotency (i.e., the potential to generate the three major neural lineages), they can markedly differ in their intrinsic propensity toward neurogenesis or gliogenesis as well as their ability to generate distinct types of neurons with respect to their transmitter identity. Here we discuss the endogenous differentiation potential of neurosphere cells derived from both embryonic and adult brain tissues and provide protocols how to direct adult SEZ neurosphere cells toward specific neuronal subtypes (i.e., GABAergic versus glutamatergic) via retroviral-mediated gene transfer of pro-neural genes.

**Key words:** Neurosphere, gene transfer, retrovirus, pro-neural genes.

## 1. Introduction

The neurosphere culture allows for the in vitro propagation of cells derived from various neural tissues at distinct developmental stages that share in common the potential to self-renew and to generate the three major neural cell types in the central nervous system (CNS), namely neurons, astrocytes, and oligodendrocytes (1, 2). These self-renewing, multipotent cells are often referred to as "neural stem cells (NSC)" (for critical review of the stemness of neurosphere-forming cells *see* **Chapter 1** by Louis and Reynolds). Interestingly, neurosphere-forming cells cannot only be isolated from the embryonic brain, but also persist in

L.C. Doering (ed.), *Protocols for Neural Cell Culture*, Springer Protocols Handbooks,
DOI 10.1007/978-1-60761-292-6_2, © Humana Press, a part of Springer Science+Business Media, LLC 1992, 1997, 2001, 2010

some brain regions throughout adulthood and have been proposed as a potential source for providing newly generated cells for brain repair (3). The focus of the present chapter will be to provide and to discuss protocols for directing neurosphere cell differentiation toward distinct neuronal phenotypes. Understanding how neurosphere cells can be differentiated into selective neuronal types is essential when considering these cells as a source for neuronal replacement (for review *see* (4)). It also serves as a model system to study intrinsic signaling mechanisms involved in neuronal determination and differentiation in vitro for being highly amenable for biochemical and genetic manipulation.

Importantly, the regional and developmental origin of the neurosphere-forming cells is likely to exert a restrictive effect on their differentiation potential. This can be easily revealed by assessing the endogenous potential of neurosphere cells derived from different brain regions and developmental stages. For example, neurosphere cells derived from cells isolated from the dorsal telencephalon (i.e., cerebral cortex) at beginning of the neurogenic period (embryonic day 10 in mice) generate more neurons than neurosphere cells derived at mid-neurogenesis (embryonic day 14) from the same region (5). This reflects most likely the changes in the progenitor potential observed in vivo and in vitro: dorsal telencephalic progenitors isolated at early corticogenesis are highly neurogenic and become gradually more gliogenic at mid- to late corticogenesis (*see e.g.*, (6) and references herein). Accordingly, both neurosphere cells derived from embryonic stem cells (ESC) or embryonic telencephalic cells generate larger number of glial cells when differentiated at progressively later passages (7), suggesting that a similar transition from neurogenesis to gliogenesis has taken place in the neurospheres. Additionally, neurosphere cells keep at least some of their regional identity even after several passages (8, 9). For instance, neurosphere cells derived from embryonic dorsal telencephalon generate large numbers of glutamatergic neurons, whereas neurosphere cells isolated from ventral telencephalon or adult subependymal zone (SEZ) give rise predominantly to GABAergic cells (10, 11). Therefore, some of the original properties of stem/progenitor cells isolated from distinct germinative zones in the developing or adult brain appear to be maintained following culturing in the neurosphere assay. On the other hand, there isevidence that some transcription factors, such as Olig2, become upregulated in cells of distinct germinative zones under neurosphere growth conditions (12, 13), while other more region-specific transcription factors, such as Pax6 and Dlx2, are downregulated (13, 14). This data indicate that some intrinsic properties of the neurosphere-forming stem/progenitor cells are lost following expansion in vitro. Importantly, this effect may be in part attributable to culturing the neurosphere cells in high concentrations of epidermal growth factor (EGF) and fibroblast growth factor-2 (FGF2) (14). For instance, in the adult SEZ infusion of EGF results in the downregulation of the transcription factor Dlx2 in rapid dividing transit amplifying cells (a type of precursor placed between the *bona fide* stem cell and fate-restricted neuroblasts), with the consequence that these cells give rise to glia at the expense of the generation of neuroblasts. This change in fate decision is consistent with the important role of Dlx2 in adult SEZ neurogenesis (15). Given that the bulk of adult SEZ neurosphere cells during expansion in vitro divide rather rapidly thereby resembling the transit amplifying cells in vivo, EGF treatment may exert a similar effect on their fate decision. Indeed, the massive upregulation of Olig2 that occurs in adult SEZ neurosphere cells (13) may be nothing but the other side of the coin of Dlx2 downregulation, reflecting the switch toward a gliogenic potential (15, 16). Thus the developmental stage, tissue origin, and passage number are obviously important determining factors of the endogenous potential of neurosphere cells and hence will strongly influence the outcome of experiments that aim at analyzing the relative propensity of neurosphere cells toward neuro- or gliogenesis as well as their propensity toward generating distinct neuronal phenotypes.

Neurosphere cells can be isolated from most embryonic murine neural tissues as well as some adult murine neural tissues, including the SEZ (1), the spinal cord (17), the hippocampal dentate gyrus (18) [but possibly of a more restricted potential as compared to SEZ neurospheres (19)], the injured, but not intact, adult cerebral cortex (20), and the adult carotid body (21). Yet rather little information is available on the specific differentiation potential of these distinct types of neurosphere cells. Here, we will provide protocols focussing on how to assess the differentiation potential of neurosphere cells derived from both the embryonic cerebral cortex and the adult SEZ and will describe a methodology how to direct adult SEZ neurosphere cells toward specific neuronal subtypes via the forced expression of neurogenic fate determinants, such as the pro-neural genes *Neurog2* and *Mash1*.

**1.1. Directing Neuronal Differentiation via Retrovirus-Mediated Transduction**

Studies of key transcription factors acting during development have provided insights into the molecular mechanisms of neuronal subtype specification (*see e.g.,* 22–24) and provide important information with regard to the attempt to drive neural stem cells/precursor cells toward specific cell fates. It has previously been reported that the expression of pro-neural basic helix loop helix transcription factors, such as the neurogenins (i.e., Neurog2) and Mash1, can effectively induce the differentiation of adult neuronal progenitors toward specific cellular fates (11, 25, 26).

Retroviral transduction is a useful method for gene delivery into neurosphere (precursors) cultures, since it is efficient and allows for stable and longlasting expression of transgenes. Nonviral gene delivery protocols such as electroporation can also be a useful method to genetically manipulate neurosphere cells. However, it has the disadvantage of being a rather aggressive method that can result in the damage of cellular integrity resulting in cell death (up to 50%). Furthermore, the introduced genes do not become faithfully transmitted to the progeny of cells continuing to proliferate and may thus become substantially diluted with each cell division. As an alternative to retroviral vectors, also lentiviral or adenoviral vectors can be exploited to efficiently transduce neural stem/precursor cells (26). One interesting difference between retroviral and other viral gene delivery methods regards the timing of gene delivery with respect to the cell division cycle as retroviral integration into the host genome requires nuclear breakdown, thus excluding nonproliferating cells from transduction. Furthermore, retroviral gene transfer may also narrow the time window for the onset of virally mediated gene expression with respect to the cell cycle. These aspects may prove important, given that the susceptibility of cells for the effect of fate determinants is very likely to depend on the cell cycle status. In the following, we will focus on the use of retroviral vectors for gene delivery into neurosphere cells, as we have most experience using this system.

*1.1.1. Retroviral Vectors*

The complete cDNAs for neurogenic transcription factors such as Neurog2, Mash1, and others have been inserted into retroviral vectors previously described, such as the pCAG-IRES-GFP (27, 25), pMXIG and pCLIG (11, 28, 29, 30). These recombinant retroviral vectors contain viral long terminal repeats at their ends, a viral packaging signal, a promoter driving expression, and a gene of interest. The transgene should desirably contain a bicistronic cassette, which will allow both transgene expression and the identification of transduced cells [for example, hooking up the gene to be delivered via an internal ribosomal entry site (IRES), a nucleotide sequence that allows for translation initiation in the middle of a messenger RNA (31, 32), to enhanced green fluorescent protein (eGFP)].

The pCAG-IRES-GFP (*see* schematic map below) has been found to be highly effective in our hands, less prone to silencing compared to other retroviral vectors (pMXIG, pCLIG), likely due that the promoter sequence being an hybrid consisting of the human cytomegalovirus immediate-early enhancer and a modified chicken beta-actin promoter (33, 34). Moreover, this retroviral vector allows for increased expression of the transgene, due to the woodchuck hepatitis post-transcriptional regulatory element (WPRE), which has been reported to stabilize transgene transcripts (35).

**Schematic Map of pCAG-IRES-GFP** *Schematic map of the plasmid pCAG-IRES-GFP.* The gene of interest is subcloned between the CAG promoter and the IRES-GFP. This strategy allows expression of both the gene of interest and reporter gene (for instance GFP) under control of the CAG promoter. The WPRE is located downstream of the transcript and helps to stabilize it.

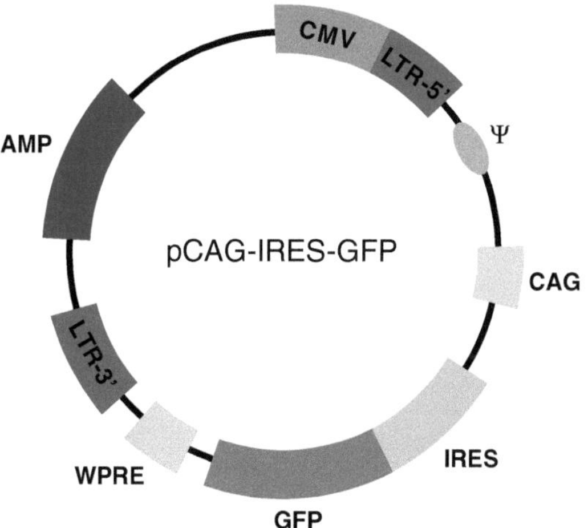


*1.1.2. Retroviral Genome*

Simple retroviral genomes contain at least three sets of genes, *gag*, *pol*, and *env*, encoding the structural, enzymatic, and envelope proteins required for virus replication and particle assembly. In recombinant retroviral vector systems, the genes required for viral replication and particle assembly are deleted from the basic retroviral vector, with the consequence that the retrovirus is replication-incompetent and the assembly of viral particles requires *trans*-complementation of the deleted genes in a viral producer cell line. Only in cells that are either transiently or stably transfected with pro-virus plasmids (separately coding for gag, pol, and envelope proteins) can the viral particles assemble correctly containing the retroviral RNA, thus enabling it to infect cells and permanently integrate its pro-viral DNA into the host genome of dividing cells. Since the genome is split, the deleted genes have to be co-transfected into packaging cell lines, such as the human kidney cell line 293 cells (HEK293T) cells. Several alternative protocols exist regarding the mode of transfection (for instance, lipofectamin2000 or calcium phosphate transfection), the use of inducible and stably transfected packaging cell lines, such as the HEK 293-GPG cells (26, 36), and the pseudotyping using different envelope proteins [here we limit us to the use of retroviruses pseudotyped using the vesicular stomatitis virus

G protein (vsv-G), which interacts with a phospholipid component in the plasma membrane giving a rather broad host range of infection].

## 2. Reagents and Equipment

### 2.1. Culture Medium Components

1. Opti-MEM (Invitrogen, 31950-062 or 31985-070)
2. Dulbecco's modified Eagle's medium – DMEM (Invitrogen, 21969035)
3. DMEM/ F12 (Invitrogen, 31331093)
4. B27 supplement (Invitrogen, 17504044)
5. Fetal bovine serum – FBS (Invitrogen, 16140-063)
6. Antibiotics: penicillin–streptomycin (Invitrogen, 15140-122)
7. Hepes buffer, 1 M (Invitrogen, 15630)
8. 45% glucose (Sigma, G 8796)

### 2.2. Dissociation of Neurosphere Cells

1. Trypsin 0.05% (Invitrogen, 25300)

### 2.3. Glass-Coated Coverslips

1. Glass coverslips (VWR, 631-0149)
2. Poly-D-lysin (Sigma, 81358)

### 2.4. Culture Dishes

1. 24-well culture dishes (Becton Dickinson Labware, 3047)
2. 10 cm cell culture dish (BD Falcon, 353003)

### 2.5. Conical Tubes

1. 15 mL (BD Falcon)
2. 50 mL ((BD Falcon))

### 2.6. Filter

1. Corning® filter-top centrifuge tubes (Sigma Aldrich, CLS430320)

### 2.7. Plasmid DNA

1. pCMV-GP [expresses the Moloney murine leukemia virus (MoMLV) *gag-pol* genes from the immediate early promoter–enhancer of the human cytomegalovirus].
2. pCMV-VSV-G (expresses the *env* gene of the vesticular stomatitis virus, VSV)
3. pCAG-IRES GFP or pMXIG (with subcloned transgenes, e.g., *Mash1*, *Neurog2*).
   **Note:** All plasmids and cDNAs encoding the transcription factors encoding Neurog2, Mash1 and Pax6 are available upon request from the authors.

| | |
|---|---|
| ***2.8. Transfection Reagent*** | 1. Lipofectamine 2000 (Invitrogen, 11668-019).<br>**Note:** Other transfection reagents can be used; however, lipofectamine 2000 gives consistent results with a relatively high transfection efficiency, which is prerequisite in generating viruses with a high titre. Calcium phosphate precipitation is also a relatively efficient and cheaper procedure for transfection of HEK293T cells. |
| ***2.9. Centrifuges*** | 1. Hettich, Rotanta 460 R (for 15 and 50 mL tubes; refrigeration system)<br>2. Beckman coulter, Optima L-100 XP (ultracentrifuge) |
| ***2.10. Laminar Flow*** | 1. HERA safe, KS 12 (1200×780×627 mm) |
| ***2.11. Incubator*** | 1. VWR; BINDER, CB 210 |

---

## 3.  Protocol

### 3.1. Preparing Media

*3.1.1. Differentiation Medium*

| Component | For 50 mL | For 10 mL | For 5 mL |
|---|---|---|---|
| DMEM/ F12 | 47.6 mL | 9.42 mL | 4.76 mL |
| B27 supplement | 1 mL | 200 μL | 100 μL |
| Antibiotics | 0.5 mL | 100 μL | 50 μL |
| Hepes buffer, 1 M | 0.4 mL | 80 μL | 40 μL |
| Glucose 45% | 0.5 mL | 100 μL | 50 μL |

*3.1.2. Dissociation Medium (DMEM/10%FBS)*

| Component | For 50 mL | For 10 mL | For 5 mL |
|---|---|---|---|
| DMEM Glutamax | 43.6 mL | 8.72 mL | 4.36 mL |
| FBS | 5 mL | 1 mL | 0.5 mL |
| Antibiotics | 0.5 mL | 100 μL | 50 μL |
| Hepes buffer, 1 M | 0.4 mL | 80 μL | 40 μL |
| Glucose 45% | 0.5 mL | 100 μL | 50 μL |

***3.2. Differentiation of Neurosphere Cultures***

Overall, two methods can be used for the differentiation of neurospheres:

(1) Whole sphere culture that allow for the identification of multipotent, bipotent, or unipotent spheres.

(2) Dissociated cells: used to determine the relative percentage of differentiated cell types generated.

**Note:** A complete protocol on the preparation of neurospheres is described by Louis and Reynolds in **Chapter 1**.

*3.2.1. Preparation of Glass-Coated Coverslips*

Pre-coat glass slides by adding a sufficient volume of PDL to completely cover the glass coverslip for a period of 2 h at 37°C. Remove PDL and wash three times (10 min each) with sterile 0.1 M phosphate buffer saline (PBS). Remove the PBS just before plating neurosphere cells.

*3.2.2. Differentiation of Whole Neurospheres*

1. Typically after 7 days in vitro, floating neurospheres reach sizes of about 150–200 μm. It is not advisable to keep them longer, as further growth in size leads to cell death (probably due to low perfusion of nutrients and $O_2$ into the center of the sphere). Transfer the neurospheres kept in neurosphere growth medium (*see* **Chapter 1**) into an appropriately sized sterile tissue culture tube. Spin at 500 rpm for 2 min.

2. Remove the neurosphere growth medium and gently resuspend the neurospheres with an appropriate volume of differentiation medium.

3. Transfer the neurosphere suspension to a 10 cm Petri dish from where individual neurospheres can be visualized and harvested, using a sterile disposable plastic pipette or a Gilson P1000 pipette.

4. Transfer 3–5 neurospheres to individual wells of a 24-well tissue culture plate with PDL-coated coverslips. Add differentiation medium up to 500 μl.

5. After 7–10 days in vitro, individual attached neurospheres disperse in such a manner so as to appear as a flattened monolayer of cells.

6. Fix the cells by the a ddition of 4% paraformaldehyde (in PBS, pH 7.2) for 5 min at room temperature and then process the adherent cells for immunocytochemistry as required. **Notes**: Neurospheres attach to the substrate quite rapidly and can be processed for immunocytochemistry after 2 h (already after this short period of time differentiating cells can be observed inside the neurospheres, indicating that neurospheres are not homogenous and comprise some fate-restricted cells). The minimum and maximum times for differentiation need to be tested for each application. For example, it may be necessary to culture the cells under differentiation conditions for longer periods of time (up to a month) to obtain fully functional neurons. On the other hand, 3 days might be enough to assess for multipotentiality of neurosphere-forming cells (**Fig. 2.1**).

*3.2.3. Differentiation of Dissociated Cells*

1. Typically after 7 days in vitro, floating neurospheres reach sizes of about 150–200 μm. As stated above, it is not advisable to expand them for longer time. Transfer the neurospheres kept in the neurosphere growth medium into

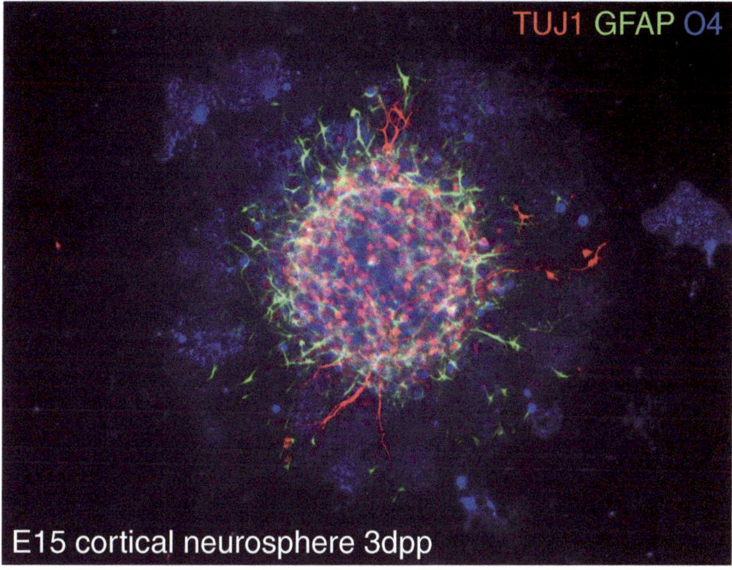


Fig. 2.1. Neurosphere cells express cell type-specific markers shortly after differentiation. The micrograph depicts a whole neurosphere grown from E15 cortical cells that was plated under adherent conditions without EGF and FGF2. Three days post-plating (dpp), neurons, astrocytes, and oligodendrocytes can be distinguished by immunocytochemistry using antibodies against class βIII-tubulin (Tuj1), glial fibrilliary acidic protein (GFAP) or O4, respectively. Note that cells have migrated little out from the neurosphere at that stage.

an appropriately sized sterile tissue culture tube. Spin at 500 rpm for 2 min.

2. Remove the neurosphere growth medium and add 1 mL trypsin 0.05%. Gently mix and incubate at 37°C for 5 min.

3. Add 1 mL of DMEM/10% FCS and dissociate neurospheres mechanically by using a fire-polished Pasteur pipette. Centrifuge the cell suspension(s) at 1000 rpm for 5 min.

4. Remove the neurosphere growth medium and then gently resuspend the cells with 1–2 mL of DMEM/10% FCS. Centrifuge the cell suspension(s) at 1000 rpm for 5 min. Repeat this step to completely remove trypsin from the cell solution.

5. Remove the neurosphere growth medium and then gently resuspend the neurospheres with an appropriate volume of differentiation medium.

6. Combine an aliquot from the cell suspension with trypan blue and count live cells.

7. Prepare the appropriate cell suspension in differentiation medium so as to seed individual wells of 24-well tissue culture plate containing a PDL-coated glass coverslip with $1–2 \times 10^5$ cells (Final volume 500 μL).

8. After 7–10 days in vitro, most cells will have differentiated sufficiently to be characterized via immunocytochemistry using antibodies against neuronal and glial antigens.

9. Fix the cells by the addition of 4% paraformaldehyde (in PBS, pH 7.2) for 5 min at room temperature and then process the adherent cells for immunocytochemistry as required.

*3.2.4. Viral Infection of Whole and Dissociated Neurosphere Cultures*

1. After plating, cells can be infected with viruses carrying a bicistronic cassette encoding for the protein of interest and a reporter gene, such as green or red fluorescent proteins (GFP and RFP, respectively) for visualization. For retroviruses, transduction needs to be performed preferentially shortly after plating as the cells have to undergo mitosis to incorporate the retroviral vector before inducing differentiation. For lentiviral gene delivery, infection can be performed at any time, as both proliferating and postmitotic cells will be transduced.
   **Note:** Viral infection must be performed under particular safety conditions in accordance with legal regulations. The medium containing pro-viral particles must be replaced/ discarded in accordance with these regulations.

2. Choosing the appropriate viral delivery system: The choice between retro- or lentiviral expression of a given gene can markedly affect the outcome of the experiment. In contrast to lentiviral vectors, retroviral vectors will integrate only into genomes of dividing cells. This means that expression driven by the construct will commence with specific timing with respect to the cell cycle which is likely to affect the functional consequences of expression of fate determinants. Our experience suggests that neurogenic fate determinants encoded by retroviral vectors are more efficient in driving adult neurosphere cells toward neurogenesis than the same genes encoded by lentiviral vectors.

3. Choosing the amount of virus to be used for transduction: Depending on the experiment, it might be desirable to obtain different numbers of transduced cells. The amount of virus used to transduce cells will be then determined by two main factors: (i) how many cells the experimenter wants to infect; this will vary depending on the experimental question: for instance, if a clonal analysis should be performed, it is desirable to infect only few clones per cover slip to avoid that nearby clones become infected simultaneously, thus suggesting the wrong interpretation of being one single large clone; (ii) the titre of the original viral stock solution (*see* below). Additionally, for retroviral vectors the proportion of transduced cells will also depend on the proportion of cells undergoing cell division.

*3.2.5. Virus Production*

This method is modified from previously described method (37). The aim is to produce a relatively large batch of highly concentrated virus by harvesting the medium from twelve 10 cm plates. However, depending on the transgene and retroviral backbone (size and other unknown factors), the number of plates required for the same quantity of virus produced will vary.

1. Unthaw HEK293T cells at 37°C and plate 5 million ($5 \times 10^6$) HEK293T cells in 5 mL Opti-MEM per 10 cm Petri dish.

   **Note:** HEK293T cells should be maintained as a stock using standard cell culture methods in DMEM with 10% FBS.

2. At 24 h after plating, add 3 mL of the DNA/lipofectamine[1] solution onto each Petri dish of 293T cell plate. Transfer dropwise while rocking the Petri dish. Incubate at 37°C for 5–6 h.

3. Replace medium with fresh DMEM/ 10% FBS in order to wash out DNA/lipofectamine solution. Incubate 48 h at 37°C.

4. Collect the supernatant of medium in 50 mL Falcon tubes (2 days after transfection).

5. Filter the supernatant through 0.22 μm filter top and transfer filtered supernatant into ultracentrifuge tubes.

6. Centrifuge at $50,000g$ at 4°C using an ultracentrifuge for 2 h.

7. Remove the supernatant with a Pasteur pipette.

8. Resuspend the precipitated virus with 1 mL Opti-MEM by pipetting up and down. Pool the resuspended viruses and fill up tubes with Opti-MEM.

9. Repeat the centrifugation step at $50,000g$ at 4°C using an ultracentrifuge for 2 h.

---

[1] *Troubleshooting: Preparing the DNA/lipofectamine solution:*

(A) Mix Opti-MEM and cDNAS (pro-viral cDNAs and backbone plasmid containing your gene of interest). Incubate at room temperature for 5 min (*Solution A*).

   For each 10 cm Petri dish:

   9 μg of retroviral back bone DNA (e.g., pCAG-IRES GFP or pMXIG with transgene)

   6 μg of CMV-GP

   3 μg of CMV-VSV-G

   1.5 mL of Opti-MEM

(B) Mix Opti-MEM and lipofectamine 2000. Incubate at room temperature for 5 min (*Solution B*).

   For each 10 cm Petri dish:

   60 μL lipofectamine 2000

   1.5 mL of Opti-MEM

(C) Mix *Solution A* and *B* by constant pipetting up and down and tapping the tube. Incubate at room temperature for 30 min.

10. Remove the supernatant with Pasteur pipette.

11. Resuspend the final pellet in a desired volume of Optimem or PBS, (recommended 50–100 μL for the first time). Let the pellet incubate in medium for 2 h at 4°C.

12. Pipette up and down until pellet is dissolved. Aliquot in small volumes as viruses do not tolerate repeated refreezing. Store at –80°C until use.

    **Note:** It is possible to harvest viruses, from the same cells transfected, a second and third time, by replacing the medium with new medium (Opti-MEM) and repeat the harvesting procedure 2–3 days later (Steps 4–12).

*3.2.6. Determining the Viral Titre*

**Note:** Determining the titreof the virus should ideally be done in the cell type used for the experimental question, as the infection rate is dependent on cell cycle kinetics and the general infectivity of cells. For simplicity, however, titration may be performed in a cell line. Here we describe determining the titre using HEK293T cells.

1. Plate 30,000 HEK293T cells per well in 24-well plate on coverslips. Add 500 μL of DMEM/10%FBS and incubate at 37°C. Seed six wells for each virus preparation.

2. Perform a serial dilution of viruses (example: mix 199 μL of medium with 1 μL of viral stock). Transfer 100, 20, 10, 2, 1, and 0.2 μL of mixture to each well. These volumes correspond to 1/2, 1/10, 1/20, 1/100, 1/200, and 1/1000 of the viral stock. The denominator of these fractions is termed *Diluting Factor (DF)*.

3. After 48 h, fix the cells by the addition of 4% paraformaldehyde (in PBS, pH 7.2) for 5 min at room temperature and then process the adherent cells for immunocytochemistry as required (for example, using anti-GFP primary antibody followed by a secondary antibody conjugated to a fluorescent protein, in order to amplify the signal for the endogenous GFP produced in a cell transduced with pCAG-IRES-GFP).
    **Note:** In some cases, the endogenous levels of fluorescent reporter proteins are high enough to allow direct visualization after fixation. However, we recommend the staining to amplify the signal always, because there might be cells expressing lower, undetectable levels of fluorescent reporter proteins and, as a consequence, the titre will be underestimated.

4. Count the number of fluorescent colonies (clusters of cells) in wells where there are not too many infected cells (i.e., per coverslip with less than 100 colonies). The virus titre per millilitre is calculated by the formula:

$$\text{Number of colonies } (n) \times \text{dilution factor (DF)} \times 10^3$$

Example: If you count 50 colonies in a well where 2 μL of the viral mixture (200 μL medium + 1 μL viral stock) had been added:

$$\text{Titre} = 50(n) \times 100(\text{DF}) \times 10^3 = 5 \times 10^6 \text{ viral particles/ml}$$

**Note:** From viral supernatants obtained from 12 plates resuspended in a volume of 50–100 μL, one should obtain titres of $10^7$–$10^8$ viral particles per millilitre.

## 4. Typical Protocol Results

### 4.1. Neurospheres Derived from Developing Cortex Generate Glutamatergic Neurons

According to the literature when neurospheres derived from mouse embryonic dorsal telencephalon are differentiated, distinct proportions of neurons, astrocytes, and oligodendrocytes are generated, depending on (1) the age when cells were obtained (5) and (2) the number of passages prior to differentiation (7). We have observed that neurons derived from E14 embryonic cortical neurospheres acquire a predominantly glutamatergic identity, in contrast to neurospheres derived from embryonic ventral telencephalon (10) or adult SEZ (11). **Figure 2.2** shows examples of neurons derived from E14 cortical neurospheres. Note that virtually all MAP2-positive neurons also express Tbr1 **(Fig. 2.2A,B)**, a transcription factor involved in the specification of glutamatergic neurons (38). Electrophysiological pair recording of neurons in these cultures confirmed their glutamatergic phenotype, as revealed by the complete block of the postsynaptic responses in the presence of CNQX, a competitive AMPA/kainate receptor antagonist **(Fig. 2.3)**. We have also observed the majority of

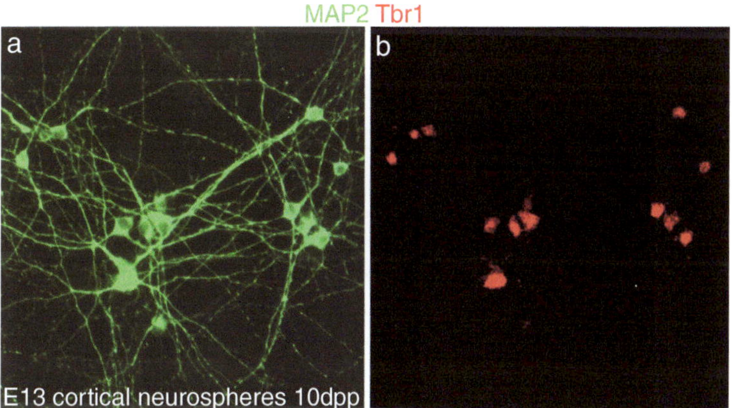


Fig. 2.2. Embryonic cortical neurospheres generate glutamatergic neurons. (**A**) The micrograph depicts a dissociated neurosphere cell culture 10 dpp, stained for the dendritic marker protein MAP2. (**B**) The micrograph depicts the same culture stained for Tbr1, a transcription factor involved in the genesis of glutamatergic neurons. Note that virtually all MAP2-positive neurons also express Tbr1.

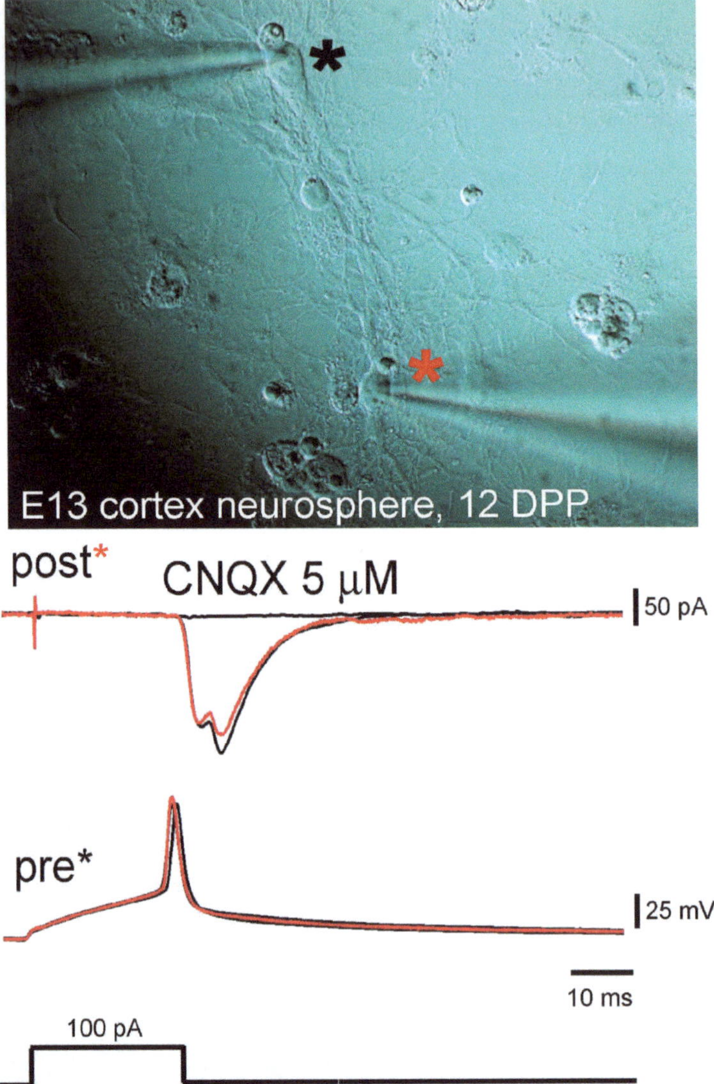



Fig. 2.3. E13 cortical neurosphere cells give rise to functional glutamatergic neurons. (**A**) Bright-field micrograph showing two neurons derived from E13 cerebral cortical neurospheres, 12 days post-plating (dpp). The neuron marked with the asterisk is presynaptic (pre* in b) to the other neuron (post in b). (**B**) In current clamp, a brief step current injection of 100 pA induces an action potential in the presynaptic neuron (pre*). The postsynaptic neuron (held in voltage clamp) exhibits a synaptic response (*black trace*), which is blocked by the AMPA/kainate receptor antagonist CNQX, indicating the glutamatergic nature of the presynaptic neuron. Following washout (*red trace*), the response returns to the control level. The response exhibits a short-latency monosynaptic and a later polysynaptic component, presumably due to recruitment of another glutamatergic neuron synapsing onto the postsynaptic neuron.

MAP2-positive neurons derived from E14 cortex neurospheres express a vesicular glutamate transporter (vGluT) up to the fifth passage (data not shown), indicating that although the overall neurogenic potential of cortical neurosphere cells becomes progressively reduced with passage number, the neurochemical properties of the neurons generated appear still to be consistent with the regional identity from which the original neurosphere-forming cell had been isolated.

## 4.2. Directing Adult SEZ Neurosphere Cells Toward Specific Transmitter Identities

### 4.2.1. Endogenous Potential of Adult SEZ Neurosphere Cells

When allowed to differentiate, adult SEZ neurosphere cells generate a certain proportion of neurons (roughly between 5 and 15%). The transmitter identity of these cells is heterogeneous: Approximately 80–90% of all neurons adopt a GABAergic identity as shown by their expression of GFP when isolated from GAD67::GFP knock-in mice, in which GFP is driven by the promoter of the GABA-synthesizing enzyme isoform glutamic acid decarboxylase (GAD) 67 (not shown). The remaining cells exhibit vesicular glutamate transporter (vGluT) immunoreactivity (11). These data are consistent with the electrophysiological analysis, which reveals that the majority of cells form GABAergic synapses and autapses (synaptic connections of a neurons onto itself), but there is a subpopulation (i.e., roughly 10–20%) of glutamatergic neurons within this culture system (11). **Figure 2.4** shows an example of such a glutamatergic neuron derived from an adult SEZ neurosphere cell. Presumably, the differences in transmitter identity are accounted for by the mosaic nature of

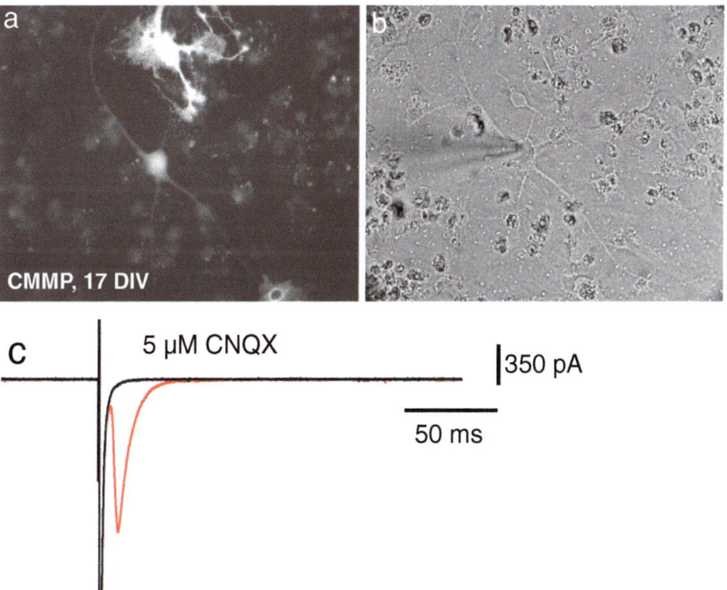


Fig. 2.4. A small percentage of control vector (CMMP encoding GFP) transduced adult SEZ neurosphere cells generate glutamatergic neurons. (**A**) Fluorescence micrograph depicting a GFP-positive neuron close to a GFP-positive astrocyte, 16 days post-infection (dpi). (**B**) Bright-field micrograph depicting the recorded cell. (**C**) Step depolarization of the neuron in voltage clamp resulted in the activation of an autaptic connection (i.e., connection of the stimulated neuron onto itself; *red trace*) that was entirely blocked by the AMPA-type glutamate receptor blocker CNQX (*black trace*).

the adult SEZ (39). There is growing evidence that there is not a single generic neural stem cell but distinct "stem cells" that differ with respect to their potential for subtype specification and hence transmitter identity. A still open question in this regard is whether neurons derived from the same neurosphere (that is of clonal origin) can acquire distinct neurotransmitter identities or whether all neurons derived from the same neurosphere-forming cell acquire the same neuronal phenotype.

*4.2.2. Forced Expression of Neurog2 Drives Adult SEZ Neurosphere Cells Toward a Glutamatergic Identity*

In order to direct adult SEZ neurospheres toward specific neuronal phenotypes, we have previously shown that expression of neurogenic fate determinants not only favors neurogenesis at the expense of gliogenesis but also permits the acquisition of a specific transmitter identity (11). Transcription factor-driven differentiation of neurosphere cells may be a valuable tool for understanding the molecular mechanism underlying subtype specification as considerable numbers of cells can be generated, thereby permitting a biochemical analysis.

Following transduction with the proneural gene *Neurog2*, >95% of the neurosphere cells adopt a neuronal identity (26, 11). Moreover, when analyzing their phenotype, these neurons exhibit the feature of typical glutamatergic telencephalic neurons **(Fig. 2.5)**. This is consistent with the fact that within the developing telencephalon Neurog2 is expressed in the dorsal domain, which gives rise to glutamatergic pyramidal neurons (40). Morphologically, Neurog2-transduced neurosphere cells resemble cultured pyramidal neurons **(Fig. 2.5)** and over time

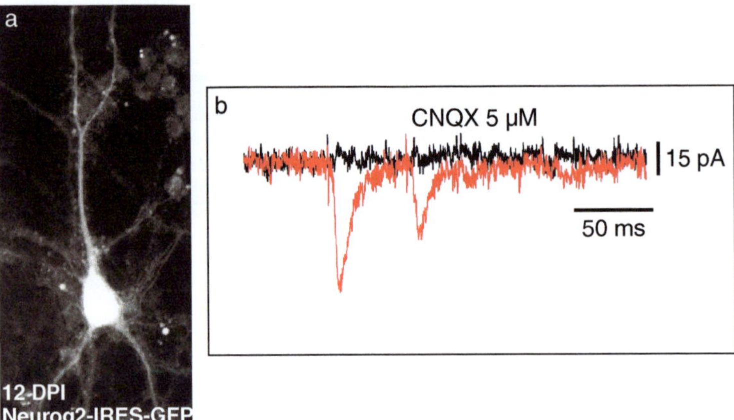


Fig. 2.5. Neurog2-transduced adult SEZ neurosphere cells give rise to glutamatergic neurons. (**A**) Fluorescence micrograph depicting a Neurog2-expressing neuron after 12 days post-infection (dpi). The neuron exhibited a pyramidal neuron-like morphology. (**B**) Dual recording from the cell depicted in (**A**) and a neuron nearby revealed that activation of the Neurog2-expressing neurosphere cell generated a postsynaptic response (*lighter trace*) in the other neuron which was blocked by CNQX (*Black trace*), indicating its glutamatergic nature.

in culture acquire dendritic spines. Massive expression of vGluT immunoreactivity (11) indicates that virtually all transduced cell have acquired a glutamatergic identity, suggesting a quantitative effect of Neurog2 expression. Conversely, there is no expression of GABAergic markers (such as GABA and GAD) in neurons derived from Neurog2-expressing neurosphere cells, indicating that these cells do not acquire a hybrid identity. Consistent with their neurochemical features, Neurog2-transduced neurosphere cells also form glutamatergic synapses and autapses (**Fig. 2.5**). Electrophysiological signs for the presence of synapses can be obtained around day 9 post-infection, indicating that the maturation of these cells does not occur at a slower pace compared to murine cortical neurons in culture. Particularly interesting is the fact that following Neurog2 expression there is also upregulation of the downstream transcription factor Tbr1 (**Fig. 2.6**), which is known to form part of the transcriptional cascade involved in specification of glutamatergic neurons (38). However, in contrast to E13 neurosphere cultures (example above), Tbr1 expression is detected in only one third of the transduced cells (**Fig. 2.6**). One explanation may be that not all Neurog2-transduced neurosphere cells acquire the same subtype specification. However, given the wide spread expression of vGluT and our evidence from electrophysiological recordings that virtually all Neurog2-expressing neurosphere cells become glutamatergic neurons, the limited expression of Tbr1 may indicate that this transcription factor is expressed only transiently. Alternatively, expression may be restricted to a subpopulation of Neurog2-transduced cells with additional subtype characteristics. For instance, while Tbr1 is believed to be expressed in virtually all glutamatergic neurons in the developing cerebral cortex, its expression is maintained only in a subpopulation of deep-layer neurons (38). These findings point to the interesting possibility that expression of neurogenic fate determinants may permit to direct not only the neu-

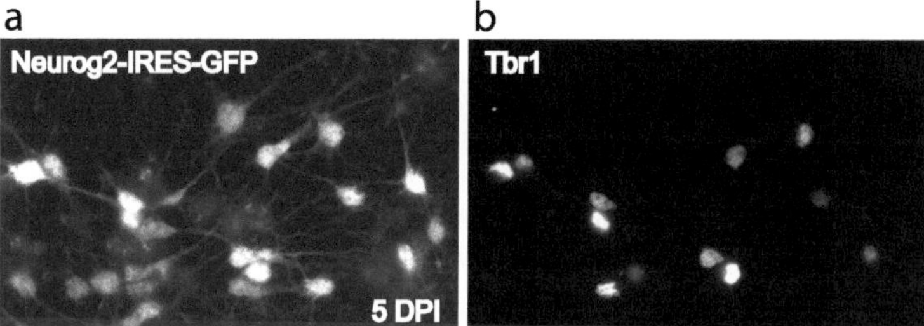


Fig. 2.6. Induction of the T-box transcription factor Tbr1 following forced expression of Neurog2 in adult SEZ neurosphere cells. (**A**) Fluorescence micrograph depicting a culture transduced with a retrovirus encoding Neurog2-IRES-GFP 5 days post-infection (dpi). The transduced cells all exhibited a neuronal morphology. (**B**) Many of the transduced cells also express Tbr1.

rotransmitter identity of adult SEZ cells but also their precise subtype. Using this approach, it may be in principle possible to selectively generate different classes of cortical neurons (i.e., callosal, corticothalamic, or subcortical projection neurons) in vitro. Future studies will have to show whether by expression of additional fate determinants in conjunction with Neurog2, such as the zinc finger protein Fezl (41), a more precise subtype specification can be achieved.

### 4.2.3. Forced Expression of Mash1 Promotes a GABAergic Identity

Given the fact that the close relative of Neurog2, Mash1, is expressed in the developing ventral telencephalon where mostly GABAergic neurons are generated (40), we hypothesized that forced expression of Mash1 may limit the differentiation potential of adult SEZ neurons to a GABAergic fate. Indeed, we found that virtually all neurons expressing Mash1 adopt a GABAergic phenotype (11) (**Fig. 2.7**). However, neurons derived from Mash1-expressing neurosphere cells mature more slowly compared to Neurog2-expressing cells (for instance, with regard to the pace of synapse formation). Thus, while principally confirming the approach, other or additional factors may be needed to differentiate adult SEZ into selective types of GABAergic neurons. Recently, we have shown that adult SEZ neurosphere cells can be

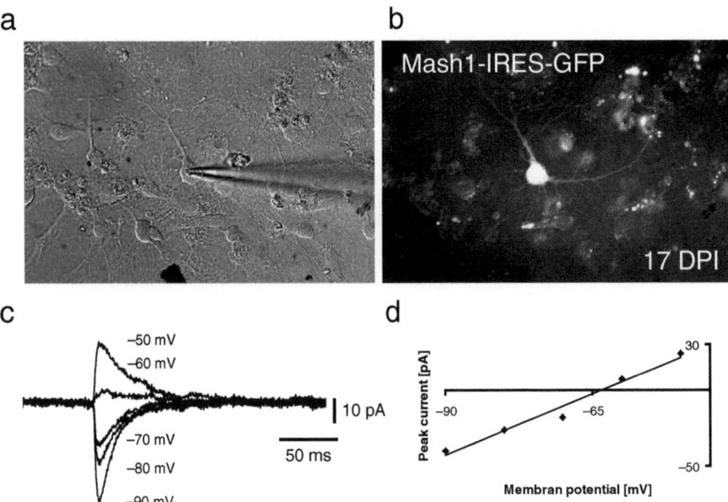


Fig. 2.7. Mash1 expressing adult SEZ neurosphere cells give rise to GABAergic neurons. (**A**) Bright-field micrograph depicting a culture 17 days post-infection (dpi). (**B**) Fluorescence micrograph depicting a Mash1-expressing neuron derived from an adult SEZ neurosphere. (**C**) Step depolarization of the Mash1-expressing neuron resulted in a postsynaptic response in a nearby neuron held at different holding membrane potential to reveal the reversal potential of this synaptic connection. The response reversed already at membrane potentials more positive than −65 mV, indicating that it is of inhibitory (i.e., GABAergic) nature. (**D**) The peak current of the synaptic response is blotted versus the holding membrane potential of the postsynaptic neuron.

driven toward neurogenesis by forced expression of Dlx2, another transcription factor in the genesis of GABAergic neurons (15). It remains to be shown whether Dlx2 or other transcription factors promote the specific generation of GABAergic neurons.

## 5. Future Directions

The selective differentiation of adult neurosphere cells as exemplified here by the forced expression of Neurog2 and Mash1 may suggest that it should be in principle possible to selectively direct neurosphere cells toward distinct neuronal sublineages. As discussed above, by combining expression of factors inducing a generic glutamatergic fate (such as Neurog2) in conjunction with transcription factors responsible for the specification of additional features such as precise location within the circuitry (e.g., layer specificity in case of cortical neurons) and projection patterns (e.g., callosal versus subcortical), it may become feasible to tailor adult SEZ-derived neurosphere cells into specific types of neurons that degenerate in different neurodegenerative diseases, as for instance corticospinal motor neurons that are lost in amyotrophic lateral sclerosis. A challenge for the future will be to test whether neurosphere cells from the adult SEZ can also be differentiated toward neuron subtypes occurring in non-telencephalic brain regions, such as midbrain dopaminergic neurons that degenerate in Parkinson's disease. This may bear therapeutic implications, but also provide insights into the fate limitations of these adult "stem cells." If it turns out that the site of origin from where the neurosphere-forming cells have been isolated poses limits on the plasticity of their progeny, it may become even more important to assess the differentiation potential of neurosphere cells derived from other regions such as the spinal cord (17).

## References

1. Reynolds BA, Weiss S. Generation of neurons and astrocytes from isolated cells of the adult mammalian central nervous system. Science 1992; 255: 1707–10.
2. Reynolds BA, Weiss S. Clonal and population analyses demonstrate that an EGF-responsive mammalian embryonic CNS precursor is a stem cell. Dev Biol 1996; 175: 1–13.
3. Gage FH. Mammalian neural stem cells. Science 2000; 287: 1433–8.
4. Berninger B, Hack MA, Gotz M. Neural stem cells: on where they hide, in which disguise, and how we may lure them out. Handb Exp Pharmacol 2006; 174: 319–60.
5. Shen Q, Wang Y, Dimos JT, Fasano CA, Phoenix TN, Lemischka IR, Ivanova NB, Stifani S, Morrisey EE, Temple S. The timing of cortical neurogenesis is encoded within lineages of individual progenitor cells. Nat Neurosci 2006; 9: 743–51.
6. Miller FD, Gauthier AS. Timing is everything: making neurons versus glia in the developing cortex. Neuron 2007; 54: 357–69.

7.  Naka H, Nakamura S, Shimazaki T, Okano H. Requirement for COUP-TFI and II in the temporal specification of neural stem cells in CNS development. Nat Neurosci 2008; 11(9): 1014–23.

8.  Parmar M, Skogh C, Bjorklund A, Campbell K. Regional specification of neurosphere cultures derived from subregions of the embryonic telencephalon. Mol Cell Neurosci 2002; 21: 645–56.

9.  Hitoshi S, Tropepe V, Ekker M, van der Kooy D. Neural stem cell lineages are regionally specified, but not committed, within distinct compartments of the developing brain. Development 2002; 129: 233–44.

10.  Eriksson C, Bjorklund A, Wictorin K. Neuronal differentiation following transplantation of expanded mouse neurosphere cultures derived from different embryonic forebrain regions. Exp Neurol 2003; 184: 615–35.

11.  Berninger B, Guillemot F, Gotz M. Directing neurotransmitter identity of neurones derived from expanded adult neural stem cells. Eur J Neurosci 2007; 25: 2581–90.

12.  Gabay L, Lowell S, Rubin LL, Anderson DJ. Deregulation of dorsoventral patterning by FGF confers trilineage differentiation capacity on CNS stem cells in vitro. Neuron 2003; 40: 485–99.

13.  Hack MA, Sugimori M, Lundberg C, Nakafuku M, Gotz M. Regionalization and fate specification in neurospheres: the role of Olig2 and Pax6. Mol Cell Neurosci 2004; 25: 664–78.

14.  Doetsch F, Petreanu L, Caille I, Garcia-Verdugo JM, Alvarez-Buylla A. EGF converts transit-amplifying neurogenic precursors in the adult brain into multipotent stem cells. Neuron 2002; 36: 1021–34.

15.  Brill MS, Snapyan M, Wohlfrom H, Ninkovic J, Jawerka M, Mastick GS, Ashery-Padan R, Saghatelyan A, Berninger B, Gotz M. A dlx2- and pax6-dependent transcriptional code for periglomerular neuron specification in the adult olfactory bulb. J Neurosci 2008; 28: 6439–52.

16.  Colak D, Mori T, Brill MS, Pfeifer A, Falk S, Deng C, Monteiro R, Mummery C, Sommer L, Gotz M. Adult neurogenesis requires Smad4-mediated bone morphogenic protein signaling in stem cells. J Neurosci 2008; 28: 434–46.

17.  Johansson CB, Momma S, Clarke DL, Risling M, Lendahl U, Frisen J. Identification of a neural stem cell in the adult mammalian central nervous system. Cell 1999; 96: 25–34.

18.  Babu H, Cheung G, Kettenmann H, Palmer TD, Kempermann G. Enriched monolayer precursor cell cultures from micro-dissected adult mouse dentate gyrus yield functional granule cell-like neurons. PLoS ONE 2007; 2: e388.

19.  Seaberg RM, van der Kooy D. Adult rodent neurogenic regions: the ventricular subependyma contains neural stem cells, but the dentate gyrus contains restricted progenitors. J Neurosci 2002; 22: 1784–93.

20.  Buffo A, Rite I, Tripathi P, Lepier A, Colak D, Horn AP, Mori T, Gotz M. Origin and progeny of reactive gliosis: A source of multipotent cells in the injured brain. Proc Natl Acad Sci USA 2008; 105: 3581–6.

21.  Pardal R, Ortega-Saenz P, Duran R, Lopez-Barneo J. Glia-like stem cells sustain physiologic neurogenesis in the adult mammalian carotid body. Cell 2007; 131: 364–77.

22.  Hoshino M. Molecular machinery governing GABAergic neuron specification in the cerebellum. Cerebellum 2006; 5: 193–8.

23.  Ang SL. Transcriptional control of midbrain dopaminergic neuron development. Development 2006; 133: 3499–506.

24.  Molyneaux BJ, Arlotta P, Menezes JR, Macklis JD. Neuronal subtype specification in the cerebral cortex. Nat Rev Neurosci 2007; 8: 427–37.

25.  Jessberger S, Toni N, Clemenson GD, Jr., Ray J, Gage FH. Directed differentiation of hippocampal stem/progenitor cells in the adult brain. Nat Neurosci 2008; 11: 888–93.

26.  Falk A, Holmstrom N, Carlen M, Cassidy R, Lundberg C, Frisen J. Gene delivery to adult neural stem cells. Exp Cell Res 2002; 279: 34–9.

27.  van Praag H, Schinder AF, Christie BR, Toni N, Palmer TD, Gage FH. Functional neurogenesis in the adult hippocampus. Nature 2002; 415: 1030–34.

28.  Hatakeyama J, Kageyama R. Retrovirus-mediated gene transfer to retinal explants. Methods 2002; 28: 387–95.

29.  Berninger B, Costa MR, Koch U, Schroeder T, Sutor B, Grothe B, Gotz M. Functional properties of neurons derived from in vitro reprogrammed postnatal astroglia. J Neurosci 2007; 27: 8654–64.

30.  Costa MR, Wen G, Lepier A, Schroeder T, Gotz M. Par-complex proteins promote proliferative progenitor divisions in the developing mouse cerebral cortex. Development 2008; 135: 11–22.

31.  Pelletier J, Sonenberg N. Internal initiation of translation of eukaryotic mRNA directed

by a sequence derived from poliovirus RNA. Nature 1988; 334: 320–5.

32. Jang SK, Krausslich HG, Nicklin MJ, Duke GM, Palmenberg AC, Wimmer E. A segment of the 5′ nontranslated region of encephalomyocarditis virus RNA directs internal entry of ribosomes during in vitro translation. J Virol 1988; 62: 2636–43.

33. Niwa H, Yamamura K, Miyazaki J. Efficient selection for high-expression transfectants with a novel eukaryotic vector. Gene 1991; 108: 193–9.

34. Xu ZL, Mizuguchi H, Ishii-Watabe A, Uchida E, Mayumi T, Hayakawa T. Optimization of transcriptional regulatory elements for constructing plasmid vectors. Gene 2001; 272: 149–56.

35. Zufferey R, Donello JE, Trono D, Hope TJ. Woodchuck hepatitis virus posttranscriptional regulatory element enhances expression of transgenes delivered by retroviral vectors. J Virol 1999; 73: 2886–92.

36. Ory DS, Neugeboren BA, Mulligan RC. A stable human-derived packaging cell line for production of high titer retrovirus/vesicular stomatitis virus G pseudotypes. Proc Natl Acad Sci USA 1996; 93: 11400–6.

37. Tashiro A, Zhao C, Gage FH. Retrovirus-mediated single-cell gene knockout technique in adult newborn neurons in vivo. Nat Protoc 2006; 1: 3049–55.

38. Hevner RF. From radial glia to pyramidal-projection neuron: transcription factor cascades in cerebral cortex development. Mol Neurobiol 2006; 33: 33–50.

39. Lledo PM, Merkle FT, Alvarez-Buylla A. Origin and function of olfactory bulb interneuron diversity. Trends Neurosci 2008; 31: 392–400.

40. Bertrand N, Castro DS, Guillemot F. Proneural genes and the specification of neural cell types. Nat Rev Neurosci 2002; 3: 517–30.

41. Molyneaux BJ, Arlotta P, Hirata T, Hibi M, Macklis JD. Fezl is required for the birth and specification of corticospinal motor neurons. Neuron 2005; 47: 817–31.

# Chapter 3

# Culture and Differentiation of Human Neural Stem Cells

## Soojung Shin and Mohan Vemuri

## Abstract

We describe the steps in detail to expand human neural stem cells, to bank and cryopreserve the stem cells. The methods described in this protocol represent the latest improvement in tissue culture and molecular biology reagents to help identify specific cell lineages. Specific tissue culture media and methods are outlined to characterize neural stem cells and to differentiate the cells to neurons/oligodendrocytes/astrocytes. The protocols could be modified to obtain specific types of neurons such as spinal cord motoneurons and dopaminergic neurons by incorporating key morphogens during lineage specification.

**Key words:** Human neural stem cell, cell lineage, differentiation, cryopreservation, RT-PCR.

## 1. Introduction

Neural stem cells (NSC) are self-renewing multipotent stem cells in the nervous system that can differentiate into neurons, oligodendrocytes, and astrocytes. Multipotent neural stem cells reside in the region of the subventricular zone (SVZ) or the hippocampus in the fetal or adult brain. NSC can be isolated from these tissues (1–3), but often there is limitation in their differentiation potential with a propensity toward their site of origin (4–6). On the other hand, NSC can be derived from embryonic stem cells (7, 8) with enhanced specification (9, 10). Because of the property to generate neurons and glial cells, both NSC are a valuable source not only for neuroscience but also for potential clinical use to treat neurodegenerative disease or neurological disorders.

Markers that define neural stem cells are still in development. The identification of neural stem cells relies on the expression of a combination of positive and negative phenotypic markers (11, 12). As a self-renewing population, NSC express the proliferation marker Ki67 or labeled by EDU (BrdU alternative).

L.C. Doering (ed.), *Protocols for Neural Cell Culture*, Springer Protocols Handbooks,
DOI 10.1007/978-1-60761-292-6_3, © Humana Press, a part of Springer Science+Business Media, LLC 1992, 1997, 2001, 2010

CD133 is a cell surface protein that can be used to sort precursor cells (13). Nestin is a class VI intermediate filament protein expressed predominantly in neural stem cells. Sox1 and Sox2 are HMG box transcription factors expressed in neural stem cells. In addition, neural stem cells do not express markers for differentiated lineages. Neurofilament, NCAM, MAP2, $\beta$ III tubulin, NeuN, HuC/D, and Dcx are expressed in neuronal progenitors or fully differentiated neurons (14, 15). GalC, NG2, O4, myelin basic protein, CNPase, and RIP are expressed in oligodendrocyte progenitors or oligodendrocytes (16). CD44, GFAP, and S-100 protein are expressed in astrocyte progenitors or fully differentiated astrocytes (17). However, upon differentiation, NSC can give rise to downstream populations of neurons, oligodendrocytes, and astrocytes.

Neural stem cells and their progenies can be stimulated to proliferate upon exposure to the right conditions in vitro (8, 18–20). Many factors contribute to the proliferation of NSC, including the source of cells, growth factors, antioxidants, nutrients, and lipids. STEMPRO NSC SFM is a specifically developed defined medium to support NSC. Growth factors of bFGF and EGF were included as pivotal mitogens as well as antioxidants and optimized macromolecules.

In this chapter, we describe the methods to culture human neural stem cells and further characterize their differentiation.

## 2. Reagents and Equipment

Make sure that all the reagents and materials are sterile. Complete media are stable for a month if stored in the refrigerator. If the consumption rate is lower than 500 mL per month, make growth factor stock solutions according to the following protocol.

**Note:** All products are Invitrogen listings unless specified otherwise.

### 2.1. Reagents

1. STEMPRO NSC SFM (A10509-01, 500 mL) is a media kit developed specifically to support the growth of neural stem cells. The kit is composed of base medium (12660, 500 mL), bFGF (PHG0024, 10 μg), EGF (PHG0314, 10 μg), and supplement (A10508-01, 10 mL). Thaw the supplement at 37°C and add 10 mL to 485 mL of the base medium. Add 5 mL GlutaMAX 1 to 495 mL solution. Using this medium, wash the FGF and EGF bottle to transfer the entire volume (500 mL medium). Add 910 μL of 2ME to make the medium complete consisting of 1× supplement, 20 ng/mL bFGF, 20 ng/mL EGF, 2 mM Glutamax 1 in base medium (Refer to **Table 3.1**).

**Table 3.1**
**Formulation of STEMPRO NSC SFM**

| Components | Cat. no. | Final con. | 100 mL | 500 mL |
| --- | --- | --- | --- | --- |
| Knockout DMEM/F12 | A10509-01 | 1× | 97 mL | 485 mL |
| Glutamax | 35050 | 2 mM | 1 mL | 5 mL |
| bFGF | A10509-01 | 20 ng/mL | 0.1 mL | 10 μg |
| β-Mercaptoethanol | 21985 | 0.1 mM | 0.182 mL | 0.91 mL |
| EGF | A10509-01 | 20 ng/mL | 0.1 mL | 10 μg |
| Stempro Neural supplement | A10509-01 | 2% | 2 mL | 10 mL |

*Note*:

(1) White precipitation on the bottom of STEMPRO Neural supplement (50×) can be observed at the time of thawing. It will dissolve as it warms up.

(2) Once medium is supplemented with growth factors, allocate amount to be used and keep the remaining medium at 4°C. (Alternatively, growth factors can be added at time of use.)

(3) Thawed supplement and complete medium are stable for 4 weeks if properly stored at 4°C.

2. GlutaMAX™-I Supplement (35050, 200 mM, 100 mL).

3. Beta mercaptoethanol (21985, 100 mL).

4. FGF2 stock solution (1000×, 20 μg/mL). If medium consumption rate is less than 500 mL per month, kit components of the growth factors should be stored as 1000× stock solution. Human recombinant fibroblast growth factor 2 is part of the kit component of A10509-01 or PHG0024. Resuspend lyophilized FGF2 in its vial with 500 μL of 0.1% BSA or HSA containing PBS (14040). Mix well and make an aliquot (50–100 μL) to store at −20°C. Thawed stock can be used up to 1 week.

5. EGF stock solution (1000×, 20 μg/mL). If medium consumption rate is less than 500 mL per month, kit components of the growth factors should be stored as 1000× stock solution. Human recombinant fibroblast growth factor 2 is part of the kit component of A10509-01 or PHG0314. Resuspend lyophilized EGF in its vial with 500 μL of 0.1% BSA or HSA containing PBS (14040). Mix well and make an aliquot (50–100 μL) to store at −20°C. Thawed EGF stock can be used up to 2 weeks.

6. dBcAMP stock solution (Sigma D0627-250 mg, 1000×, 100 mM). Make 100 mM stock solution by adding 5 mL

PBS (14190) to 250 mg dBcAMP. Filter solution using a 0.22 μm syringe filter and store the stock solution as single-working aliquots at −20°C.

7. 3,3,5-Triiodo thyronine (T3) (Sigma T6397). Make 30 μg/mL stock solution in PBS (14190) and filter solution using a 0.22 μm syringe filter. Store the stock solution in single-working aliquots at −20°C.

8. Polyornithin (Sigma P3655 50 mg). Make 10 mg/mL stock in distilled water (15230). Store the stock solution in aliquots at −20°C.

9. Laminin (IVGN 23017 1 mg). Thaw (1 mg/mL) and store the stock solution in aliquots at −20°C.

10. Geltrex (12760-021)

11. PBS (14040) with calcium and magnesium

12. PBS (14190) without calcium and magnesium

13. CELLStart (A10142)

14. DMSO (Sigma D2650)

15. Paraformaldehyde (Sigma P6148). Measure and make a 4% solution under the fume hood. Add 4 g PFA in 90 mL MQ water. Add three drops of 10 N NaOH and heat the solution up to 70°C with constant stirring. Once the solution is transparent, slowly add 10 mL of 10× PBS (14080). If the solution becomes cloudy, check the pH. If the pH is higher than 7.4, make the solution again from the beginning. Chill the solution and adjust pH up to 7.0–7.4 using 10 N NaOH. Filter aliquot and store at −80°C. Frozen solution is stable for more than 6 months. *Note*: Do not overheat the solution and keep it under 70°C.

16. Accutase (A1110501). Make aliquots (5–10 mL) for storage at −20°C.

17. TrypLE™ Express (12604-013)

18. Synth-a-freeze (R005-50, 50 mL). A defined, protein-free, sterile cryopreservation medium containing 10% DMSO. Store at −20°C.

19. Blocking solution (serum containing PBS). Add 5% of serum (decide serum kind according to second antibody host) to PBS (14040). Add 0.1% triton X to permeabilize tissue for intercellular antigen staining.

20. DAPI (D3571, 10 mg). Resuspend in vial with 1 mL of DMSO to make 10 mg/mL stock solution. Stock can be stored at −20°C and is stable for 1 year. Thaw stock and dilute 1:10,000 in PBS. Diluted working solution can be stored at 4°C and used for a month.

21. Dynabeads mRNA DIRECT Micro Kit (610.21)

22. Dynal MPC-S (159-14D)

23. TRIzol® reagent and PureLink™ Micro-to-Midi™ total RNA purification (cat. no. 12183)

24. Phase Lock Gel 2 ml Heavy (Eppendorf 0032 005.152)

25. Chloroform (Sigma C2432)

26. 96–100% ethanol and 70% ethanol (in RNase-free water)

27. SuperScript III First-Strand Synthesis SuperMix (11752-050)

28. Platinum Blue PCR SuperMix (12580-015)

**2.2. Equipment**

1. Biosafety cabinet

2. 5% $CO_2$ 37°C incubator

3. Water bath (shaking) set at 37°C

4. 60 mm style polystyrene tissue culture dishes

5. 24 well (multi-well) tissue culture plates

6. 1.5 mL RNase-free microcentrifuge tubes and RNase-free pipette tips

7. 5 mL polypropylene round-bottom tubes (Falcon cat. no. 352063)

8. 15 mL polystyrene conical tubes

9. 50 mL polypropylene conical tubes

10. CultureWell™ chambered coverglass, 16 wells (C37000)

11. Inverted fluorescence microscope

12. Table top centrifuge

13. Temperature-controlled microcentrifuge capable of 12,000$g$

14. E-Gel 2% Starter kit (G6000-02)

15. Tissue culture dishes

16. Pipettes and tips

17. Thermocycler

18. Gel imaging system

# 3. Protocol

**3.1. CELLstart Coating of Culture Flask**

1. Dilute CELLstart (A10142) 1:100 in DPBS (i.e., 50 μL of CELLstart into 5 mL of DPBS). Pipette gently to mix. DO NOT VORTEX. Coat T25 culture flasks by adding 5 mL of the CELLstart solution to each flask.

2. Place culture flasks with CELLstart in the incubator at 36–38°C in a humidified atmosphere of 4–6% $CO_2$ in air for 60 min.

3. After incubation, remove flasks from the incubator and store at 2–8°C. Do not remove CELLstart solution until just prior to use.

    *Note*:

    (1) For optimal cell attachment, CELLstart must be diluted in DPBS with calcium and magnesium (14040).

    (2) Ensure that plates do not dry out.

    (3) There is no washing step needed and use the dish directly after the CELLstart is aspirated.

    (4) Coated flasks are stable for 2 weeks if properly stored at 4°C in refrigerator.

***3.2. Recovery of Cryopreserved Human NSC***

1. Rapidly thaw frozen vial of cells in a 37°C water bath.

2. Pipette the entire contents of the cryovial into a 50 mL conical tube.

3. Carefully add 4 mL of pre-warmed (37°C) STEMPRO NSC SFM complete medium to conical tube at an approximate rate of 1 drop per second while swirling the tube. Then add 5 mL of pre-warmed medium.

4. To remove cryoprotectant, spin down the tube at $200g$ for 4 min and remove the supernatant.

5. Resuspend cells gently in pre-warmed STEMPRO NSC SFM and transfer the entire contents of the tube into a CELLstart-coated tissue culture flask.

6. Incubate at 36–38°C in a humidified atmosphere of 4–6% $CO_2$ in air.

7. Change the STEMPRO NSC SFM after 24 h.

    *Note*:

    (1) For recovery of cells grown in STEMPRO NSC SFM, it is recommended to seed cells at $\geq 1 \times 10^5$ cells/$cm^2$ for the initial recovery passage.

    (2) In general, the viability of thawed cells will be around 80% if you follow the freezing protocol described in this protocol.

***3.3. Adherent Subculture of NSC***

1. Observe stock culture flask (cells growing in current medium formulation or in STEMPRO NSC SFM) under the microscope and confirm that the cells are ready to be passed (~90% confluent).

2. Pre-warm cell dissociation reagent (TrypLE™ or Accutase™) and complete STEMPRO NSC SFM to 37°C before use.

3. Remove medium from the flask with a pipette.

4. Wash cell surface with 3 mL DPBS, remove, and discard.

5. Add 1.0 mL of the cell dissociation reagent to the T25 flask and tilt flask in all directions to evenly distribute. Incubate for 2–5 min at room temperature.

6. Once cells lift (detached cells will move with tilting of the flask), gently pipette up and down to break clumps into a single cell suspension. Stop cell dissociation reaction by diluting cell dissociation reagent with 9 mL of complete medium. Transfer cell suspension to a conical tube.

7. Centrifuge tube at $200g$ for 4 min.

8. Resuspend cells in a minimal volume of warmed complete STEMPRO NSC SFM. Take a sample from the cell suspension for cell-counting using preferred counting method (i.e., hemocytometer, automated cell counter).

9. Following the CELLstart coating procedure, remove CELLstart coating solution from each coated flask and discard. Add optimum volume of STEMPRO NSC SFM to each flask (Refer to **Table 3.2**).

10. Add enough cell suspension to each CELLstart-coated flask and mix or swirl cell suspension to ensure even distribution.

11. Place culture flask in the incubator at 36–38°C in a humidified atmosphere of 4–6% $CO_2$ in air.

12. For optimal performance and cell growth, cultures should be re-fed every 2 days with fresh complete STEMPRO NSC SFM.

    *Note*:

    (1) Cells will be lifted off from the culture dish right after application of dissociation reagent (~30 s).

    (2) Do not wait more than 3 min before the stop of dissociation reagent action by adding medium.

    (3) Cells are ready to be passaged when the culture reaches to ~90% confluent.

    (4) $0.1–1 \times 10^5/cm^2$ of cell Con. or 1:4 split ratio can be used for passage.

***3.4. Suspension Subculture of NSC***

1. Transfer medium with neurospheres into 15–50 mL conical tube.

2. Leave the tube at room temperature to allow neurospheres to settle. (Alternatively, you can centrifuge spheres at 500 rpm = $200g$ for 2 min.)

3. Aspirate the supernatant carefully to leave the neurospheres in a minimum volume of medium.

## Table 3.2
## Tissue culture vessel information

| Culture vessel | Vendor | SKU | Surface area (cm$^2$) | Plating density ($\sim 5 \times 10^4$ cells/cm$^2$) | Working volume (mL) |
|---|---|---|---|---|---|
| *ICC chamber* | | | | | |
| 16-well culture well coverglass | Molecular Probes | C37000 | 0.32 | $1.5 \times 10^4$ | 0.1–0.2 |
| 96-well imaging plate | BD Falcon | 353219 | 0.31 | $1.5 \times 10^4$ | 0.1–0.2 |
| 8-well culture slide | Nalge Nunc | 177445 | 0.81 | $4 \times 10^4$ | 0.2–0.4 |
| 4-well culture slide | Nalge Nunc | 177437 | 1.8 | $9 \times 10^4$ | 0.5–0.9 |
| *Tissue culture plate* | | | | | |
| 96-well TC plate | BD Falcon | 353072 | 0.32 | $1.5 \times 10^4$ | 0.1–0.2 |
| 48-well TC plate | BD Falcon | 353078 | 0.75 | $3.75 \times 10^4$ | 0.2–0.4 |
| 24-well plate | BD Falcon | 353047 | 2 | $1.0 \times 10^5$ | 0.5 |
| 12-well plate | BD Falcon | 353043 | 3.8 | $1.9 \times 10^5$ | 1 |
| 6-well plate | BD Falcon | 353046 | 9.6 | $5 \times 10^5$ | 2 |
| *Tissue culture dish* | | | | | |
| 35 × 10 mm dish | BD Falcon | 353001 | 11.78 | $5 \times 10^5$ | 2.0–3.0 |
| 60 × 15 mm dish | BD Falcon | 353004 | 19.5 | $1 \times 10^6$ | 5.0–7.0 |
| 100 × 20 mm dish | BD Falcon | 353003 | 58.95 | $3 \times 10^6$ | 10.0–15.0 |
| *Tissue culture flask* | | | | | |
| T25 cm$^2$ flask | BD Falcon | 353109 | 25 | $1.25 \times 10^6$ | 5.0–7.5 |
| T75 cm$^2$ flask | BD Falcon | 353136 | 75 | $3.75 \times 10^6$ | 15–20 |
| T175 cm$^2$ flask | BD Falcon | 353112 | 175 | $8.75 \times 10^6$ | 25–35 |

4. Wash neurospheres with 10 mL PBS (cat. No. 14190) and leave a minimal volume of PBS.

5. Add 1 mL of TrypLE$^{\mathrm{TM}}$ to the spheres and gently triturate neurospheres using a Pasteur pipette to generate a single cell suspension.

6. Stop the treatment by adding 4 mL of medium.

7. Spin down cells (1200 rpm, 4 min).

8. Aspirate supernatant and resuspend cells in formulated STEMPRO NSC SFM.

9. Count cell number with a hematocytometer.

10. Seed cells in fresh medium in suspension to a tissue culture dish (noncoated flask can be used).

*Note*:

(1) Neurospheres are ready to be passaged when the diameter exceeds 0.5–1 mm.

(2) 200,000 cells/mL of cell density can be used.

**3.5. Cryopreservation of NSC**

1. Prepare NSC cryopreservation solution by thawing Synth-a-freeze at 4°C.

2. Chill the isopropanol chamber to 4°C.

3. Determine the viable cell density and calculate the required volume of cryopreservation medium to give a desired final cell density (i.e., $2.0 \times 10^6$ cells/mL; 1 mL/vial).

4. Centrifuge tubes at $200g$ for 4 min and aspirate supernatant.

5. Resuspend the pellet in chilled Synth-a-freeze solution and dispense aliquots of this suspension into cryovials.

6. Put isopropanol chamber in −80°C freezer for a controlled freezing rate of 1°C decrease per minute.

7. Next day, transfer frozen cells to liquid nitrogen.
*Note*:

(1) Avoid repeated freezing and thawing of Synth-a-freeze solution.

**3.6. Differentiation of NSC to Neurons, Oligodendrocytes, and Astrocytes**

*3.6.1. ECM Preparation (Poly-l-Ornithine/Laminin)*

1. Dilute polyornithine stock in distilled water (1:500 dilution for final con. of 20 μg/mL).

2. Add 2 mL of 20 μg/mL polyornithine solution to a 35 mm dish (0.5 ml for 4 well dish or slide, 0.25 mL for 8 well slide).

3. Leave it at least 1 h in the incubator.

4. Rinse once with distilled water.

5. Dilute Laminin stock in distilled water (1:1000 dilution for final con. of 1 μg/mL).

6. Add 2 mL of 1 μg/mL Laminin solution to a 35 mm dish.

7. Leave it at least 1 h in the incubator and store at 4°C for further use.

*3.6.2. ECM Preparation (Geltrex 12760)*

1. Thaw bottle of Geltrex at 4°C overnight to prevent polymerization.

2. Next day, add DMEM/F12 medium to make 10 mg/mL stock solution (50×). Use ice bucket to keep the bottle at 4°C.

3. Store the stock solution in single-working aliquots at −20°C.

4. At time of dish coating, thaw one tube slowly at 4°C and dilute 50 fold in DMEM/F12 and use this working solution

to cover whole surface of the culture dish (1.5 mL for 35 mm dish, 3 mL for 60 mm dish).

5. Seal each dish with parafilm to prevent dish from drying out and store at 4°C.

6. At time of use, aspirate Geltrex solution and use directly without a washing step.

7. Avoid any drying of the surface (aspirate solution just before use).

*Note*:

(1) ECM plays an important role in differentiating NSC. We have found that laminin for neurons, CellStart for oligodendrocytes, and Geltrex for astrocytes can be selected as the preferable extracellular matrix for respective lineages.

(2) Dish can be used up to 1 month if stored at 4°C.

(3) Prevent dish from drying out before use (use parafilm or closed box).

*3.6.3. Differentiation to Neurons*

1. Plate neural stem cells on polyornithine- and laminin-coated plates in STEMPRO NSC SFM at $5 \times 10^4$ cell/cm$^2$.

2. After 2 days, change medium to neural differentiation medium (*see* **Table 3.3**).

3. Change medium every 3 days.

*3.6.4. Differentiation to Oligodendrocytes*

1. Plate neural stem cells on CellStart-coated plates in fresh medium at $2.5 \times 10^4$ cell/cm$^2$.

2. After 2 days, change medium to oligodendrocyte differentiation medium (*see* **Table 3.3**).

3. Change medium every 3 days.

*3.6.5. Differentiation to Astrocytes*

1. Plate neural stem cells on Geltrex-coated plates in fresh medium at $2.5 \times 10^4$ cell/cm$^2$.

2. After 2 days, change medium to astrocyte differentiation medium (*see* **Table 3.3**).

3. Change medium every 3 days.

***3.7. Characterization of NSC by Phenotype Marker Expression (Live Staining of Cell Surface Antigens)***

1. Coat the slide chamber or imaging plate according to the CELLStart-coating protocol.

2. Harvest NSC according to the protocol described in adherent subculture of NSC.

3. Make cell suspension in complete medium at a cell density of 50,000 cells/cm$^2$.

**Table 3.3**
**Formulation of differentiation medium**

| Components | Cat. no. | Final concentration | 100 mL |
| --- | --- | --- | --- |
| *Neuronal differentiation* | | | |
| Neurobasal | 21103 | 1× | 97 mL |
| B27 | 17504 | 2% | 2 mL |
| Glutamax | 35050 | 2 mM | 1 mL |
| β-Mercaptoethanol | 21985 | 0.1 mM | 0.182 mL |
| BDNF | PHC7074 | 10 ng/ml | 0.1 mL |
| cAMP | Sigma D0627 | 100 μM | 0.1 mL |
| *Oligodendrocyte differentiation* | | | |
| Neurobasal | 21103 | 1× | 97 mL |
| B27 | 17504 | 2% | 2 mL |
| Glutamax | 35050 | 2 mM | 1 mL |
| T3 | Sigma D6397 | 30 ng/ml | 0.1 mL |
| β-Mercaptoethanol | 21985 | 0.1 mM | 0.182 mL |
| *Astrocyte differentiation* | | | |
| DMEM | 11995 | 1× | 97 mL |
| N2 | 17502 | 1% | 1 mL |
| Glutamax | 35050 | 2 mM | 1 mL |
| FBS | 16141 | 1% | 1 mL |
| β-Mercaptoethanol | 21985 | 0.1 mM | 0.182 mL |

4. Aspirate CELLStart and dispense cell suspension into each well.

5. After 2 days, cells are ready to be characterized. Dilute primary antibody (**Table 3.4**) in complete medium and change the spent medium with antibody-containing medium. Place in the incubator.

6. After 30 min of incubation, aspirate primary antibody solution and add complete medium.

7. Wash one more time by aspirating the medium and adding new medium. (Cells can be fixed at this stage if desired.)

8. Dilute secondary antibody (**Table 3.5**) in complete medium and replace spent medium with secondary antibody-containing medium.

**Table 3.4**
**Immunocytochemical markers used to characterize NSC and differentiated progenies**

| Category | Antigen | Cat. no. | Type | ICC dilution |
| --- | --- | --- | --- | --- |
| Neural stem cells | Sox1 | Millipore AB15766 | Rabbit IgG | 1:200 |
|  | Sox2 | R&D system | Mouse IgG | 2 µg/mL |
|  | Nestin | BD Science 611658 | Mouse IgG | 1:500 |
|  | CD133 | Abcam 16518 | Rabbit IgG | 1:100 |
| Neuronal progenitors | MAP2 | 13-1500 | Mouse IgG | 1:200 |
| Neurons | HuC/D | A21271 | Mouse IgG | 20 µg/ml |
|  | NF | 18-0171Z | Mouse IgG | 1:100 |
|  | NCAM | MHCD5600 | Mouse IgG | 1:50 |
|  | βIII tubulin | Sigma T8660 | Mouse IgG | 1:2000 |
|  | Dcx | 48-1200 | Rabbit IgG | 1:200 |
| Astrocyte progenitors | CD44 | MHCD4400 | Mouse IgG | 1:50 |
| Astrocytes | GFAP | 13-0300 | Rat IgG | 1:100 |
|  | GFAP | 18-0063 | Rabbit IgG | 1:500 |
| Oligodendrocyte progenitors | GalC | Millipore AB15766 | Mouse IgG | 1:200 |
| Oligodendrocytes | NG2 | Abcam ab20156 | Mouse IgG | 1:200 |
|  | O4 | Millipore MAB345 | Mouse IgM | 1:50 |
| Proliferation | Ki67 | 18-0191Z | Rabbit IgG | 1:50 |
|  | EDU | C35002 | Chemicals | 1:1000 |
| Isotype control | Mouse | 08-6599 | IgM and IgG | Use as it is |
|  | Rabbit | 08-6199 | IgG | Use as it is |
|  | Rat | R2C00 | IgM and IgG | 1:50 |

9. After 30 min of incubation, aspirate the antibody solution and add complete medium.

10. Wash one more time with PBS and observe staining under a fluorescence microscope.

11. If desired, cells can be fixed and processed for detection of intracellular antigen.
    *Note*: If the target antigen is present on cell surface, avoid Triton X, which can wash out and abolish the expression of the antigen.

***3.8. Characterization of NSC by Phenotype Marker Expression (Fixed Cells)***

1. Prepare cells as described in **Section 3.7**.

2. Aspirate spent medium and add 4% PFA solution.

3. After 20 min at room temperature, remove PFA and wash cells by adding PBS (14040).

**Table 3.5**
**Secondary antibody information**

| Ex/Emission (color) | Alexa no. | Second host | Second against | Cat. no. | Dilution rate |
|---|---|---|---|---|---|
| 346/442 (Blue) | 350 | Goat | Mouse IgM | A31552 | 1:1000 |
| | | Goat | Mouse IgG | A21049 | 1:1000 |
| | | Goat | Rat IgG | A21093 | 1:1000 |
| | | Goat | Rabbit IgG | A21068 | 1:1000 |
| | | Donkey | Goat IgG | A21081 | 1:1000 |
| 495/519 (Green) | 488 | Goat | Mouse IgM | A21042 | 1:1000 |
| | | Goat | Mouse IgG | A11029 | 1:1000 |
| | | Goat | Rat IgM | A21212 | 1:1000 |
| | | Goat | Rat IgG | A11006 | 1:1000 |
| | | Goat | Rabbit IgG | A11034 | 1:1000 |
| | | Donkey | Goat IgG | A11055 | 1:1000 |
| 590/617 (Red) | 594 | Goat | Mouse IgM | A21044 | 1:1000 |
| | | Goat | Mouse IgG | A11029 | 1:1000 |
| | | Goat | Rat IgM | A21213 | 1:1000 |
| | | Goat | Rat IgG | A11007 | 1:1000 |
| | | Goat | Rabbit IgG | A11037 | 1:1000 |
| | | Donkey | Goat IgG | A11058 | 1:1000 |
| 496, 536, 565/576 (Red) | NA | Goat | Mouse IgM | M31504 | 1:500 |
| | | Goat | Mouse IgG | P852 | 1:1000 |
| | | Goat | Rabbit IgG | P2771MP | 1:1000 |

4. Wash twice by replacing spent PBS with new PBS solution.

5. Depending on the expression site of the antigen, block cells by adding the respective blocking solution and incubate for 30 min at room temperature.

6. Dilute primary antibody (**Table 3.4**) in PBS (14040) and apply on cells either 1 h at room temperature or overnight at 4°C.

7. Remove primary antibody and wash samples with PBS three times.

8. Dilute secondary antibody (**Table 3.5**) in blocking solution (without Triton X) and apply to cells for 30 min at room temperature.

9. Remove secondary antibody and wash samples with PBS three times.

10. If nuclear staining is desired, replace secondary PBS washing step with DAPI solution and incubate for 5 min at room temperature and proceed with the third PBS washing step.

11. Samples are ready to be observed by fluorescence microscopy.

    *Note*:

    (1) If cells tend to lift with differentiation, perform two-step fixations.

    **Step 1.** Keep the temperature of PFA at room temperature and add equal volume of PFA on top of medium to leave cells in 2% PFA solution for 5 min.

    **Step 2.** Aspirate the 2% PFA and add 4% PFA and leave cells for 15 min at room temperature.

    (2) For cell surface antigens, use the blocking solution without Triton X to avoid the destruction of the antigen. If antigen expression is nuclear or cytoplasmic, permeabilize cells with Triton X in the blocking solution.

    (3) If the antibody expression is weak, perform an overnight incubation at 4°C with the primary antibody or for 1 h at room temperature.

    (4) To evaluate nonspecific binding, negative control should be prepared by replicating the primary antibody incubation step with an isotype control (depending on the subtype of the primary antibody, etc., mouse IgG for primary antibody raised in mouse with an IgG isotype) followed by secondary antibody staining.

    (5) If cells did not permeablize at the time of DAPI staining, add 0.1% Triton X to DAPI solution to permeabilize cells.

### 3.9. Characterization of NSC by Gene Expression (RT-PCR)

*3.9.1. RNA Extraction (Sample Size Is Less than $5\times10^4$, Dynabeads mRNA Direct Micro Kit, Cat. No. 610.21)*

1. Resuspend Dynabeads oligo(dT)25 thoroughly before use and transfer required amount of Dynabeads to an Rnase-free 1.5 mL microcentrifuge tube (20 μL for one reaction).

2. Place the tube in a magnetic field using Dynal MPC-S.

3. Remove the supernatant when the suspension is clear (~30s).

4. Remove the tube from the magnet and pre-wash the Dynabeads oligo(dT)25 by suspending in lysis/binding buffer to the original volume.

5. Place the tube on the magnet and remove the supernatant.

6. Remove the tube from the magnet and resuspend the beads in lysis/binding buffer to the original volume and aliquot 20 μL to each microcentrifuge tube.

7. If cells are adherent, wash once with PBS before the lysis step. If cells are in suspension, spin them down to make a cell pellet.

8. Aspirate PBS and add 100 μL lysis/binding buffer to lyse the cells.

9. Transfer the clear lysate to the tube containing 20 μL prewashed Dynabeads oligo(dT)25.

10. Mix by pipetting up and down a few times.

11. Place the tube on a roller for 5 min at room temperature.

12. Place the sample tube on the magnet and discard the supernatant when the solution clears.

13. Remove the sample tube from the magnet and resuspend the Dynabeads–mRNA complex in 100 μL washing buffer A.

14. Place the sample tube on the magnet and discard the supernatant.

15. Repeat Steps 13 and 14 once.

16. Resuspend the Dynabeads–mRNA complex in 100 μL washing buffer B and transfer the suspension to a new microcentrifuge tube.

17. Place the tube on the magnet and discard the supernatant.

18. Remove the sample tube from the magnet and resuspend the Dynabeads–mRNA complex in 100 μL washing buffer B.

19. Place the sample tube on the magnet and discard the supernatant.

20. Remove the sample tube from the magnet and resuspend in 100 μL of ice-cold 10 mM Tris–HCl.

21. Place the sample on the magnetic field on ice and aspirate the supernatant.

22. Resuspend bead–mRNA in 10–20 μL 10 mM Tris–HCl and incubate at 80–90°C for 2 min to elute. Then place the tube on the magnet and immediately transfer the supernatant to a new microcentrifuge tube (eluted mRNA).

23. Measure mRNA concentration by measuring the absorbance of eluted mRNA at 260 nM.

*Note*:

(1) All buffers except the 10 mM Tris–HCl should be brought to room temperature prior to use.

(2) Do not leave the Dynabeads oligo(dT)25 dry for long periods, as this may lower their capacity.

(3) Lysate can be stored at −80°C for later use.

(4) Use the mRNA–Dynabeads immediately for the cDNA reaction.

*3.9.2. RNA Extraction (Sample Size Is More than 5×10⁴, Trizol/Safelock Gel/Purelink Purification System)*

1. If cells are adherent, wash once with PBS before Trizol application. If cells are suspended, spin them down to obtain a cell pellet.

2. Add Trizol solution directly on the cell plate (adherent) or cell pellet to lyse the cells (1 mL for 3 mm dish or 1 mL for 2 million cells).

3. Add 60 mL 96–100% ethanol to wash buffer II. Check the box on the wash buffer II label to indicate that ethanol was added. Store wash buffer II with ethanol at room temperature.

4. Incubate the lysate with the Trizol reagent at room temperature for 5 min to allow complete dissociation of nucleoprotein complexes.

5. Pre-spin phase lock gel-heavy tubes briefly to collect the gel on the tube bottoms ($1500g$ for 30 s).

6. Transfer lysate with Trizol reagent to the phase lock gel-heavy tubes and incubate 5 min at room temperature.

7. Add 0.2 mL chloroform per 1 mL TRIzol® reagent, cap the tubes securely, and shake the tube vigorously by hand for 15 s.

8. Centrifuge the sample at $12,000g$ for 10 min at 4°C.

9. After centrifugation, examine and locate the clear aqueous phase atop the phase lock gel.

10. Transfer the colorless, upper phase containing RNA to a fresh tube.

11. Add an equal volume of 70% ethanol to obtain a final ethanol concentration of 35% and mix well by vortexing.

12. Transfer up to 650 μL of the sample from Step 11 above to the RNA spin cartridge supplied with the kit.

13. Centrifuge at $12,000g$ for 15 s at room temperature and discard the flow-through.

14. Transfer any remaining sample to the RNA spin cartridge and centrifuge at $12,000g$ for 15 s at room temperature.

15. Discard the flow-through and add 700 μL of wash buffer I to the spin cartridge.

16. Centrifuge at 12,000*g* for 15 s at room temperature and discard the collection tube and place spin cartridge into a clear RNA wash tube supplied with the kit.

17. Add 500 μL wash buffer II with ethanol to the RNA spin cartridge and centrifuge at 12,000*g* for 15 s at room temperature.

18. Discard the flow-through, repeat Step 17 once and then discard the flow-through.

19. Centrifuge the spin cartridge at 12,000*g* for 1 min at room temperature to dry the membrane with attached RNA.

20. Discard the collection tube and insert the cartridge into an RNA recovery tube supplied with the kit.

21. To elute the RNA, add 30–100 μL of RNase-free water to the center of the spin cartridge and incubate at room temperature for 1 min.

22. Centrifuge the spin cartridge for 2 min at $\geq$12,000*g* at room temperature.

    *Note*:

    (1) Lysate in Trizol can be stored at $-80°C$ for later use.

    (2) Vortexing the sample at Step 7 may increase DNA contamination of your RNA sample. Avoid vortexing if your downstream application is sensitive to the presence of DNA.

    (3) If no clear aqueous phase is observed at Step 9, add another 0.2 mL chloroform and shake vigorously, repeat centrifugation and re-examine the phases.

    (4) Eluted RNA can be stored on ice for a few hours. For long-term storage, store at $-80°C$.

### 3.10. First-Strand cDNA Synthesis

1. Kit Component:
   RT enzyme mix (includes SuperScript™ III RT and RNaseOUT™),
   2× RT reaction mix (includes oligo(dT)20 (2.5 μM), random hexamers (2.5 ng/μL), 10 mM MgCl2, and dNTPs)),
   *E. coli* RNase H

2. Transfer desired RNA to PCR tube on ice and add RNase/Dnase-free water up to 8 μL.

3. Make master mix of RT enzyme mix (2 μL each for reaction) and 2× RT reaction mix (10 μL each for reaction) and add 12 μL of master mix to RNA on ice (total volume will be 20 μL = 8 μL RNA + 12 μL master mix).

4. Incubate in a thermal cycler at 25°C for 10 min.

5. Incubate tube at 50°C for 30 min.

6. Terminate the reaction at 85°C for 5 min and chill on ice.

7. Add 1 μl (2 U) of *E. coli* RNase H and incubate at 37°C for 20 min.

   *Note*:

   (1) Protocol is designed to convert 0.1 pg to 5 μg of total RNA or 0.1 pg to 500 ng of poly(A) + RNA into first-strand cDNA.

   (2) Use diluted or undiluted cDNA in PCR or store at −20°C until use.

***3.11. Polymerase Chain Reaction***

1. Add the following component to each reaction tube or to a 96 well plate.

   | | |
   |---|---|
   | Platinum Blue PCR SuperMix | 18 μL |
   | Forward and backward primer | 1 μL |
   | cDNA | 1 μL |

2. Mix the contents and briefly spin to gather contents on the bottom of the tube.

3. Load the tube in the thermocycler to get following reaction cycle.

   Initial denature at 94°C for 2 min.

   | Cycle (30–40 cycles) | Denature at 94°C for 30 s |
   |---|---|
   | | Anneal at 55°C for 30 s |
   | | Extend at 72°C for 30 s |

   Final extension at 72°C for 7 min.
   Hold at 4°C.

4. Take out E-Gel (precast buffer-free agarose gel) from the plastic bag and insert gel cassette into the apparatus right edge first and push on the top and bottom to seat the gel nicely in the base.

5. Press and hold bottom in the apparatus until the red light turns to a flashing green light (Prerun).

6. Wait ~2 min of prerun, which will change flashing green light to flashing red light with beeping when it finishes the run.

7. Press button once to stop the beeping and remove the comb from the gel cassette.

8. Load PCR amplicon from Step 3 and press button to start electrophoresis. The light will change from red to green.

9. The run will stop automatically when the programmed time has elapsed (beeps).

10. Press button once to stop the beeping and take out gel cast to observe under UV box or imaging system.

    *Note*:

    (1) Supermix is composed of 22 U/mL complexed recombinant Taq DNA polymerase with Platinum® Taq antibody, 22 mM Tris–HCl (pH 8.4), 55 mM KCl, 1.65 mM MgCl2, 220 μM dGTP, 220 μM dATP,

220 μM dTTP, 220 μM dCTP, stabilizers, glycerol, and blue tracking dye.

(2) For multiple reactions with common components, prepare a master mix of the components common to all reactions to reduce pipetting errors.

(3) Primer stock includes 4 μM of forward and 4 μM of backward primers, which will result in 200 nM final concentration per primer in each reaction.

(4) Design primer to flank intron so that amplicon size from contaminating DNA is different from that of mRNA (**Table 3.6**). Amplicon from genomic DNA will include the intron size.

## Table 3.6
## PCR primers used to characterize NSC and differentiated progenies

| Target | Primer | Sequence | Tm | Amplicon size | Intron size |
| --- | --- | --- | --- | --- | --- |
| NSC | SOX1-F | GCGGAAAGCG TTTTCTTG | 53.0 | 406 | No intron |
| | SOX1-B | TAATCTGACT TCTCCTCCC | 50.2 | | |
| | SOX2-F | ATGCACCGCTA CGACGTGA | 59.3 | 59.3 | No intron |
| | SOX2-B | CTTTTGCACCC CTCCCATTT | 56.0 | | |
| | NESTIN-F | CAGCGTTGGAAC AGAGGTTGG | 58.6 | 389 | 1142 |
| | NESTIN-B | TGGCACAGGTGT CTCAAGGGTAG | 60.7 | | |
| Oligodendrocytes | MAG-F | TCTGGATTATG ATTTCAGCC | 49.7 | 366 | 159 |
| | MAG-B | GCTCTGAGAAGG TGTACTGG | 54.7 | | |
| | OSP-F | ACTGCTGCTGA CTGTTCTTC | 55.1 | 283 | 5714 |
| | OSP-B | GTAGAAACGGT TTTCACCAA | 50.8 | | |
| Astrocytes | ALDH1L1-F | TCACAGAAGTCT AACCTGCC | 55.5 | 398 | 21837 |
| | ALDH1L1-B | AGTGACGGGTG ATAGATGAT | 54.4 | | |
| | GFAP-F | GTACCAGGAC CTGCTCAAT | 55.0 | 321 | 2989 |
| | GFAP-R | CAACTATCCTGC TTCTGCTC | 55.3 | | |

(continued)

## Table 3.6 (continued)

| Target | Primer | Sequence | Tm | Amplicon size | Intron size |
|---|---|---|---|---|---|
| Neurons | 55.3 | CCACCTGAGATT AAGGATCA | 55.1 | 482 | 11798 |
| | MAP2-B | GGCTTACTTTG CTTCTCTGA | 55.0 | | |
| | CHAT-F | ACTGGGTGTCT GAGTACTGG | 55.0 | 451 | 7692 |
| | CHAT-B | TTGGAAGCCATT TTGACTAT | 54.9 | | |
| Endogenous control | ACTB-F | ACCATGGATGA TGATATCGC | 58.2 | 281 | 135 |
| | ACTB-B | TCATTGTAGAAG GTGTGGTG | 54.4 | | |

(5) Electrophoresis apparatus has two buttons. Left button is for double-comb gel and right button is for single-comb gel.

## 4. Critical Steps and Troubleshooting

1. Culturing NSC: A low yield of NSC is often a result of (1) poor adhesion of NSC by using the wrong version of PBS (calcium- and magnesium-free); (2) improper storage of STEMPRO NSC SFM complete medium (warming and cooling medium repeatedly compromises growth factors in the medium); (3) prolonged treatment with enzymes results in low viability. Make sure to use right buffer solution for CellStart and make an aliquot of the complete medium for the warming for the day. Expose NSC to enzyme solution just to initiate disaggregation.

2. Clustering of cells at time of differentiation: Clustering is often a result of the following: (1) coating is compromised if dish dries out; (2) plating density is too high; (3) cells do not dissociate well at time of passage.

3. Detachment of cells after the fixation: Detachment is often a result of the following: (1) sudden temperature change to cells; (2) fixative (4% PFA) goes bad; (3) bacterial contamination. If cells are loosely attached to the substrate, apply the two-step fixation in protocol. If open handling (nonsterile environment) is desired, add preservative in staining solution.

## 5. Typical Protocol Results

Neural stem cells cultured in STEMPRO NSC SFM typically have 60–80 h of doubling time (**Fig. 3.1**). Synth-a-freeze freezing method typically results in more than 80% cell viability after the thawing. Characterization of neural stem cells includes proliferation and phenotype marker expressions (**Fig. 3.2**). During differentiation to neurons, a portion of cell death is typical. One week of differentiation gives rise to immature neurons, oligodendrocytes, and astrocytes, whose phenotype can be labeled with phenotype markers (**Fig. 3.3**).

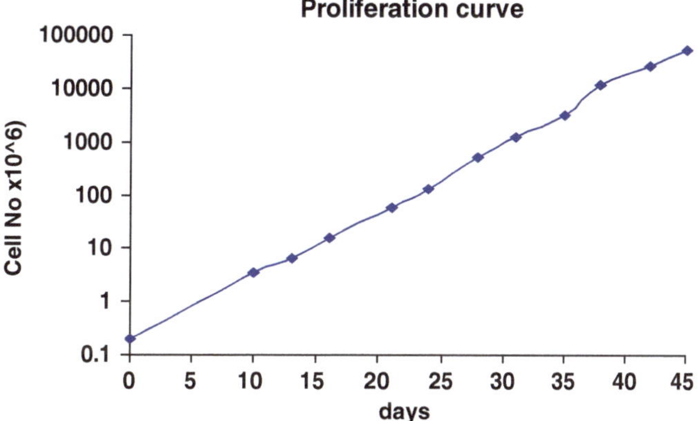


Fig. 3.1. Growth curve of neural stem cell in STEMPRO NSC SFM. Cumulative cell number over passage has been recorded and shown over time.

Fig. 3.2. Phenotype marker expression of NSC after culture in STEMPRO NSC SFM for five passages. NSC-expressed NSC phenotype markers (Nestin-*left*, Sox2-*middle*) and proliferation marker (Ki67-*right*). Inset at the right corner of each panel represents low-power image of nuclei staining with DAPI.

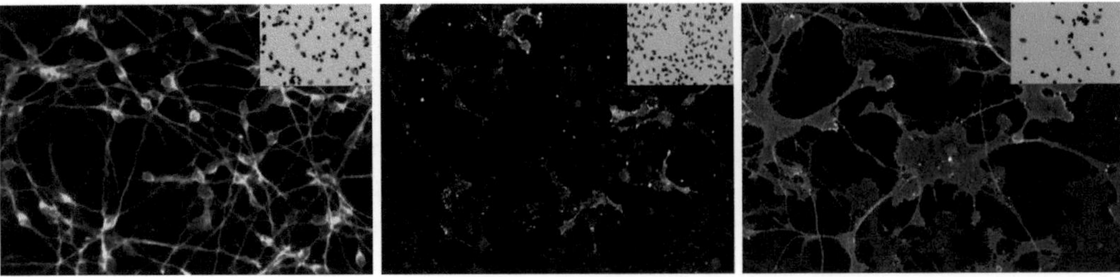

Fig. 3.3. Differentiation of NSC cultured in STEMPRO NSC SFM. NSC cultured in STEMPRO NSC SFM differentiated to generate neurons *(left)* and glial cells (oligodendrocytes-*middle*, astrocytes-*right)*. Neurons were labeled with Dcx *(left)*; cells whose lineage to oligodendrocytes were labeled with GalC *(middle)* and cells whose lineage to astrocytes were labeled with CD44 *(right)*. Cell nuclei were labeled with DAPI, and lower-power image is shown at the corner of each panel.

## Acknowledgments

We are indebted to Dr. Mahendra Rao for the support and ideas. We also thank the lab members of the Department of Stem Cell and Regenerative Medicine for useful discussions.

## References


1. Wu YY, Mujtaba T, Rao MS. Isolation of stem and precursor cells from fetal tissue. Methods Mol Biol 2002;198:29–40.
2. Bjorklund A, Lindvall O. Cell replacement therapies for central nervous system disorders. Nat Neurosci 2000;3(6):537–44.
3. Gage FH. Mammalian neural stem cells. Science 2000;287(5457):1433–8.
4. Studer L, Tabar V, McKay RD. Transplantation of expanded mesencephalic precursors leads to recovery in parkinsonian rats. Nat Neurosci 1998;1(4):290–5.
5. Caldwell MA, He X, Wilkie N, et al. Growth factors regulate the survival and fate of cells derived from human neurospheres. Nat Biotechnol 2001;19(5):475–9.
6. Jain M, Armstrong RJ, Tyers P, Barker RA, Rosser AE. GABAergic immunoreactivity is predominant in neurons derived from expanded human neural precursor cells in vitro. Exp Neurol 2003;182(1): 113–23.
7. Reubinoff BE, Itsykson P, Turetsky T, et al. Neural progenitors from human embryonic stem cells. Nat Biotechnol 2001; 19(12):1134–40.
8. Shin S, Mitalipova M, Noggle S, et al. Long-term proliferation of human embryonic stem cell-derived neuroepithelial cells using defined adherent culture conditions. Stem Cells 2006;24(1):125–38.
9. Colombo E, Giannelli SG, Galli R, et al. Embryonic stem-derived versus somatic neural stem cells: a comparative analysis of their developmental potential and molecular phenotype. Stem Cells 2006;24(4): 825–34.
10. Watanabe K, Kamiya D, Nishiyama A, et al. Directed differentiation of telencephalic precursors from embryonic stem cells. Nat Neurosci 2005;8(3):288–96.
11. Mayer-Proschel M, Kalyani AJ, Mujtaba T, Rao MS. Isolation of lineage-restricted neuronal precursors from multipotent neuroepithelial stem cells. Neuron 1997;19(4): 773–85.
12. Cai J, Wu Y, Mirua T, et al. Properties of a fetal multipotent neural stem cell (NEP cell). Dev Biol 2002;251(2):221–40.
13. Uchida N, Buck DW, He D, et al. Direct isolation of human central nervous system stem cells. Proc Natl Acad Sci USA 2000;97(26):14720–5.
14. Gleeson JG, Lin PT, Flanagan LA, Walsh CA. Doublecortin is a microtubule-associated protein and is expressed widely by migrating neurons. Neuron 1999;23(2): 257–71.

15. Barami K, Iversen K, Furneaux H, Goldman SA. Hu protein as an early marker of neuronal phenotypic differentiation by subependymal zone cells of the adult songbird forebrain. J Neurobiol 1995;28(1):82–101.

16. Zhang SC. Defining glial cells during CNS development. Nat Rev Neurosci 2001; 2(11):840–3.

17. Liu Y, Han SS, Wu Y, et al. CD44 expression identifies astrocyte-restricted precursor cells. Dev Biol 2004;276(1):31–46.

18. Conti L, Pollard SM, Gorba T, et al. Niche-independent symmetrical self-renewal of a mammalian tissue stem cell. PLoS Biol 2005;3(9):e283.

19. Wachs FP, Couillard-Despres S, Engelhardt M, et al. High efficacy of clonal growth and expansion of adult neural stem cells. Lab Invest 2003;83(7):949–62.

20. Ostenfeld T, Svendsen CN. Requirement for neurogenesis to proceed through the division of neuronal progenitors following differentiation of epidermal growth factor and fibroblast growth factor-2-responsive human neural stem cells. Stem Cells 2004;22(5): 798–811.

# Chapter 4

# Neural Differentiation of Human Embryonic Stem Cells

## Mirella Dottori, Alice Pébay, and Martin F. Pera

## Abstract

Human embryonic stem cells (hESC) and their neural derivatives hold great potential for developing therapies for treating disorders and injuries of the nervous system. It is therefore imperative that the methods used for differentiating hESC to neurons and glia be robust and well characterized. This chapter describes the efficient generation of neural progenitors from hESC using the BMP inhibitor protein, Noggin, their expansion as neurospheres, and subsequent differentiation into neurons and glia. This protocol has been used as an in vitro model for understanding neural differentiation of hESC as well as for deriving neural progenitors for treating neurodegenerative disorders. Since the overall method can be divided into defined stages of neural differentiation, it allows flexibility for introducing modifications to generate specific cell lineages. This feature of the protocol, together with its efficiency of generating neural progenitors, makes it attractive for other researchers to use for their own studies.

**Key words:** Neural differentiation, embryonic stem cells, human, neurospheres, noggin, neural progenitors.

## 1. Introduction

The defining characteristic of embryonic stem (ES) cells is their pluripotency. This unique feature can be viewed as a double-edged sword by neurobiologists. Working with ES cells is ideal for deriving any cell type of the central and peripheral nervous system, and although this is advantageous, the challenge is then to develop and refine methods for directing their differentiation toward desired lineages. For this, we need to appreciate and understand how ES cells undergo neural induction, followed by the generation of precursor and progenitor cells and ultimately mature neural and glial lineages.

L.C. Doering (ed.), *Protocols for Neural Cell Culture*, Springer Protocols Handbooks,
DOI 10.1007/978-1-60761-292-6_4, © Humana Press, a part of Springer Science+Business Media, LLC 1992, 1997, 2001, 2010

In our laboratory we have developed an efficient protocol for differentiating human ES cells (hESC) to mature neurons and glia. Our method can be divided into three main stages of neural differentiation (**Fig. 4.**1): (a) neural induction of hESC, (b) expansion of neural progenitors, and (c) differentiation to post-mitotic neurons and glia. Our previous studies showed that treatment of hESC with the BMP inhibitor protein, Noggin, results in their differentiation toward neuroectoderm only, as shown by the expression of Sox2, Pax6, and nestin, and a lack expression of early mesoderm or endoderm markers (1, 2). Noggin-treated hESC colonies can then be mechanically dissected and harvested and cultured in suspension in neural basal media (NBM) supplemented with fibroblast growth factor 2 (FGF2) and epidermal growth factor (EGF). Under these conditions, noggin-treated cells aggregate to form a spherical-like cluster resembling neurosphere cultures derived from the fetal or adult brain. These culture conditions do not support the growth of undifferentiated ES cells, and thus the neurospheres consist of a heterogeneous population of neural progenitor (NP) cell types with no contaminating hESC present (2). Neurospheres can be differentiated to give rise to neurons and glia when plated on laminin and fibronectin substrates, respectively (3).

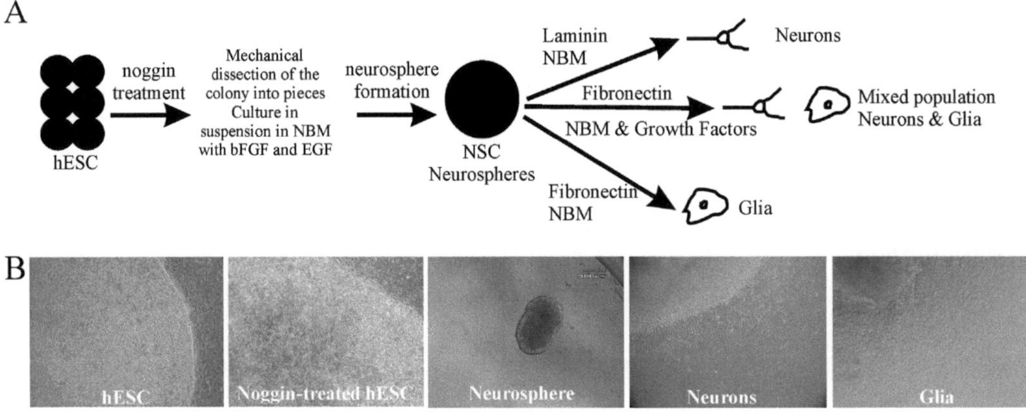


Fig. 4.1. Schematic outline (**A**) and phase-contrast images (**B**) showing multistage neural differentiation protocol.

This neural induction protocol is highly efficient and robust with no differentiation of other germ lineages (1). In addition, each stage of neural differentiation can be monitored and studied in isolation, and the protocol is therefore very useful for understanding how to direct neural differentiation toward a specific lineage as well as examining the critical factors/mechanisms involved at each stage of differentiation. For example, we found that Wnt proteins promote the expansion and survival of hESC-derived neural progenitors when cultured as neurospheres (2).

We also found that lysophosphatidic acid inhibits differentiation of hESC-derived neural progenitors to mature neurons (4). This neural induction protocol was adopted to generate dopaminergic neurons in vitro as well as neurospheres for transplantation into the brains of parkinsonian rodents showing in vivo differentiation into neural and glial cells (5). Overall, this protocol of hESC neural differentiation is efficient and well characterized with both in vitro and in vivo studies and thus gives neuroscientists a good working model for their research.

## 2. Reagents and Equipment

### 2.1. General Reagents for Culture Media and Solutions

The culture media for MEF, hESC, and neurospheres share some common reagents. The abbreviations and source of these reagents are listed below:

1. Stereomicroscope (Leica MZ6, Leica Microsystems) and microscope stage (Leica Microsystems).

2. Mouse embryonic fibroblasts are expanded in T75 flask (75cm$^2$ Blue Plug Seal Cap Tissue Culture Treated, BD Falcon #353135).

3. hESC and NSC are cultivated in different formats depending on the experiments to perform. In all cases, we culture hESC with a feeder layer of mouse embryonic fibroblasts (MEF) supplemented with 20% fetal calf serum or 20% knockout serum replacement (Invitrogen #10828-028) plus 4ng/mL bFGF (R&D #233-FB-025/CF).

4. For hESC maintenance and Noggin induction, we culture hESC in centre-well organ culture dishes, 60 × 15mm style (35mm culture dishes, Falcon #353037).

5. For neurosphere formation, we culture cells in 96-well plate, ultra-low attachment (Costar ultra-low cluster plate, 96 well with lid, flat bottom, ultra-low attachment, sterile, polystyrene, Corning #3474).

6. For neurosphere differentiation toward neuronal and glial lineages, we culture cells on tissue culture-treated glass slides (8 chamber polystyrene vessel, BD Falcon #354108, #354118).

7. Distilled water (Invitrogen #15230-204).

8. Phosphate buffered saline solution without calcium and magnesium (PBS$^-$) 10×: 1.37M NaCl, 27mM KCl, 100mM Na$_2$HPO$_4$, 18mM KH$_2$PO$_4$, pH 7.4, or from Invitrogen #Gibco14190).

9. PBS with calcium and magnesium (PBS$^+$, Invitrogen #14040).

10. Dispase (Invitrogen #17105-041).

11. Dulbecco's modified eagle medium high glucose (DMEM, Invitrogen #11960).

12. Fetal calf serum (FCS).
    **Note:** Fetal calf serum for hESC and MEF cultures need to be batch tested for support of hESC maintenance over at least four passages.

13. Insulin–transferrin–selenium Solution (ITS, Invitrogen #41400).
    **Note:** ITS-A supplement consists of 1g/L insulin, 0.67mg/L sodium selenite, 0.55g/L transferrin, and 11g/L sodium pyruvate.

14. L-Glutamine (L-Glut, Invitrogen #25030).

15. Non-essential amino acids solution (NEAA, Invitrogen #11140-050).

16. Penicillin–streptomycin (Pen/Strep, Invitrogen #15070).
    **Note:** Pen/Strep solution consists of 5000 U/mL penicillin and 5000μg/mL streptomycin.

17. Mitomycin C from *S. Caespitosus* (Sigma-Aldrich #M0503-10 × 2mg). Mitomycin C is cytotoxic, mutagenic, and carcinogenic; appropriate safety measures must be adhered to for use and disposal of this agent.

18. β-mercaptoethanol (Sigma #M7154).

19. B-27 supplement (Invitrogen, #17504-044).

20. N2 supplement (Invitrogen, #390424).

21. 0.2μm filters (Sartorius).

22. Gelatin (type A from porcine skin, Sigma #G1890). Stocks of 1% gelatin can be made in distilled $H_2O$ and stored at room temperature.

23. Poly-D-lysine (BD Bioscience #354210).

24. Mouse laminin (BD Bioscience #356008).

25. Human plasma fibronectin (Invitrogen #33016-015).

26. Noggin (R&D Systems #3344-NG).

27. Neurobasal media (NBM, Invitrogen # 21103).

28. Human recombinant epidermal growth factor (EGF; R&D Systems #236-EG or Preprotech Inc. #100-15).

29. Human recombinant basic fibroblast growth factor (bFGF; Preprotech Inc. #100-18B).

30. Human recombinant platelet-derived growth factor-AA (PDGF-AA; Peprotech Inc. #100-13A).

31. Trypsin/EDTA (Invitrogen #25200).

### 2.2. Preparation of Mouse Embryonic Feeder Culture Media (F-DMEM)

Supplement 1 × DMEM with 10% heat-inactivated FCS, 2mM L-Glut, and 0.5% Pen/Strep. Mix well and sterilize using a 0.22μm filter.

### 2.3. Preparation of Mitomycin C Solution

1. Dissolve 2mg Mitomycin C in 4 mL of distilled $H_2O$ (0.5mg/mL final).
2. Filter sterilize using 0.22μm filter.
3. Keep solution protected from light and store at 4°C.
4. When ready to use, dilute to 10μg/mL in F-DMEM.

### 2.4. Preparation of Trypsin–EDTA Solution

1. A stock solution of 0.4% ethylenediaminetetraacetic acid (EDTA) and 2.5% trypsin (Gibco/BRL) is made in PBS⁻.
2. Filter-sterilize, aliquote, and store at –20°C. Once thawed, an aliquot can remain active at 4°C for up to a week.
3. Prepare a working solution (0.5% trypsin–EDTA) from stock aliquot with PBS⁻.

### 2.5. Preparation of Human Embryonic Stem Cell Culture Media

1. Supplement 1× DMEM with 20% heat-inactivated FCS, 0.1mM NEAA (**Note:** Non-essential amino acids solution consists of 750mg/L glycine, 890mg/L L-alanine, 1320mg/L L-asparagine, 1330mg/L L-aspartic acid, 1470mg/L L-glutamic acid, 1150mg/L L-proline, and 1050mg/L L-serine), 2mM L-glut, 0.5% pen/strep, 0.1mM 2β-mercaptoethanol, and 1% ITS.
2. Mix well and sterilize using a 0.22μm filter.

### 2.6. Preparation of Dispase Solution

1. A 10mg/mL solution of Dispase is made in pre-warmed hESC culture media.
2. Incubate at room temperature for 5min, gently invert, and leave for at least 15min at room temperature.
3. Sterilize the solution using a 0.22μm filter.
4. The solution can be used or stored at 4°C for up to 2 days.

### 2.7. Preparation of Neurobasal Medium

1. Supplement 1 × neurobasal A (**Note:** The components of neurobasal A and B-27 supplements are proprietary.) with 2% B-27 supplement (**Note:** The components of neurobasal A and B-27 supplements are proprietary), 1% ITS, 1% $N_2$ supplement (**Note:** $N_2$ supplement consists of 0.5mg/mL recombinant insulin, 1mM human transferrin, 10mM putrescine, 3 nM selenite, and 2 nM progesterone), 2mM L-glut, and 0.5% pen/strep.
2. Mix well and sterilize using a 0.22μm filter.

### 2.8. Preparation of Adhesive Substrates

#### 2.8.1. Gelatinized Plates for Culturing MEFs

1. Stock solutions of gelatin are diluted to 0.1% using distilled $H_2O$ and filter-sterilized.
2. Coat plates with 0.1% gelatin and keep at room temperature for at least 1h.
3. Aspirate the gelatin solution and plates are ready to use for plating cells.

#### 2.8.2. Laminin-Coated Plates for Neuronal Differentiation

1. Cover plates with poly-D-lysine solution (10μg/mL in PBS$^+$), keep at room temperature for at least 30min.
2. Aspirate the solution, wash the plates three times with PBS$^+$.
3. Coat the plates with mouse laminin solution (5μg/mL in PBS$^+$) and incubate overnight at 4°C.
4. The next day remove the laminin solution, wash three times with PBS$^+$.
5. Culture media can be added for plating cells/neurospheres.

#### 2.8.3. Fibronectin-Coated Plates for Glial Differentiation

1. Cover plates with poly-D-lysine solution (10μg/mL in PBS$^+$), and keep at room temperature for at least 30min.
2. Aspirate the solution and wash the plates three times with PBS$^+$.
3. Coat the plates with fibronectin solution (10μg/mL in PBS$^+$) and incubate overnight at 4°C.
4. The next day remove the fibronectin solution; wash three times with PBS$^+$.
5. Culture media can be added for plating cells/neurospheres.

---

## 3. Methods

### 3.1. Culture of Human Embryonic Stem Cells

hESC cells are cultured as colonies on mitotically inactivated mouse embryonic fibroblast (MEF) feeder layer, in a gelatin-coated organ tissue culture dish. The MEF feeder layer density is $6.0 \times 10^4$ cells/cm$^2$ ($1.7 \times 10^5$ cells/organ culture dish). Isolation of MEF, their inactivation by mitomycin C treatment, and plating onto gelatinized organ culture dishes are described below (**Section 3.1.1**). hESC are passaged weekly by mechanical slicing of the colonies using either glass-pulled pipettes or a 27-gauge needle (**Section 3.1.2**). hESC are maintained at 37°C with 5% $CO_2$ in a humidified incubator. All tissue culture procedures are performed using aseptic techniques in class II biological safety cabinets.

*3.1.1. Preparation of MEF*

3.1.1.1. Derivation of MEF

MEF are derived from E12.5 to 14 embryos. Usually, about three embryos are used to obtain a confluent T75 flask of fibroblasts.

1. Extract embryos from placental tissue.

2. Remove head and visceral tissue (including fetal liver).

3. Transfer remaining fetal tissue to a clean dish and wash three times with PBS$^-$ by transferring tissue from dish to dish.

4. Mince the tissue using either scissors or scalpel blade.

5. Add 2 mL of 0.5% trypsin and continue mincing.

6. Add an additional 5 mL of trypsin and incubate for 20min at 37°C.

7. Pipette the embryos in the trypsin until few chunks remain and then incubate for further 10min at 37°C. The aim is to get a single cell suspension.

8. Add 10 mL F-DMEM to neutralize the trypsin and transfer contents to a 50 mL tube.

9. Mix well and transfer all to T75 flask.
**Note:** In primary cultures of mouse embryo fibroblast preparations, blood cells, as well as large intact fragments of tissue, may be present along with fibroblasts. Generally, these contaminating cell types disappear following subculture as the fibroblasts overgrow them.

10. The next day change media to remove cell debris and dead cells.

11. When the flask is reaching 70–80% confluency, freeze down the cells and label as passage 0. At this stage the cells grow very vigorously and so can be expanded 1:5 (freeze at 1:5). They cannot be used past passage 4 as their growth rate begins to slow down significantly, they may develop morphological changes, and their suitability as feeder cells declines.

To use these cells as feeder layers in hESC culture, it is necessary to passage them a few times prior to use to minimize contamination with residual non-fibroblast cells. Thus, to use these cells as feeder layers, one needs to:

1. Thaw passage 0.

2. Culture for 2 days until confluent, then trypsinize and expand (becomes p1).

3. Culture p1 cells for 2 days, then trypsinize and expand again (p2).

4. Culture p2 cells for 2 days, then trypsinize and freeze at passage 3 (p3).

5. Cells at p3 are used for hESC feeder layer. The cells are mitotically inactivated either by mitomycin C treatment or irradiation.

*3.1.1.2. Mitomycin C Treatment of MEF*

1. Aspirate existing media from flask containing MEF cultures.

2. Add mitomycin C ($10\mu g/mL$) made in warm F-DMEM to each flask.

3. Incubate at $37°C$ for 2.5h.

4. Aspirate media containing mitomycin C.

5. Wash flask once with F-DMEM and twice with PBS$^-$.

6. Add 0.5% trypsin–EDTA solution (**Note:** The volume of trypsin/EDTA solution used for a T75 ($75cm^2$)–T175 ($175cm^2$) flask is 2–5 mL, respectively) and incubate at $37°C$ for 3min.

7. Slightly agitate the flask until cells start to detach.

8. Collect the cells with F-DMEM and transfer into a 10–50 mL tube.

9. Spin the cells for 2min at 2000rpm.

10. Aspirate and resuspend the cells with 10–15 mL of F-DMEM.

11. Obtain a 10 $\mu$L sample and perform a cell count.

12. Plate cells as required on gelatinized tissue culture plates.

*3.1.2. Passage of hESC Colonies*

hESC are cultured as colonies on a feeder layer of mitotically inactivated MEF and are routinely passaged once a week. The method used for hESC passage is by mechanical dissociation of colonies, combined with dispase treatment, to dissect the colony into "pieces." Dissection of colony is performed using a stereomicroscope with a $37°C$ warm microscope stage. Each colony "piece" is then transferred onto a fresh feeder layer. The detailed description of this method is as follows.

1. Prepare 2 × 6cm Petri dishes with PBS$^+$ for two washes.

2. Make dispase solution in hESC media and filter-sterilize (*see* **Section 2.1** , **Step 6**).

3. Change the media of hESC culture to PBS$^+$.

4. Working with a stereomicroscope, cut hESC colonies into pieces using fine glass pulled pipettes or 27-gauge needles. Avoid cutting differentiated portions of partially differentiated colonies.

   **Note:** By 7 days after passage, hESC colonies show some regions of cell differentiation, particularly within the centre of the colony. These regions are morphologically distinguished by cystic-like structures extending up from the colony.

5. Aspirate the PBS⁺.

6. Add dispase solution and leave for 3min at 37°C on heating stage, or until edges of the cut colony pieces start to detach.

7. Using a p20 pipette, nudge the corner of the colony piece until it complete detaches, collect and transfer to the PBS⁺ wash dish.

8. Once all the colony pieces have been collected and transferred to the wash dish, transfer them to the second PBS⁺ wash dish.

9. Finally, transfer the pieces to an organ culture dish containing a fresh feeder layer of mitotically inactivated MEFs and hESC media. Usually, about eight pieces are placed per organ culture dish, evenly spaced and neatly arranged in a circle.

### 3.2. Neural Induction of hESC by Noggin

1. Add 500ng/mL recombinant Noggin to hESC media at the time of hESC colony transfer onto a fresh feeder layer of MEFs (day 0).
**Note:** Growth factors (bFGF, EGF, and Noggin) are reconstituted according to manufacturer's recommendations to stock concentrations of 50–100µg/mL. Long-term storage of factors is at –80°C. Stocks can be kept short term (up to 1–3 months) at 4°C.

2. Cells are cultured for 14 days without passage and Noggin media is replaced every other day.
**Note:** After 14 days of Noggin treatment, immunostaining and PCR expression analyses show that hESC colonies consist of a mixed population of hESC and neural stem cells, the latter usually found more within the central regions of the colony (*5; Dottori unpublished data*).

### 3.3. Expansion of Neural Progenitors as Neurosphere Cultures

After 14 days of Noggin treatment, colonies are dissected into pieces (as performed for normal hESC passage, *see* **Section 3.1.2**) and then transferred to individual wells in a non-adherent 96-well plate to allow neurosphere formation. Neurospheres are cultured in suspension in NBM supplemented with 10ng/mL human recombinant EGF and 10ng/mL human recombinant bFGF. Noggin-treated colonies do still contain some undifferentiated hESC even after 14 days of treatment.
**Note:** After 14 days of Noggin treatment, immunostaining and PCR expression analyses show that hESC colonies consist of a mixed population of hESC and neural stem cells, the latter usually found more within the central regions of the colony (*5; Dottori unpublished data*).

Contaminating hESC that may be transferred to the neurosphere culture conditions either do not survive or form cystic-like structures. The latter are easily identified and removed from culture dish. A more detailed protocol of transfer to neurosphere cultures is:

1. Rehydrate the surface of the 96 well with NBM without any growth factors for at least 15min in the incubator.

2. Aspirate and add NBM with 20ng/mL bFGF and EGF (100–200 µL/well), and leave in the incubator until ready to plate the pieces.

3. Prepare a Petri dish with PBS⁺.

4. Wash the day14 Noggin-treated hESC cultures with PBS⁺.

5. Aspirate and add fresh PBS⁺.

6. Using either a pulled glass pipette or 27-gauge needle, cut the center region of the colony.
   **Note:** After 14 days of Noggin treatment, immunostaining and PCR expression analyses show that hESC colonies consist of a mixed population of hESC and neural stem cells, the latter usually found more within the central regions of the colony (*5; Dottori unpublished data*). ˜˜˜**Note:** After 14 days of culture, the colonies are easily detached by mechanical dissociation from the feeder layer, and thus additional enzymatic treatment is not required.

7. Using a p20 pipette, pick up pieces and transfer to the PBS dish.

8. Transfer one piece per well.

9. By 3–5 days in culture, the pieces will form a smooth sphere indicating neurosphere formation. Those that do not form neurospheres will either disintegrate or form cysts.

10. Maintain neurosphere cultures at 37°C with 5% $CO_2$ in a humidified incubator. Change media every 2–3 days.

11. Neurospheres may be subcultured by dissection using glass needles in neurobasal medium. Dissected neurosphere fragments are returned to 96-well plates and grown as described above.

***3.4. Differentiation of Neurospheres to Neurons and Glia***

For neuronal differentiation, whole neurospheres are plated onto laminin substrates (*see* **Section 2.3**, **Step 2**) in NBM without bFGF and EGF supplements. For differentiation biased toward glial lineages, whole neurospheres are plated onto fibronectin substrates (*see* **Section 2.3**, **Step 3**) in NBM with 20ng/mL bFGF, 20ng/mL EGF, and 20ng/mL PDGF-AA supplements for 1 week, followed by NBM without supplements. Cells are maintained at 37°C with 5% $CO_2$ in a humidified incubator. Media is changed every other day.

**Note:** On laminin substrates, neuronal differentiation with axonal outgrowths can be observed usually within 5 days of culturing. Expression of early neuronal markers, such as beta-tubulin III, can be observed at this time. On fibronectin substrates, migrating glia expressing glial fibrillary acidic protein (GFAP) can be observed after 2 weeks of culture.

## 4. Conclusions

This method of generating neural progenitors from hESC is based upon neural induction by Noggin treatment. While some groups have reported Noggin treatment helps to maintain hESC undifferentiated (6, 7), other groups have also reported similar results to ours showing a bias toward neural differentiation (8, 9). Neural progenitors derived by Noggin induction are tripotent, as shown by their differentiation to neurons, glia, and oligodendrocytes, and give rise to mature functional neurons, as demonstrated by their electrophysiological properties (8). Noggin treatment has also been combined with other neural-inducing methods, such as using stromal cell co-culture systems, to obtain high yields of midbrain dopaminergic neurons from hESC (10). Although the exact mechanism of neural induction by Noggin is not known, our studies and others suggest a model whereby Noggin is suppressing paracrine BMP signaling in hESC (1, 9). Further work is required to determine specific signaling pathways involved and find common ground with all procedures.

In summary, the strengths of this protocol are that it is robust and efficient and that different stages of neural differentiation can be monitored and studied separately. Our studies from using this current in vitro model of hESC neuronal differentiation will compliment future work as research in stem cell biology advances.


## References

1. Pera, M. F., Andrade, J., Houssami, S., Reubinoff, B., Trounson, A., Stanley, E. G., Ward-van Oostwaard, D., and Mummery, C. Regulation of human embryonic stem cell differentiation by BMP-2 and its antagonist noggin. *J Cell Sci* 2004; 117: 1269–80.

2. Davidson, K. C., Jamshidi, P., Daly, R., Hearn, M. T., Pera, M. F., and Dottori, M. Wnt3a regulates survival, expansion, and maintenance of neural progenitors derived from human embryonic stem cells. *Mol Cell Neurosci* 2007; 36: 408–15.

3. Reubinoff, B. E., Itsykson, P., Turetsky, T., Pera, M. F., Reinhartz, E., Itzik, A., and Ben-Hur, T. T. Neural progenitors from human embryonic stem cells. *Nat Biotechnol* 2001; 19: 1134–40.

4. Dottori, M., Leung, J., Turnley, A. M., and Pebay, A. Lysophosphatidic acid inhibits neuronal differentiation of neural stem/progenitor cells derived from human embryonic stem cells. *Stem Cells* 2008; 26: 1146–54.

5. Ben-Hur, T., Idelson, M., Khaner, H., Pera, M., Reinhartz, E., Itzik, A., and Reubi-

noff, B. E. Transplantation of human embryonic stem cell-derived neural progenitors improves behavioral deficit in Parkinsonian rats. *Stem Cells* 2004; 22: 1246–55.

6. Xu, R. H., Peck, R. M., Li, D. S., Feng, X., Ludwig, T., and Thomson, J. A. Basic FGF and suppression of BMP signaling sustain undifferentiated proliferation of human ES cells. *Nat Methods* 2005; 2: 185–90.

7. Wang, G., Zhang, H., Zhao, Y., Li, J., Cai, J., Wang, P., Meng, S., Feng, J., Miao, C., Ding, M., Li, D., and Deng, H. Noggin and bFGF cooperate to maintain the pluripotency of human embryonic stem cells in the absence of feeder layer. *Biochem Biophys Res Commun* 2005; 330: 934–42.

8. Itsykson, P., Ilouz, N., Turetsky, T., Goldstein, R. S., Pera, M. F., Fishbein, I., Segal, M., and Reubinoff, B. E. Derivation of neural precursors from human embryonic stem cells in the presence of noggin. *Mol Cell Neurosci* 2005; 30: 24–36.

9. Yao, S., Chen, S., Clark, J., Hao, E., Beattie, G. M., Hayek, A., and Ding, S. Longterm self-renewal and directed differentiation of human embryonic stem cells in chemically defined conditions. *Proc Natl Acad Sci U S A* 2006; 103: 6907–12.

10. Sonntag, K. C., Pruszak, J., Yoshizaki, T., van Arensbergen, J., Sanchez-Pernaute, R., and Isacson, O. Enhanced yield of neuroepithelial precursors and midbrain-like dopaminergic neurons from human embryonic stem cells using the bone morphogenic protein antagonist noggin. *Stem Cells* 2007; 25: 411–8.

# Chapter 5

# Isolation and Culture of Primary Human CNS Neural Cells

**Manon Blain, Veronique E. Miron, Caroline Lambert, Peter J. Darlington, Qiao-Ling Cui, Philippe Saikali, V. Wee Yong, and Jack P. Antel**

## Abstract

We present our current methods for isolating and culturing cells from the adult and fetal human CNS. The cell isolation procedures used are also well suited to obtain samples for immediate RNA- and protein-based analyses, particularly as techniques to minimize the amount of material needed become increasingly available. Further tissue culture techniques commonly used for rodent material can be adapted to human adult and fetal tissues such as the hippocampal slice culture and the neurosphere assay (Dorr et al. J Neuroimmunol 167:204–209, 2005; Chojnacki et al. Ann Neurol 64:127–142, 2008).

**Key words:** Human, adult, fetal, neural progenitors.

## 1. Introduction

The access to surgically derived human central nervous system (CNS) tissue provides the opportunity to isolate and culture the various cell types present in such tissues. Once in culture, these cells can be used to understand the molecular mechanisms that account for their phenotypic properties observed in situ under physiologic and pathologic conditions. Most surgically derived adult specimens come from resections carried out to ameliorate intractable epilepsy. Most commonly, the samples are from tissue that needs to be resected (deep white matter) as an approach to the site of actual pathology (usually the hippocampal region), as illustrated in **Fig. 5.1**. Other tissue samples include peri-tumoral tissues, but concerns exist as to what the source (tumor vs. non-tumor) of derived cells may be.

L.C. Doering (ed.), *Protocols for Neural Cell Culture*, Springer Protocols Handbooks,
DOI 10.1007/978-1-60761-292-6_5, © Humana Press, a part of Springer Science+Business Media, LLC 1992, 1997, 2001, 2010

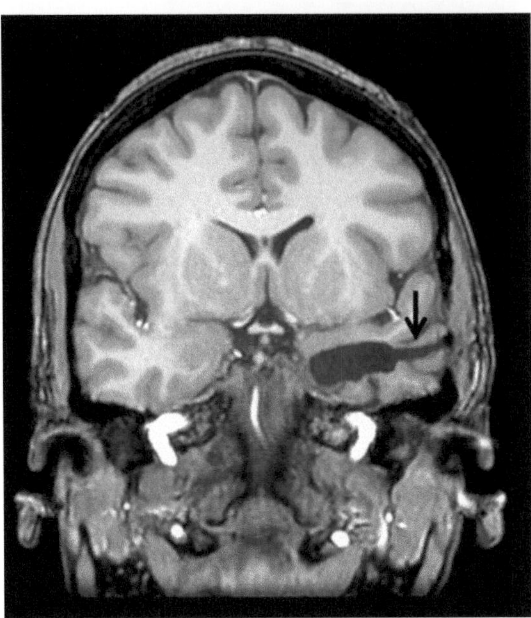

Fig. 5.1. Post-operative magnetic resonance imaging (MRI) scan indicating site of surgical resection of adult human CNS tissue (*black arrow* indicates areas where deep white matter was removed).

The cells we have been able to extract from the tissues are mature and progenitor oligodendrocytes, and microglial cells (1–3). Human brain endothelial cells can also be derived from this material but are not covered in this chapter (4, 5). Biopsy-derived human brain specimens offer several advantages over autopsy-derived materials. Being relatively fresh, the cell viability and yield are high (1–5 million cells/g wet weight of tissue; cell yield for autopsy specimens is at least an order of magnitude less). The success rate of bulk-isolating viable cells from biopsy specimens has been in the over 90% range. The material we obtain is removed by Cavitron$^{TM}$ ultrasonic aspiration (CUSA), which fragments the tissue into cubes of 2 mm, on average. Other centers use en bloc dissections; our limited experience with such tissue is that cell viability is reduced. Another advantage offered by biopsy specimens is that blood samples can be procured from the living patients, allowing analysis of autologous immune–neural interactions.

The human fetal tissue is also obtained from surgically removed tissues, usually from 16- to 24-week-old fetuses. As with the adult tissues, all tissues are obtained using Canadian Institutes of Health Research (CIHR) guidelines and institutional review board approved protocols. The human fetal tissue serves as a source of astrocytes and neurons, cell types that either cannot be obtained as a purified preparation or maintained in long-term culture, respectively, from the adult CNS (6, 7). Since progenitor cells and microglia can be obtained from adult and fetal tissues, changes in properties with development and aging can be addressed (1, 3). Human CNS-derived cells from different age sources also allow comparisons with animal-based studies using tissues of different ages. Of note, survival of human adult cells, especially oligodendrocytes and microglia cells, under the relatively basal cell culture conditions described later seem to exceed that observed using corresponding cell types derived from the rodent CNS.

## 2. Adult Human Preparations

### 2.1. Reagents and Equipment

#### 2.1.1. For Mixed Cell Population

1. Human adult brain tissue
2. Incubator at 37°C and 5% $CO_2$
3. Tissue culture hood
4. Pipette aid and 10 mL disposable plastic pipettes with wide mouth (Fisher)
5. 9 in. glass Pasteur pipettes (2 mL) and rubber bulb
6. Aspirator
7. Phosphate-buffered saline (PBS), pH 7.4 (calcium- and magnesium-free)
8. Percoll (GE Healthcare)
9. 70% ethanol
10. Glass bottles with cap (100 mL, 500 mL)
11. Trypsin final concentration 0.25% (2.5% stock, Invitrogen)
12. DNase I final concentration 100 μg/mL (1 mg/mL stock, Roche Diagnostics)
13. Shaking water bath set at 37°C
14. Fetal calf serum (FCS)
15. Plastic 2 piece Buchnel funnel with perforated plate removed, and containing a 132 μm nylon mesh (industrial fabrics) (refer to **Fig. 5.2** for assembly instructions)
16. 20 mL sterile plastic syringe plunger
17. Sterile 50 mL plastic conical tubes
18. Polycarbonate tubes (40 mL, Nalgene)
19. Beckman Coulter Allegra 6 with GH 3.8 swing bucket rotors (02U 36300 or 05U 43749)
20. Beckman Coulter high-speed fixed-angle rotor centrifuge (J2-21 M/E)
21. Uncoated plastic tissue culture Falcon flasks (25 cm$^2$)
22. Hemocytometer

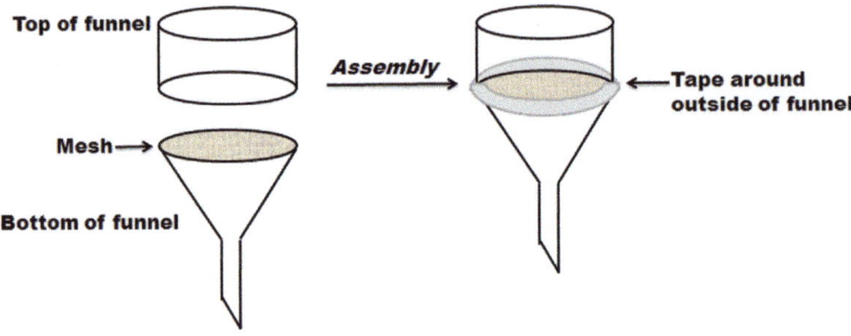



Fig. 5.2. Mechanical dissociation of human adult brain preparations. Nylon mesh (132 $\mu$m pore size) tightly spread over the surface of the bottom portion of the funnel. The top portion of the filter is fitted over the filter (edges of filter mesh should stick out). The assembly is held together with autoclavable tape wrapped around the funnel.

23. 10 and 200 $\mu$L micropipettes

24. Bleach

25. EDTA

*2.1.2. For Oligodendrocytes*

1. 16-well glass chamber slides coated with 10 $\mu$g/mL poly-L-lysine and air-dried for 6–24 h

2. Minimal essential medium (MEM) supplemented with 0.1% glucose, 5% FCS, 1% penicillin–streptomycin, 1% glutamine

*2.1.3. For Microglia*

1. Minimal essential medium (MEM) supplemented with 0.1% glucose, 5% FCS, 1% penicillin–streptomycin, 1% glutamine

2. Plastic tissue culture vessel (flasks or plates)

**2.2. Protocol**

All of the following procedures are carried out in a tissue culture hood with ethanol-sterilized gloves and a lab coat. All discarded solutions that have come into contact with the brain preparation are incubated with 10% bleach for a minimum of 24 h. Buchnel funnel nylon mesh, glass bottles, glass Pasteur pipettes and polycarbonate tubes are autoclaved in bags. All other sterile disposable materials are purchased from manufacturers.

*2.2.1. Brain Mechanical and Chemical Dissociation*

1. Pour the CUSA bag contents into 50 mL conical tubes.

2. Wash the CUSA bag with PBS to remove any brain tissue that may have stuck to the sides of the bag.

3. Once brain tissue has settled to bottom of the tube, slowly pour out as much blood as possible without losing brain tissue.

4. Complete with PBS to 40 mL and let tissue settle. Discard liquid and repeat until contaminating blood is cleared.

5. Pool tissue into one tube and remove capillaries and blood clots with a glass Pasteur pipette.

6. Note the amount of tissue (mL).

7. Pour the brain tissue into a 100 mL sterile glass bottle.

8. Rinse the conical tube with PBS and add to bottle. Total volume should be approximately 65 mL. Add the Trypsin and DNase. For less than 15 mL of tissue, add 0.25% trypsin and 100 μg/mL DNAse I. For more than 15 mL of tissue, concentrations can be doubled.

9. Parafilm the lid of the bottle and incubate in a shaking water bath at 37°C for 30 min at 180 rpm. Alternatively, add a sterilized stirrer into the bottle and stir the tissue suspension gently using a magnetic stirrer set at 37°C.

10. Add FCS (10% of the total volume) to inhibit the trypsin, and swirl the bottle to distribute the serum.

11. Place funnel with mesh onto sterile 500 mL glass bottle.

12. Pour brain preparation onto mesh 10 mL at a time. Use plunger of a 20 mL plastic syringe to grind the tissue through the mesh in a circular motion. Use PBS to wash the bottle and add to the mesh, and add PBS directly to the mesh for final wash. Total volume in the bottle should amount to a maximum of 50 mL per 1 mL tissue.

13. Distribute the brain preparation into 50 mL conical tubes and centrifuge at 234*g* with high brakes (with a Beckman Coulter Allegra 6 centrifuge with GH 3.8 swing bucket rotors, this is equivalent to 1200 rpm). *See* **Fig. 5.2**

*2.2.2. Thirty Percentage of Percoll Density Gradient and Isolation of Neural Cells/Removal of Myelin*

1. Prepare "*n*" amount of polycarbonate tubes with 9 mL of Percoll per tube (where "*n*" is equivalent to volume in millilitre of wet tissue noted at **Step 6**).

2. Discard supernatant and resuspend each conical tube with 21 mL PBS.

3. Carefully pipette 21 mL of brain suspension into each polycarbonate tube. Use PBS to supplement the volume if required.

4. Weigh the tubes and match tubes in pairs within 0.05*g* for balancing purposes.

5. Centrifuge in fixed-angle rotor high-speed centrifuge for 30 min at 15,000rpm at 4°C, without brakes.

6. Using a glass Pasteur pipette connected to an aspirator, remove the top layer and second layer (myelin is white in appearance, outlined in **Fig. 5.3**).

7. Set pipette aid on "slow" withdrawal speed and remove the cellular layer without touching or aspirating the red blood cell layer. Transfer to a 50 mL conical tube (each cellular layer is transferred to one conical tube).

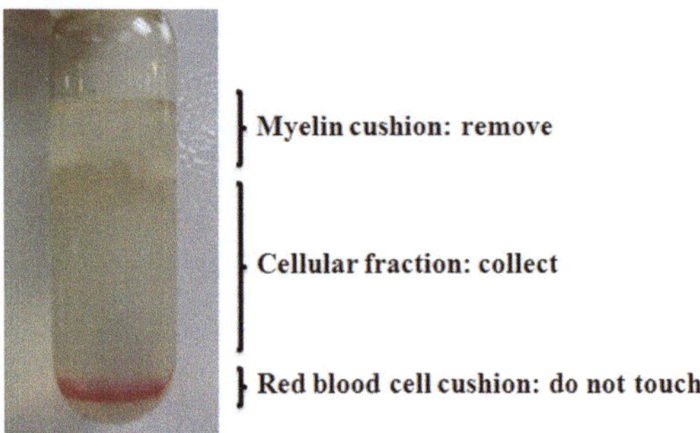


Fig. 5.3. Percoll density gradient.

8. Complete to 50 mL with PBS, mix well, and centrifuge for 10 min at $650g$ with high brakes (equivalent to 2000 rpm).

9. Discard supernatant and resuspend pellets with MEM supplemented with 5% FCS and 0.1% glucose. Wash the conical tubes with media and add to resuspended pellet. Complete to 50 mL with media and centrifuge for 10 min at $234g$ (equivalent to 1200 rpm) with low brakes.

10. Carefully discard supernatant and resuspend pellet in 5 mL of media.

11. Count cells on hemocytometer and plate cells in uncoated tissue culture flask (25 cm$^2$) at $2 \times 10^6$ per mL with 6 mL per flask ($0.5 \times 10^6$ cells/cm$^2$). Cell suspension consists mainly of mature oligodendrocytes and microglia, as well as a few astrocytes and oligodendrocyte progenitor cells.

*2.2.3. Oligodendrocyte and Microglia Isolation and Culture*

*2.2.3.1. Oligodendrocytes*

Oligodendrocytes (and contaminating astrocytes/progenitors) are separated from microglia by differential adhesion to tissue culture flasks. Microglia adhere to the plastic whereas oligodendrocytes remain in the floating fraction.

1. On the day following processing of brain tissue, remove floating fraction from flask and place in a 15 mL plastic tube.

2. Wash flasks vigorously with media twice and pool with floating fraction. Centrifuge at $234g$ (equivalent to 1200 rpm) for 10 min. Add media to flask for microglia culture.

3. Discard supernatant and resuspend with media. Re-plate in one tissue culture flask.

4. Repeat **Steps 1–2**. Plate the floating fraction in poly-L-lysine-coated (10 µg/mL) 16-well glass chamber slides at 100,000 cells/well, leaving the wells at each extremity of the chamber slide empty. Fill the empty wells with PBS to reduce evaporation. Change the media every 3 days.

2.2.3.2. Microglia

1. Flask(s) from **Step 2** routinely contain(s) above 85% microglial cells adhered to the plastic. The other cells are a variable number of astrocytes, oligodendrocytes, and fibroblasts. Flasks obtained upon the second round of differential adhesion (**Step 4**) may also contain microglia, but in much sparser numbers.

2. Microglia can be used as plated in these flasks or collected, counted, and replated into different formats of tissue culture vessels. To detach microglia, first prepare a solution of 0.25% trypsin with 2 mM EDTA and warm up required medium.

3. Remove the medium from the microglia-containing flask(s) with suction and wash once with PBS. Use suction to remove all the PBS from the flask(s).

4. Add sufficient amount of trypsin/EDTA solution to cover the bottom of the flask (about 1 mL). Leave the flask at room temperature for approximately 30 s and aspirate leaving just a thin film of trypsin over the cells.

5. Place the flask(s) in the incubator for 5 min. During this time, dispense a small volume of medium into a 15 mL conical tube.

6. The next steps need to be performed promptly. Remove flask form incubator, quickly and vigorously hit the flask(s) 2–3 consecutive times against a flat surface and immediately verify under the microscope that most of the cells have been dislodged.

7. Place the flask(s) under the culture hood and flush the cells with approximately 5 mL of PBS. Collect this fraction and transfer it into the 15 mL conical tube containing medium.

8. Add PBS to the flask(s) and assess by microscopy that almost all the cells have been detached. If many cells are still adhered, repeat **Steps 6–8**, collecting the cells into the same tube.

9. Centrifuge the cells at 234$g$ (equivalent to 1200 rpm).

10. Count cells and plate at desired density. We usually allow the cells to recover from trypsinization for 48 h prior to subjecting them to experiments.

*2.2.4. Astrocyte Culture*

In the floating oligodendrocyte fraction, a few astrocytes may be identified immunocytochemically by expression of glial fibrillary acid protein (GFAP). However, we have not developed a technique to isolate these cells for culture purposes.

*2.2.5. Oligodendrocyte Progenitor Cell Isolation and Culture*

1. Oligodendrocyte progenitor cells can be isolated from the adult human brain tissue floating fraction following the two rounds of differential adhesion, by immunomagnetic bead selection against the surface ganglioside A2B5. The protocol is outlined in detail in the subsequent **Section 3** on human fetal brain preparations.

2. One should expect only 3–5% of total brain cells to be A2B5 immunopositive. Due to excessive cell loss during the magnetic bead separation procedure, mature oligodendrocytes and progenitor cells cannot be concurrently isolated from a brain preparation; although few can be obtained from the immunonegative fraction, the physical stress associated with the column separation decreases viability and ability to re-extend processes.

3. Progenitors can be plated on poly-L-lysine (10 μg/mL) coated 16-well glass chamber slides in DMEM F-12 supplemented with N1 and the growth factors PDGF (20 ng/mL, Sigma), bFGF (20 ng/mL, Sigma), and NT3 (2 ng/mL, R&D).

## 3. Fetal Human Preparation

### 3.1. Reagents and Equipment

*3.1.1. For Mixed Cell Population*

1. Human fetal brain tissue 16–24 weeks of gestation
2. Incubator at 37°C and 5% $CO_2$
3. Tissue culture hood
4. Pipette gun and 10 mL disposable plastic pipettes with wide mouth (Fisher)
5. Glass Petri dish, 100 mm
6. PBS, pH 7.4 (calcium- and magnesium-free)
7. Bunsen burner
8. 70% ethanol
9. Scalpels and disposable blades (2)
10. Glass bottles with cap (100 mL, 500 mL)
11. Trypsin (2.5% stock, Invitrogen)
12. DNase I final concentration 100 μg/mL (1 mg/mL stock, Roche Diagnostics)
13. Shaking water bath set at 37°C
14. FCS
15. Buchnel funnel with perforated plate removed, and containing a 132 μm nylon mesh (industrial fabrics)
16. 20 mL sterile plastic syringe plunger

17. Sterile 50 mL plastic conical tubes

18. Beckman Coulter Allegra 6 with GH 3.8 swing bucket rotors

19. Hemocytometer

20. 10 and 200 μL micropipettes

21. Bleach

*3.1.2. For Oligodendrocyte Progenitor Cell Isolation*

1. A2B5 IgM antibody (200 μL; derived from hybridoma)

2. Microbead-conjugated rat anti-mouse IgM secondary antibody (Miltenyi Biotech)

3. Dulbecco's modified medium (DMEM) F-12 (Gibco) supplemented with 1% penicillin–streptomycin (Invitrogen), 1% glutamine (Invitrogen), N1 supplement 1× (Sigma)

4. Growth factors PDGF-AA (20 ng/mL, Sigma), bFGF (20 ng/mL, Sigma), T3 (2 ng/mL, Sigma)

5. MACS buffer (0.5% FCS, 2 mM EDTA, in PBS) on ice

6. Miltenyi MS columns (3) and magnets (2)

*3.1.3. For Neuron Culture*

1. Minimal essential medium (MEM) (Gibco) supplemented with 1% penicillin–streptomycin (Invitrogen), 1% glutamax (Invitrogen), 5% FCS, 1% w/v glucose.

2. 5-fluoro-2′-deoxyuridine (DFU, Sigma)

3. 40 μm cell strainer (BD falcon)

4. Thermanox culture coverslips (NUNC)

5. Poly-L-lysine 10 μg/mL (Sigma)

*3.1.4. For Astrocyte Culture*

1. Dulbecco's modified eagle medium (MEM) (Gibco) supplemented with 1% penicillin–streptomycin (Invitrogen), 1% glutamine (Invitrogen), 10% FCS, 1% w/v glucose.

2. Poly-L-lysine 10 μg/mL (Sigma)

3. Trypsin (2.5% stock, Invitrogen)

*3.1.5. For Microglia Culture*

1. Dulbecco's modified eagle medium (DMEM) supplemented with 0.1% glucose, 5% FCS, 1% penicillin–streptomycin, 1% glutamine

2. Large falcon tissue culture flasks (175 cm$^2$) and plastic tissue culture vessel (flasks or plates)

**3.2. Protocol**

All of the following procedures are carried out in a tissue culture hood with ethanol-sterilized gloves and a lab coat. All discarded solutions that have come into contact with the brain preparation are incubated with bleach for a minimum of 24 h. Cultures are incubated in humidified incubators at 37°C pulsed with an atmosphere of 5% $CO_2$ and 95% air.

*3.2.1. Brain Mechanical and Chemical Dissociation*

1. Place brain in 100 mm Petri dish.

2. Remove excess shipment media with 10 mL pipette.

3. Add sterile PBS to brain to wash away excess media until solution is clear.

4. Flame ethanol-doused forceps under a Bunsen burner flame and remove blood clots, meninges, and surface blood vessels (small blood vessels within the tissue cannot be removed).

5. Remove almost all the liquid from the dish leaving behind the brain tissue pieces. Flame ethanol-doused scalpels under a Bunsen burner flame and cut up tissue into <1 mm fragments using both scalpels simultaneously.

6. Collect the brain pieces with a 10 mL pipette and put into a sterile 100 mL glass bottle.

7. Use PBS to wash the Petri dish and collect left over brain pieces.

8. Complete the volume in the bottle up to 100 mL.

9. Add trypsin (final concentration 0.25%) and DNAse I (final concentration 25 μg/mL) to the bottle. For large preparations (>22 gestational weeks), the concentrations of trypsin and DNAse I can be doubled.

10. Parafilm the lid of the bottle and place in a shaking water bath set at 37°C set at 180 rpm for 30 min. The incubation time may be reduced to 15–20 min for smaller preparations.

11. Add 4 mL of FCS to the preparation to inactivate the trypsin. Swirl the bottle to distribute the serum within the bottle.

12. Place funnel with mesh onto sterile 500 mL glass bottle.

13. Pour brain preparation onto mesh 10 mL at a time. Use plunger of a 20 mL plastic syringe to grind the tissue through the mesh in a circular motion. Use PBS to wash the bottle and add to the mesh, and add PBS directly to the mesh for final wash. Total volume in the bottle should amount to a maximum of 300 mL.

14. Transfer the brain preparation to six 50 mL plastic conical tubes.

15. Centrifuge at 365$g$ (with a Beckman Coulter Allegra 6 with GH 3.8 swing bucket rotors, this is equivalent to 1500 rpm) for 10 min with high brakes.

16. Slowly pour out supernatant into bleach (the pellet is soft and will easily be lifted if the tubes are not handled gently).

17. Resuspend pellet in one tube using 5 ml of PBS, transfer the amount to next tube and resuspend the pellet. Continue until the brain preparation is pooled into one conical tube.

18. Complete to 50 mL with PBS and centrifuge at 1200 rpm ($234g$) for 10 min with high brakes.

19. Discard supernatant into bleach and resuspend the pellet either in 5 mL of PBS (if cell suspension will be divided for culture of various cell types) or 5 mL of cell-type specific media.

20. Remove small aliquot (10 µL) to perform a live cell count using a hemocytometer using a 1:10 dilution. Exclude red blood cells from the count. Total number may vary between $3 \times 10^6$ and over $1.5 \times 10^9$ cells.

*3.2.2. Oligodendrocyte Progenitor Cell Isolation and Culture*

1. Immunomagnetic bead selection requires an initial amount of $7 \times 10^6$–$1 \times 10^9$ total cells.

2. Centrifuge the cell suspension at 1200 rpm ($234g$) for 10 min with high brakes.

3. Remove as much supernatant as possible, using a pipetteman to remove excess supernatant.

4. Resuspend the pellet with 200 µL A2B5 IgM antibody and place on ice for 30 min. As cells have the tendency to reaggregate, repeat the resuspension every 10 min during the incubation period.

5. Complete to 50 mL with MACS buffer and centrifuge at 1200 rpm for 10 min with high brakes.

6. Discard supernatant and resuspend pellet with 8 mL of MACS buffer and 2 mL of microbead-conjugated rat anti-mouse IgM secondary antibody (according to the manufacturer's instructions). Incubate at 4°C for 20 min.

7. Complete to 50 mL with MACS buffer and centrifuge at 1200 rpm ($234g$) for 10 min with high brakes.

8. Discard supernatant and resuspend pellet thoroughly with 5 mL of MACS buffer, attempting to eliminate any clumps that may clog the columns.

9. Insert 3 MS columns into the Miltenyi magnets and rehydrate with 3 mL of MACS buffer per column. Discard flow-through.

10. Place 15 mL plastic polypropylene tube beneath each magnet to collect the negative fraction (A2B5 immunonegative).

11. Place a 200 µL pipetteman tip on the end of a 10 mL pipette to eliminate the possibility of tissue aggregates from

entering the pipette. Distribute the resuspended brain tissue amongst the columns.

12. Once the preparation has run through, wash the 50 mL conical tube with MACS buffer and distribute amongst the columns. Repeat.

13. Add 3 mL of MACS buffer directly to each column.

14. Remove the columns from the magnet and place in three new 15 mL tubes.

15. Add 5 mL of MACS buffer to each column and quickly press down with plunger to remove the A2B5 immunopositive cells from the column.

16. Centrifuge positive fraction at 1200 rpm ($234g$) for 10 min with high brakes.

17. Resuspend with DMEM-F/12 supplemented with N1, PDGF-AA (20 ng/mL), basicFGF (20 ng/mL), and T3 (2 ng/mL).

18. For survival purposes, cells must be plated on a poly-lysine-coated (10 μg/mL) plastic coverslips, in a 24-well plastic plate, with an underlying confluent bed of human fetal astrocytes (derived from a previous brain preparation) that have been lysed with sterile water and washed thrice with PBS. Astrocytes (used at third passage) become confluent over the course of 2 days post-plating, and may be stored in the lysed format for 2 weeks at 4°C.

19. Plate A2B5 positive cells on top of lysed human fetal astrocytes at $2 \times 10^5$ cells per well. Media is changed with fresh supplementation of growth factors every 2–3 days. Cells will begin to differentiate into mature neural cell types following 6 days of culture. For fixation for immunochemistry, cells are first washed in PBS and then fixed for 45 min in 2% paraformaldehyde, and subsequently washed thrice in PBS and stored at 4°C for up to 1 week before immunostaining.

*3.2.3. Neuron Isolation and Culture*

Before isolation, coat all plasticware with poly-L-lysine at 10 μg/mL. Poly-L-lysine is diluted in sterile water for 2 h, washed once with water and dried for several hours or overnight in a tissue culture hood. Poly-lysine-coated plasticware may be stored for 1 month at 4°C before use. The coated plasticware is then sterilized for 30 min under ultraviolet light in a tissue culture hood to reduce possibility of contamination. Thermanox coverslips may be added to wells of a 24-well plate prior to poly-lysine coating (the coverslip allows removal of culture later on for small-volume staining and mounting on thin glass for confocal analysis). Plastic or glass chamber slides are not recommended as neurons grow unevenly and lift easily.

1. Following mechanical and chemical dissociation, aliquot the desired number of cells (*see* below for plating density), take about 20% more than necessary as cells are lost in the subsequent steps.

2. Wash cells once in ~5 mL of MEM supplemented with 5% FCS and 0.1% glucose. Resuspend pellet in media at $2.5 \times 10^6$/mL.

3. Pour through a 40 μm cell strainer into a 50 mL conical tube. Recount cells and adjust to $2.2 \times 106$/mL.

4. Pipette 1 mL/well of the $2.2 \times 10^6$/mL suspension in poly-L-lysine-coated 24-well plates. Only fill the center eight wells, and fill the surrounding empty wells with PBS to prevent evaporation.

5. Incubate for 4 days. Change the media with MEM supplemented with 5% FCS and 0.1% glucose containing 1 mM DFU every 3 days for a total of three changes. After the last change the cells can be used. Pre-warm all media to 37°C, and pipette slowly to avoid lifting neurons. Expect to attain $\sim 0.5 \times 10^6$ neurons per well, with 10–20% contaminating astrocytes. They will last at least a week before the neurons begin to degenerate.

6. The experiment is carried out in the 24-well plate without disturbing neurons. Trypsinization is not recommended as it reduces viability and quality of neurons.

*3.2.4. Astrocyte Isolation and Culture*

1. Following mechanical and chemical dissociation, aliquot the desired number of cells. Wash in ~5 mL of DMEM supplemented with 10% FCS.

2. Put $30–50 \times 10^6$ cells in 30 mL of media, transfer to a tissue culture Falcon flasks (75 cm$^2$) coated in advance with poly-L-lysine (*see* previous **Section 3.2.3** on neurons for poly-L-lysine coating procedure).

3. Incubate for 1 week, and then passage using trypsin (0.25%) diluted in PBS from a 2.5% trypsin stock. Wash once quickly with trypsin, then trypsinize at room temperature for approximately 10 min until astrocytes detach. Avoid banging flask as this reduces cell viability. Split into two flasks with 30 mL media per flask. Repeat for two passages, increasing the number of flasks each time if many astrocytes are required. On passage (P) 3 the astrocytes do not require poly-L-lysine coating. They can be maintained in the flask without passage for weeks to months with occasional media change. One expects to have 90–95% pure astrocytes by

P3. The contaminating cells are neurons. The astrocytes may be passaged for up to 5 or 6 times if higher purity desired.

4. A few days prior to the experiment, astrocytes are trypsinized and plated. They grow in any size glass or plasticware (with or without poly-L-lysine), and will grow in a variety of media's including DMEM supplemented with 10% FCS or MEM supplemented with 5% FCS and 0.1% glucose. The astrocytes divide rapidly; to achieve a confluent monolayer seed at approximately 50% confluency 2–3 days prior to the onset of the experiment. When confluency is reached, astrocytes will stop growing. In 24 well plates, $0.2 \times 10^6$ will yield a confluency of approximately 50%.

*3.2.5. Microglia Isolation and Culture*

1. Immediately following dissociation, plate the cells at approximately 6 million cells/ml in 45 mL of DMEM 5% in as many T175 as needed (cell density = approximately $1.5 \times 10^6$ cells/cm$^2$). Astrocytes will adhere to the bottom of the flask, whereas microglia will tend to free-float above the astrocytes/neuron layer.

2. One week after the isolation, proceed to change the media as follows:

    (a) Prepare equal number of 50 mL conical tubes as there are Falcon 175 cm$^2$ flasks.

    (b) To empty T175 flask: carefully pour out medium of each flask into a separate 50 mL conical tube.

    (c) Add 25 mL of medium to flask (expel liquid on the surface opposite to the cell layer). Place in the incubator.

    (d) Centrifuge the tubes at 234$g$ (equivalent to 1200 rpm) for 10 min.

    (e) Pour out supernatant, resuspend the pellet in 25 mL of medium.

    (f) Add back to the original flasks (do not expel directly on cell layer or astrocytes will detach).

3. One week later, or as soon as the astrocyte adherent layer becomes too confluent and threatens to detach:

    (a) Carefully collect the floating cells, take care not to make jerky movements or hit the flask as this will likely trigger the astrocyte layer to detach in one sheet.

    (b) Centrifuge at 234$g$ and resuspend pellet in DMEM 5% FCS.

    (c) Count cells and plate at the desired density. Fetal microglia will adhere to plastic.

## 4. Critical Steps and Troubleshooting

### 4.1. Human Adult Preparations

1. If adult tissue is obtained from block dissections, the results differ. Our tissue originates from CUSA bags where the tissue is in small sections of $\sim$2 mm$^3$. For block sections, we would suggest obtaining the tissue quickly following removal from the brain and cutting into small pieces with sterile scalpels.

2. When growing adult mature oligodendrocytes for periods of less than 1 week, the density should be at 100,000 cells/well, whereas for longer culture periods (>2 weeks) cells should be plated at minimum 150,000 cells/well.

3. When plated oligodendrocyte cultures have much myelin debris, wash the wells slowly in a circular manner and change the media.

4. If there are many small blot clots in the brain tissue, a rapid trypsinization may help to solubilize them. If this is the case, begin by performing **Steps 1–6** (section 3.2.1) to remove as much blood and clots as much as possible by pouring and pipetting larger clots. Then, transfer tissue to a 500 mL glass bottle, add PBS so that the final volume is about 100 mL. Add trypsin (0.025%) and DNAse (10 μg/mL). Swirl the bottle to distribute the enzymes and keep observing until most of the clots have been released into the PBS. As soon as this happens, add more PBS, allow the tissue to settle at the bottom of the bottle, decant the liquid, and continue with **Step 7**.

5. Microglial cells prefer a plastic growth surface to adhere to and will survive for several weeks. Microglia will not thrive well on uncoated glass, and their survival is also more limited on poly-L-lysine-coated glass.

6. Microglial cells: although microglial cells can grow in a wide range of cell densities, we usually plate cells at no more than 62 500 cells/cm$^2$ as these cultures tend to be too crowded.

### 4.2. Human Fetal Preparations

1. If brain is large (>21 gestational weeks), use double the amount of trypsin and DNAse I during the dissociation steps.

2. During the progenitor isolation procedure, to avoid clogging of the columns with aggregates of tissue that cannot be dissociated, add a 200 μL micropipette tip to the end of a 10 mL plastic pipette to eliminate the possibility of clumps

entering the pipette prior to addition of brain tissue to the column.

3. The immunomagnetic bead selection results in an estimated loss of 50% of total cells. This has to be taken into account with regards to the size of the starting material and the minimum amount of cells required for the immunomagnetic bead selection.

4. Neurons: the neuron culture does not grow well in reduced sized formats such as 16-well circle or 8-well square chamber slides. They can be grown in larger formats such as 35 mm plastic or glass dish or flasks, as long as poly-L-lysine coating is done.

5. Neurons: neuronal cultures always contain 10–20% contaminating astrocytes regardless of the use of a mitotic inhibitor. Serum-free media may be used to limit astrocyte growth. However, highly purified human neurons (obtained by cell sorting via flow cytometry) do not survive well in the absence of the astrocytes; astrocyte-conditioned media does not promote neuron survival.

6. Microglia: floating fetal microglia will undergo proliferation growing over the layer of astrocytes. Hence, it is advisable to wait as long as possible (up to 2 weeks as described above) before harvesting them in order to maximize yield.

7. Microglia: once fetal microglia are harvested from the astrocyte layer and plated on their own, they will adhere to plastic. A heterogeneous population coexists of cells that remain round and cells that display bipolar or multipolar morphology. They typically contain numerous vesicles, display autofluorescence and high Fc-binding capacity. This is important to consider when staining these cells for flow cytometry or immunofluorescence. Adequate blocking of Fc receptors and assessment of the background signal must be performed.

8. Microglia: closely examine enriched microglial cultures for possible astrocyte contamination as those will proliferate at a much faster rate and take over the culture. Astrocyte contamination can occur if some astrocytes detach when the floating fraction is collected, so it is recommended to proceed with this step as gently as possible to avoid contamination with astrocytes. Astrocytes will grow in patches of phase-dark flat cells sometimes associated with a few neurons on top of them.

9. Microglia: Like human adult microglia, fetal microglia prefer plastic surfaces to adhere to and grow. They also survive well over weeks of culture.

## 5. Typical Protocol Results

*See* **Figs. 5.4 and 5.5** for further details.

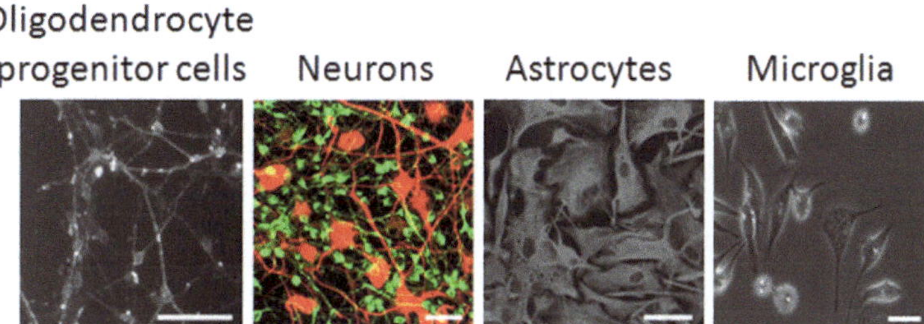


Fig. 5.4. Human fetal CNS cells. Representative image of oligodendrocyte progenitor cells labeled with A2B5; they are also immunopositive for PDGFR$\alpha$ and are typically bipolar. Neurons are labeled with $\beta$TubIII (*green*) and grow on top of GFAP-positive astrocytes (*red*); neurons are also immunopositive for MAP-2 and neurofilaments. Astrocytes are labeled with GFAP. Microglia are demonstrated as a bright-field image, and are typically immunopositive for CD68, MHC-II, CD11c, and CD45. Scale bars, 50 $\mu$m.

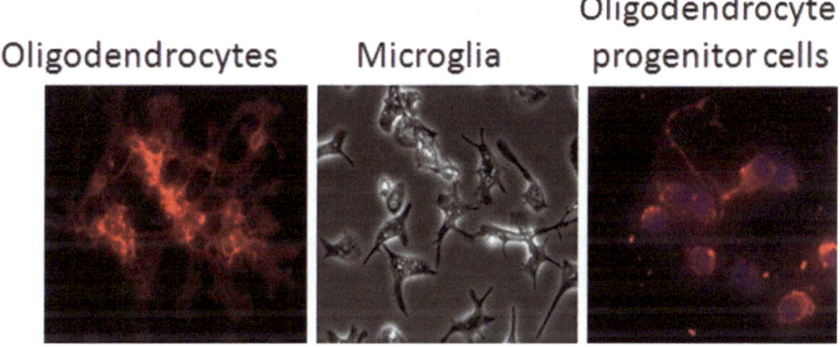


Fig. 5.5. Human adult CNS cells. Representative image of mature oligodendrocytes labeled with MAG; these cells are also immunopositive for MBP, MOG, NogoA, and GalC. Microglia are represented as a bright-field image and are immunopositive for CD68, MHC-II, CD11c, and CD45. Oligodendrocyte progenitor cells are labeled with A2B5 and are also immunopositive for another progenitor marker, PDGFR$\alpha$; a subset of these cells is immunopositive for GalC, MAG, and NogoA. Scale bars, 50 $\mu$m.

## Acknowledgments

We thank all the students, post-doctoral fellows, and technicians who have worked in our Neuroimmunology Unit and participated in the development and refinement of the tissue culture protocols presented here.

## References


1. Ruffini F, Arbour N, Blain M, Olivier A, Antel JP (2004) Distinctive properties of human adult brain-derived myelin progenitor cells. Am J Pathol 165:2167–2175.
2. Miron VE, Hall JA, Kennedy TE, Soliven B, Antel JP (2008) Cyclical and dose-dependent responses of adult human mature oligodendrocytes to fingolimod. Am J Pathol 173:1143–1152.
3. Jack CS, Arbour N, Blain M, Meier UC, Prat A, Antel JP (2007) Th1 polarization of CD4+ T cells by Toll-like receptor 3-activated human microglia. J Neuropathol Exp Neurol 66:848–859.
4. Biernacki K, Prat A, Blain M, Antel JP (2001) Regulation of Th1 and Th2 lymphocyte migration by human adult brain endothelial cells. J Neuropathol Exp Neurol 60:1127–1136.
5. Prat A, Biernacki K, Wosik K, Antel JP (2001) Glial cell influence on the human blood-brain barrier. Glia 36:145–155.
6. Darlington PJ, Goldman JS, Cui QL, Antel JP, Kennedy TE (2008) Widespread immunoreactivity for neuronal nuclei in cultured human and rodent astrocytes. J Neurochem 104:1201–1209.
7. Darlington PJ, Podjaski C, Horn KE, Costantino S, Blain M, Saikali P, Chen Z, Baker KA, Newcombe J, Freedman M, Wiseman PW, Bar-Or A, Kennedy TE, Antel JP (2008) Innate immune-mediated neuronal injury consequent to loss of astrocytes. J Neuropathol Exp Neurol 67:590–599.
8. Dorr J, Roth K, Zurbuchen U, Deisz R, Bechmann I, Lehmann TN, Meier S, Nitsch R, Zipp F (2005) Tumor-necrosis-factor-related apoptosis-inducing-ligand (TRAIL)-mediated death of neurons in living human brain tissue is inhibited by flupirtine-maleate. J Neuroimmunol 167:204–209.
9. Chojnacki A, Kelly JJ, Hader W, Weiss S (2008) Distinctions between fetal and adult human platelet-derived growth factor-responsive neural precursors. Ann Neurol 64:127–142.

# Chapter 6

# Bioengineering Protocols for Neural Precursor Cell Expansion

Behnam A. Baghbaderani, Arindom Sen, Michael S. Kallos, and Leo A. Behie

## Abstract

Neural precursor cells (NPCs) isolated from different regions of the developing or adult CNS may represent a new source of cells that may have utility in future cell replacement therapies aimed at treating neurodegenerative disorders. Moreover, these cells may have applications in a number of non-clinical areas from basic biological research to gene and drug delivery. However, their sparse concentration within the CNS means that they can only be isolated in small quantities, and thus need to be expanded to numbers that are sufficient for clinical and non-clinical applications. This chapter discusses the cell-handling protocols relevant to the expansion of murine NPCs and human NPCs in both standard tissue culture flasks (as effective means for handling the small quantities of stem cells following isolation from primary tissue) and suspension bioreactors (as highly efficient mode of culture for production of large number of cells in a standardized, controlled manner). Considering that mammalian NPCs can grow under serum-free conditions as neurospheres, or adhere to the culture flask surface, passaging protocols concerning aggregate dissociation techniques, cell sampling and inoculation procedures are described for both floating aggregates of cells and adherent cultures. Special attention is paid to important design considerations (i.e., mass transfer and culture hydrodynamics) as well as process control techniques as crucial parameters to achieve a reproducible and productive cell expansion process in suspension bioreactors.

**Key words:** Bioreactor expansion, cell therapy, long term expansion, neural precursor cells, serum-free medium, scale-up, suspension culture.

## 1. Introduction

Neural precursor cells (NPCs) represent a new source of cells that may have utility in future cell replacement therapies aimed at treating neurodegenerative disorders such as Parkinson's disease (PD) and Huntington's disease (HD). Moreover, these cells may have applications in a number of non-clinical areas from basic biological research to gene and drug delivery. NPCs can be isolated

L.C. Doering (ed.), *Protocols for Neural Cell Culture*, Springer Protocols Handbooks,
DOI 10.1007/978-1-60761-292-6_6, © Humana Press, a part of Springer Science+Business Media, LLC 1992, 1997, 2001, 2010

from different regions of the developing or adult mammalian central nervous system (CNS), including the forebrain, ventral mesencephalon, brain stem, and spinal cord (1–6). However, their sparse concentration within the CNS means that they can only be isolated in small quantities, and thus need to be expanded to numbers that are sufficient for clinical and non-clinical applications. The expansion of NPC populations in culture requires two major steps: (1) the development of a cell growth medium that can support the expansion of NPCs while allowing them to maintain their defining cell characteristics and (2) the establishment of protocols for the efficient and reproducible serial expansion of the cells in culture. The growth medium may be developed by performing growth trials to study the impact of individual or groups of major medium components including basal media; growth factors, such as epidermal growth factor (EGF), basic fibroblast growth factor (bFGF), and human leukemia inhibitor factor (hLIF); hormones; and buffer solutions. A typical starting point is to choose an existing growth medium that had been developed previously for the expansion of other mammalian cells and then modifying it appropriately to support the expansion of mammalian NPCs (7).

Serial subculturing of NPCs may be conducted in stationary culture (i.e., tissue culture flasks) or suspension bioreactors. The use of static vessels in stationary culture is advantageous since they are not technically challenging to use, are effective for handling the small quantities of precursor cells following isolation from primary tissue, and can be used to efficiently investigate the impact of several experimental conditions using small culture volumes. However, the production of large quantities of NPCs in tissue culture flasks is not feasible for several reasons. First, the production of large numbers of cells would require the use of large numbers of flasks. This would be very labor-intensive, and since each flask would be handled in a different manner, and possibly by different individuals, reproducibility would be compromised. Second, static culture vessels are typically operated only in batch mode. Third, static culture flasks are not equipped with process control devices and, therefore, important physiological parameters such as culture pH, temperature, and oxygen concentration may detrimentally fluctuate during the cell expansion process (7, 8). Alternatively, standard suspension bioreactors are amenable to scale-up, and a single bioreactor vessel can be used effectively to produce the same number of cells as hundreds of tissue culture flasks. Each bioreactor vessel can be cost-effectively operated by a single individual, and computer control systems can continuously monitor the culture to ensure that optimum and homogeneous conditions are maintained.

This chapter describes protocols that can be used for expanding populations of NPCs growing as adherent cells or floating neurospheres in culture. Special focus has been made on the expansion of these cells in suspension bioreactors.

## 2. Reagents and Equipment

1. Portable pipette-aid (BD Falcon, Cat# 357590)
2. 5 mL plastic pipettes (BD Falcon, Cat# 357543)
3. 10 mL plastic pipettes (BD Falcon, Cat# 357551)
4. P20 pipetteman (Gilson, Cat #G10534G)
5. P200 pipetteman (Gilson, Cat #G11310C)
6. P-1000 pipetteman (Eppendorf, Cat #2244020-9)
7. T-75 tissue culture flasks (Nunc, Cat#156499)
8. T-25 tissue culture flasks (Nunc, Cat#156367)
9. Standard 125 mL spinner flasks (Corning, Cat # 26502-125) with paddle impeller (**Fig. 6.1**)
10. Thermolyne Cellgro™ system (Thermolyne, Cat# 14511102)
11. Wheaton 500 mL suspension bioreactor (Wheaton Science Products, Cat#356922) (**Fig. 6.2**)
12. Wheaton control unit (Wheaton Science Products)
13. Micro stir model II single stir plate (Wheaton, Cat# 902-400)
14. 0.25% trypsin–EDTA (GIBCO, Cat#25200072)

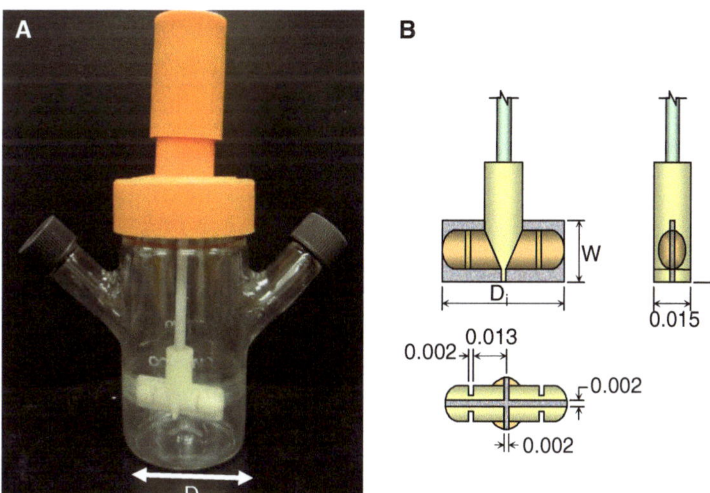


Fig. 6.1. Geometries of 125 mL spinner flasks and impellers configurations. Shown are 125 mL Corning spinner flask (**A**), and paddle impeller configuration for 125 mL spinner flasks (**B**). Standard 125 mL spinner flasks (Corning) with paddle impeller are constructed of cell culture grade glass, and the vessel bottoms are rounded and contain small raised area immediately below the impeller shaft to minimize mixing dead zones. $D_T$, $D_i$, and $W$ are 0.065, 0.052, and 0.018 m, respectively. The working volume for 125 mL bioreactors is 100 mL.

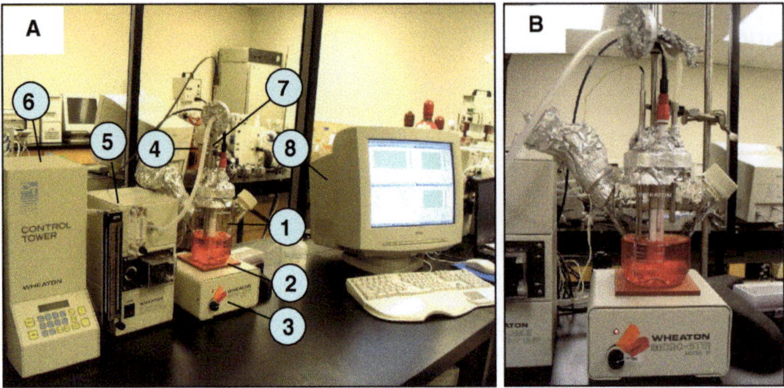

Fig. 6.2. Photographs of 500 mL computer-controlled suspension bioreactor. (**A**) Experimental setup for the 500 mL suspension bioreactor showing: (1) 500 mL suspension bioreactor, (2) heating pad, (3) stirrer plate, (4) gas inlets to the bioreactor connected to $O_2$, $CO_2$, $N_2$, and air gas cylinders via Wheaton pump, (5) Wheaton pump, (6) the control tower (Wheaton) connected to the dissolved oxygen (DO), pH, and temperature probes to monitor the level of each parameter, (7) DO, pH, and temperature probes, and (8) data acquisition computer. (**B**) A closer view of the 500 mL suspension bioreactor and measuring probes. Adapted from Baghbaderani et al. (11).

15. 0.25 mg/mL soybean trypsin-inhibitor solution (Invitrogen, Cat# 17075029)

16. Sigmacote (Sigma, Cat#SL-2)

17. Hexane (Omnisolv)

18. Dimethyl sulfoxide (DMSO) (Sigma, Cat#D-5879)

## 3. Protocols for Cell Propagation in Stationary Culture

Depending on the anatomical region of isolation, mammalian NPCs can grow under serum-free conditions as floating cell aggregates known as neurospheres, or adhered to the culture flask surface. Human NPCs isolated from the forebrain, brain stem, or spinal cord of the CNS typically form floating neurospheres in serum-free growth medium whereas human ventral mesencephalon-derived NPCs adhere to the surface of tissue culture flasks in serum-free medium (**Fig. 6.3**) (2). After a period of time in batch culture, it may be necessary to move some of the cells to a new culture vessel with fresh medium to avoid the detrimental effects associated with nutrient depletion, toxic metabolite build-up, and the formation of extremely large aggregates. This movement of cells is called passaging or subculturing, and repeatedly carrying out this procedure is referred to as serial subculturing.

The following describes the serial subculturing protocols used for NPCs grown in the form of suspended aggregates or adhered cells in stationary culture.

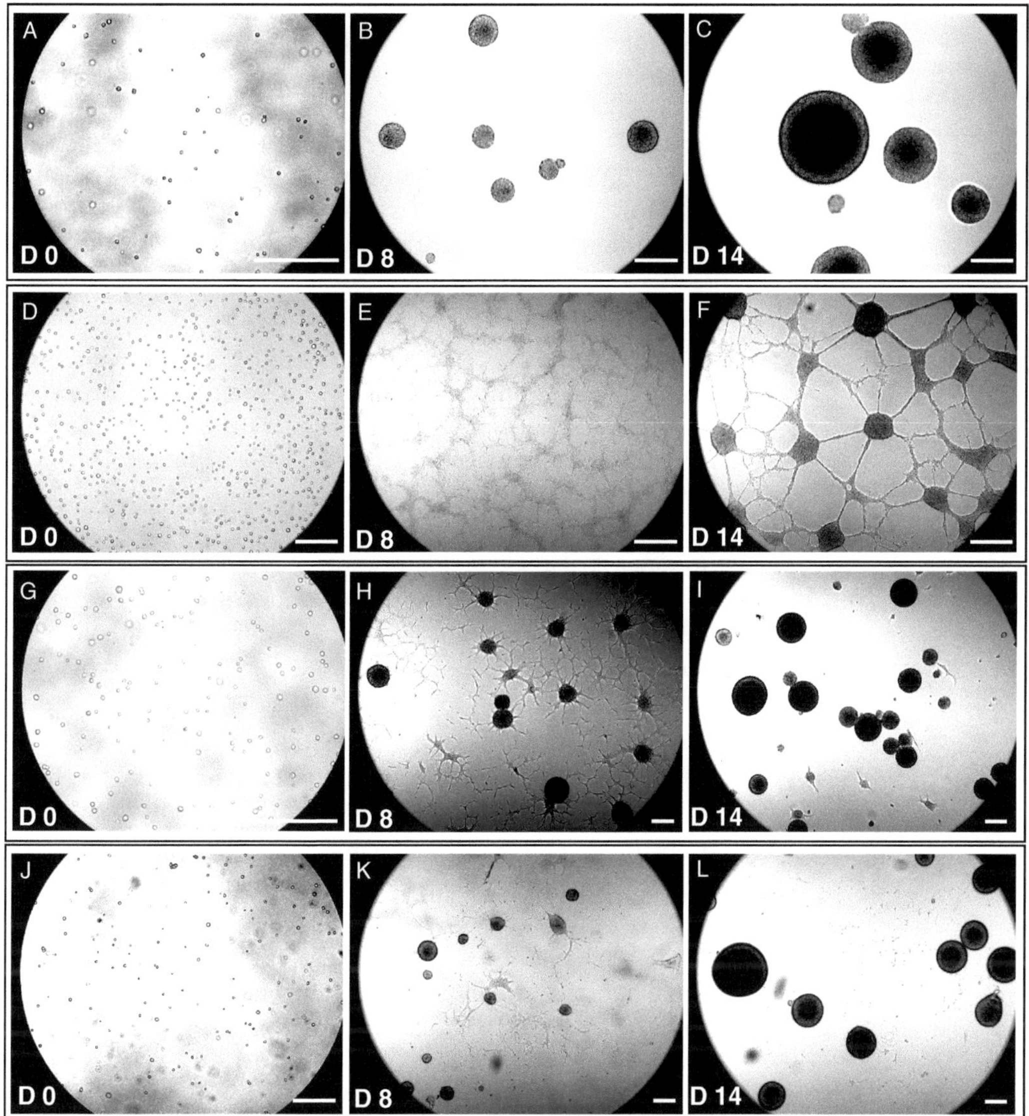

Fig. 6.3. Photomicrographs of human NPCs obtained from different regions of the central nervous system and grown in stationary culture. Shown are telencephalon-derived human NPCs immediately after inoculation (**A**), 8 days post-inoculation (**B**), and 14 days post-inoculation (**C**); ventral-mesencephalon-derived human NPCs immediately after inoculation (**D**), 8 days post-inoculation (**E**), and 14 days post-inoculation (**F**); brain-stem-derived human NPCs immediately after inoculation (**G**), 8 days post-inoculation (**H**), 14 days post-inoculation (**I**); and spinal cord-derived human NPCs immediately after inoculation (**J**), 8 days post-inoculation (**K**), and 14 days post-inoculation (**L**). Scale bar = 250 μm. Adapted from Mukhida et al. (2).

### 3.1. Serial Subculturing of Suspended Aggregates

Oxygen and nutrient concentrations at the center of multicellular aggregates are inversely related to the size of the aggregates. Cells at the centre of aggregates that grow beyond a critical diameter experience inadequate oxygen and nutrient levels, and may experience a build-up of toxic metabolic by-products, which can cause them to die and form a necrotic core. This can be avoided by dissociating the cell aggregates into a single cell suspension before they reach the critical diameter, and then placing the single cells into fresh growth medium so they can proliferate and form new aggregates. For murine NPCs, the critical aggregate size limit is about 150 μm (9, 10), which is usually attained after 4–5 days in culture. However, human NPCs can form aggregates as large as 600 μm without negatively affecting the viability of the culture (2, 6, 11, 12).

### 3.1.1. Aggregate Dissociation Procedure

Three methods have been described in the literature to dissociate NPC aggregates into single cells: mechanical dissociation, enzymatic dissociation, and chemical dissociation (1, 6, 13, 14). Here mechanical and enzymatic dissociation methods are described.

#### 3.1.1.1. Mechanical Dissociation

1. Remove T-flasks containing murine NPC aggregates (after 4–5 days of growth in culture) or human NPC aggregates (after 14–20 days of growth in culture) from the humidified incubator (37°C, 5% $CO_2$) and place in a sterile biosafety cabinet.

2. Using a 5 mL plastic pipette, aspirate the contents of each flask once to ensure it is well mixed, and carefully place 5.0 mL of the flask contents into one 15 mL centrifuge tube.

3. Centrifuge the sample for 10 min at 140$g$.

4. Remove the supernatant from the sample with the exception of approximately 200 μL (including the pellet). Triturate the remaining volume using a P200 pipetteman set to a value 30 μL less than the measured volume of the pellet and supernatant. The pipette should be placed inside the 15 mL centrifuge tube such that the tip is proximal to the bottom surface of the tube at an angle of approximately 85° from horizontal, thereby creating a small gap through which the cells can pass. Trituration is performed by repeatedly drawing the neurospheres into the pipette tip and expelling them while holding the tip against the bottom surface of the centrifuge tube. To attain a single cell suspension, murine cell aggregates need to be triturated approximately 30–40 times whereas human cell aggregates require 50–70 triturations. Using a P20 pipetteman, aliquot two 10 μL samples from the single cell suspension to perform cell counts.

#### 3.1.1.2. Enzymatic Dissociation

1–3. Harvest the cells according to Steps (1)–(3) as described in **Section 3.1.1.1**

4. Remove all of the supernatant from the sample, leaving only the cell pellet.

5. Add 1.0 mL of 0.25% Trypsin–EDTA solution to the pellet.

6. Using a P-1000 pipetteman, aspirate the pellet to resuspend the cell aggregates in the enzymatic solution and incubate the mixture at 37°C for 10 min. Remove the sample from the incubator, and in a biosafety cabinet, carefully add 5.0 mL of growth medium to the centrifuge tube to dilute the trypsin.

7. Centrifuge the sample for 10 min at 140$g$

8. Remove the supernatant and add 1.0 mL of fresh growth medium to the pellet.

9. Resuspend the pellet in the medium and aspirate the cell suspension 25–30 times using a P-1000 pipetteman set to a volume of 850 μL to achieve a single cell suspension.

10. Using a P20 pipetteman, aliquot two 10 μL samples from the single cell suspension to perform cell counts.

*3.1.2. Cell Inoculation and Propagation*

The inoculation procedure used to introduce mammalian cells into a fresh culture vessel is important as it can have a significant impact on the productivity of the culture. The following is a description of an inoculation protocol, which can be used to serially subculture mammalian NPCs.

1. Prepare each new static culture vessel ahead of time by adding fresh medium at a rate of approximately 0.2 mL/cm$^2$, and then placing the vessel into an incubator (37°C, 5% $CO_2$) for approximately 1 h in order to allow the medium to equilibrate to the appropriate temperature and pH prior to inoculation.

2. Remove a tissue culture flask containing the actively growing NPC aggregates from the incubator and place it into a biosafety cabinet. Isolate the cell aggregates and dissociate them into a single cell suspension using one of the methods described in **Section 3.1.1.**

3. Upon acquiring a single cell suspension, perform cell counts and then inoculate the single cell suspension at a density of 75,000 cells/mL (for murine NPCs) or 100,000 cells/mL (for human NPCs) into a new tissue culture flasks containing fresh medium (equilibrated as in Step (1)).

4. Swirl the T-flask gently to distribute the cells throughout the medium, and then incubate the T-flask in a humidified chamber at 37°C and 5% $CO_2$ for 4–6 days (for murine NPCs) or 14–20 days (for human NPCs).

5. Due to the longer culture period for human NPCs at a given passage level, it is necessary to feed the cultures every 4–5 days by replacing 40% of the spent medium with fresh growth medium. To minimize cell loss during feeding, tilt

the tissue culture flask to a near vertical position, and after waiting for approximately 20 s to allow the cells to settle, carefully remove the spent medium from the top portion of the vessel contents. After feeding, place the T-flask back into a humidified incubator at 37°C and 5% $CO_2$.

### 3.2. Serial Subculturing of Adherent Cells

As mentioned earlier, some of the mammalian NPC cell lines such as human ventral mesencephalon-derived NPCs grow only when adhered to a surface. This adherence is mediated by divalent cations (e.g., $Ca^{2+}$) and non-specific proteins that form a layer between the cells and the substratum (7). Given the nature of the attachment, when the cells need to be harvested, they can be detached from the surface by utilizing a combination of a proteolytic enzyme such as trypsin and a chelating agent such as EDTA. The following procedure describes protocols that can be used to harvest adherent mammalian NPCs and serially subculture them in stationary culture.

### 3.2.1. Cell Harvesting, Inoculation, and Propagation Procedure

1. Remove a T-flask containing adherent cells from the incubator and place it into a biosafety cabinet. Tilt the T-flask and carefully withdraw the cell culture medium from the T-flask using a sterile 5 mL plastic pipette.

2. Add 0.25% trypsin–EDTA to the tissue culture flask (3.0 mL for a T-25 flask and 7.0 mL for a T-75 flask). Incubate the T-flask at 37°C for 10 min.

3. After 10 min, transfer the T-flask to the biosafety cabinet and add an equal volume of soybean trypsin-inhibitor solution (0.25 mg/mL) to the culture to stop trypsin activity.

4. Aspirate the contents of the culture twice using a 5 mL plastic pipette to lift the cells from the surface of the vessel, and transfer all of the cell suspension into a 15 mL centrifuge tube. Centrifuge the cell suspension for 10 min at 140$g$.

5. Remove the supernatant, and add 1.0 mL of fresh growth medium to the cell pellet.

6. Resuspend the pellet in the growth medium by gently aspirating (10 times) the cell suspension using a P1000 pipetteman set to a volume of 850 μL.

7. Upon acquiring a single cell suspension, use a P20 pipetteman to aliquot two 10 μL samples and perform cell counts.

8. After performing cell counts, inoculate the cells into new tissue culture vessels containing fresh medium at a rate of 100,000 cells/mL (for human VM-derived NPCs). Note that as described in **Section 3.1.2**, the fresh medium should be preincubated at 37°C and 5% $CO_2$ for a period of at least 1 h prior to being inoculated.

9. Swirl the T-flask gently to distribute the cells throughout the medium, and then incubate the T-flask in a humidified chamber at 37°C and 5% $CO_2$ for 14–20 days (for human VM-derived NPCs).

10. Every 4–5 days, feed the culture by replacing 40% of the spent medium with fresh growth medium.

## 4. Protocols for Cell Propagation in Small-Scale Suspension Bioreactors

Small-scale suspension bioreactors (such as 125 mL spinner flasks) are an important link when scaling up cell production from small-scale stationary culture to large-scale computer-controlled suspension bioreactors. Not only can they provide the volume and homogenous environment required to efficiently generate the quantities of cells needed to seed large-scale bioreactors, but they can also be used to conduct a number of important preliminary studies such as determining the impact of shear stress on the cells and evaluating mass transfer characteristics under agitated conditions.

### 4.1. Important Design Considerations

Maintaining mass transfer rates and culture hydrodynamics at acceptable levels are very important when developing suspension bioreactor protocols for expanding mammalian cells. In particular, efforts should be made to maintain dissolved oxygen concentrations at optimum levels, despite the low solubility of oxygen in cell culture media (0.22 mM at 37°C in an air-saturated aqueous solution). Surface aeration (gas diffusion through the culture surface) and sparging (direct aeration within the growth medium) are the most common methods of supplying oxygen to the cells grown in culture. Due to simplicity, surface aeration is often the method of choice to meet the oxygen demand for cultures of less than 1 L in volume (7). Direct aeration by sparging is an alternative method that may be used to provide a sufficient oxygen supply to the cells in large-scale bioreactors. However, the foam produced at the culture surface as a result of sparging is known to cause damage to shear-sensitive mammalian cells through film drainage and by gas bubbles bursting. Thus sparging is more commonly used in traditional bacterial fermentation (15).

In small-scale suspension bioreactors, the cells within the culture are maintained in suspension by rotating a magnetic impeller. Whereas this provides well-mixed, homogeneous conditions within the vessel, the cells grown in these bioreactors are subjected to shear stresses created by the movement of the growth medium. The maximum shear ($\tau_{max}$) can be found at the tip of the impeller (10, 16). We have shown that maintaining the maximum shear stress at the impeller tip between 0.3 and 0.75 Pa can control the size of mammalian NPC aggregates while minimizing damage to the cells in suspension culture (9, 11, 17). Whereas increasing aggregate size is indicative of cell proliferation and growth in culture, maintaining the size of aggregates at an acceptable level is essential to avoid the development of necrotic cores as a result of oxygen and nutrient transfer limitations. Our studies in 125 mL suspension bioreactors (Corning spinner flasks with paddle impeller) have shown that surface aeration and an agitation rate of 100 rpm (equivalent to a maximum shear of 0.58 Pa) can be used to maintain the mean diameter of murine NPC aggregates to below 150 µm and human NPC aggregates to below 600 µm without compromising cell growth kinetics. Standard 125 mL spinner flasks (Corning) with paddle impeller (**Fig. 6.1**) have been widely used in the literature to grow different mammalian cell lines, including baby hamster kidney cells (18–21), islet-like structures (22), murine neural precursor cells (9, 10, 17, 23–26), mammary epithelial stem cells (27), breast cancer stem cells (28), and human NPCs (2, 11). The following procedure describes serial subculturing of mammalian NPCs in standard 125 mL suspension bioreactors.

### 4.2. Preparation of Small-Scale Suspension Bioreactors

The following procedure is used to prepare 125 mL bioreactors for cell expansion.

1. Siliconize the inner surface of the glass bioreactors and outer surface of the impellers using a 1:9 ratio of Sigmacote in hexane, and allow the vessels to dry at room temperature. This is an essential step to prevent the cells from sticking to the inner surface of the bioreactor and outer surface of the impeller during expansion.

2. Wash the inner surface of the glass bioreactors and outer surface of the impellers with double distilled water.

3. Sterilize the vessels in an autoclave and then transfer the vessels to a biosafety cabinet, and allow the vessel to cool down to room temperature.

4. Aseptically add 90.0 mL of fresh growth medium to each bioreactor.

5. Place the bioreactors containing fresh growth medium in a humidified incubator at 37°C and 5% $CO_2$ before cell inoculation.

### 4.3. Preparation of Single Cell Suspension and Inoculation into Bioreactors

We have shown than inoculation cell densities of 75,000 cells/mL (murine) (23) and 100,000 cells/mL (human) (11) can be used for successful expansion of NPCs in suspension bioreactors. The cells required for inoculation into suspension bioreactors may be generated in multiple tissue culture flasks or in suspension bioreactors. The protocol described in **Section 3** can be used to prepare cell inoculum from stationary culture. The cell sampling procedure described in **Section 4.4** is used to take NPC aggregates from suspension bioreactors, and dissociate into a single cell suspension which can be used as inoculum. The protocol described here is to prepare inoculum for a single 125 mL spinner flask bioreactor.

1. Remove the previously prepared spinner flask bioreactor (contains 90 mL of fresh medium) from the incubator and place it on a Thermolyne Cellgro magnetic stir plate that is located in a biosafety cabinet and set to operate at 100 rpm.

2. Place $7.5 \times 10^6$ murine or $1 \times 10^7$ human NPCs in a concentrated single cell suspension into a sterile 15 mL centrifuge tube. Adjust the total volume in the tube to 10.0 mL using fresh growth medium.

3. Using a 10 mL plastic pipette, gently aspirate the cell suspension in the tube and then transfer the cell suspension to the spinner flask bioreactor.

4. Transfer the spinner flask to a humidified chamber ($37^\circ$C and 5% $CO_2$). The flask should be placed on a Thermolyne Cellgro$^{TM}$ system or equivalent spinner plate set to operate at 100 rpm. Murine NPCs reach a maximum cell concentration after a period of 4–5 days, and human NPCs reach a maximum cell concentration after 14–16 days in suspension culture.

5. Human NPCs need to be fed every 4–5 days by replacing 40% of the spent medium with fresh growth medium. To replace the medium, transfer the spinner flask bioreactor into a biosafety cabinet, on a Thermolyne Cellgro$^{TM}$ system or equivalent spinner plate set to operate at 100 rpm. Stop the agitation for 20 s to allow the cells and aggregates to settle, and withdraw 40% of the spent medium from the upper layers of the culture volume using a 10 mL plastic pipette. Resume agitation at 100 rpm. Add a quantity of fresh medium to the flask equal in volume to the spent medium removed. Return the bioreactor to the incubator.

***4.4. Cell Sampling Protocol***

In order to evaluate growth characteristics of the mammalian NPCs grown in suspension bioreactors, it is necessary to frequently take samples from each bioreactor. Moreover, samples may be taken from each bioreactor to produce the single cell suspension required to inoculate other bioreactors. The following protocol can be used to take samples from a spinner flask bioreactor.

1. Remove a spinner flask bioreactor from an incubator and place it on a Thermolyne Cellgro$^{TM}$ magnetic stir plate

or equivalent magnetic stir plate (set to 100 rpm) in the biosafety cabinet.

2. While maintaining agitation, withdraw an appropriate volume of the culture contents from the bioreactor using a 5 mL plastic pipette. For cell counts, a 3.0 mL sample is recommended to be taken from each 125 mL bioreactor. To generate inoculum for other bioreactors, the sample volume taken from each bioreactor depends on the number of bioreactors that need to be inoculated.

3. Using a sterile 5 mL plastic pipette, replace the volume taken from the bioreactor with an equal volume of fresh growth medium. Return the bioreactor to the incubator.

## 5. Protocols for Cell Propagation in Large-Scale Computer-Controlled Bioreactors

In order for neural precursor cells to be approved for use in clinical settings, they need be generated in a reproducible manner under controlled, standard conditions. The small volume of standard 125 mL spinner flask bioreactors makes it difficult to incorporate the measuring probes necessary to monitor and properly control environmental parameters of the cell culture. Therefore, it is necessary to scale-up expansion of the cells from small-scale bioreactors to larger-scale computer-controlled bioreactors. The use of these larger bioreactors permits small quantities of NPCs isolated from a single fetus to be efficiently expanded to clinical quantities which would be sufficient to treat multiple patients. As discussed earlier, maintaining oxygen supply and culture hydrodynamics at acceptable levels in suspension bioreactors are crucial to achieve a successful cell culture process. Thus, these parameters need specific attention in the scale-up process. Mass transfer rates and culture hydrodynamics in suspension bioreactors are dramatically influenced by vessel geometry, impeller design, and the presence of measuring probes. Therefore, it is crucial to take these parameters into consideration when scaling up a cell production process (for further discussion on effect of vessel geometry, impeller design, and the presence of measuring probes on scale up of mammalian NPC production, the reader is referred to Gilbertson et al. (17) and Baghbaderani et al. (11)).

Prior to describing the protocols for expansion of mammalian NPCs, it is necessary to briefly discuss process control techniques that are used in computer-controlled bioreactors.

### 5.1. Process Control Techniques

In order to ensure batch-to-batch consistency when generating cells, critical parameters of the process must be controlled and maintained at optimal levels. Temperature, oxygen concentration, and pH are among the most important culture parameters that can be regulated using process control techniques (7, 15). Each culture parameter, which is also considered a process variable (PV), can be measured by a sensor or sterilizeable probe connected to a computer control system. **Figure 6.4** illustrates a process control system and the main components of the control system, which can be used to control pH. Temperature and dissolved oxygen may be controlled in a bioreactor in similar manner. Proportional control, integral control, and derivative control are three techniques that can be used by a controller to control the process variables (29). The control parameters and type of controller used are usually recommended by the bioreactor manufacturer. **Table 6.1** summarizes the control

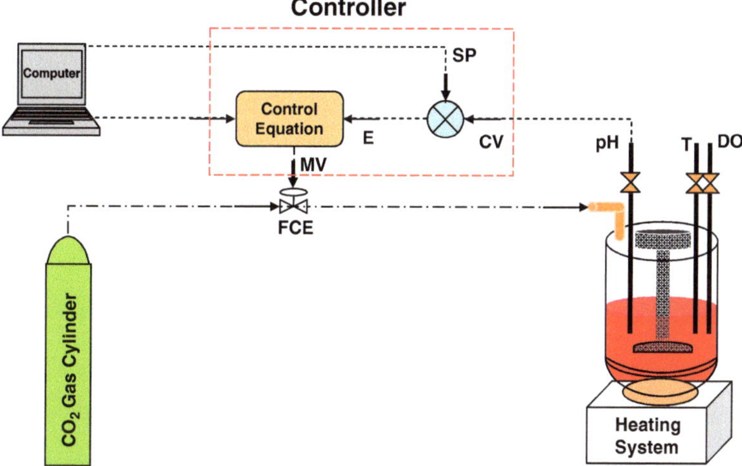


Fig. 6.4. A Schematic diagram of the main components of a process control system. A pH sensor incorporated into the bioreactor measures the pH (Controlled Variable, CV) and transmits the information to the controller where the CV value is compared with a predetermined set-point (SP) value. Depending on the deviation from the SP value, corrective feedback is provided through a final control element (FCE) to shift the culture parameter toward the optimal condition predetermined by SP value. For instance, if the pH of the culture has increased, the FCE reads the controller output and increases the CO$_2$ flow to the headspace of the bioreactor, which in turn results in lowering of the pH of the culture. The reaction provided by FCE is termed manipulated variable (MV). E denotes the error or deviation of the CV from the SP.

## Table 6.1
### The control parameters recommended by Baghbaderani et al. (11) for the 500 mL computer-controlled bioreactor (Wheaton Instruction Manual)

| Process variable | Set-point | Gain ($K_C$) | Integral ($t_i$) | Derivative ($t_d$) |
| --- | --- | --- | --- | --- |
| Temperature | 37°C | 50 | 0.03 | 1.50 |
| pH | 7.3 | 30 | 0.00 | 0.00 |
| Dissolved oxygen | 14.7%[a] | 45 | 0.00 | 0.00 |

[a]14.7% dissolved oxygen is equivalent to 70% air saturation.

parameters we recommend (11) for expansion of human NPCs in a 500 mL computer-controlled Wheaton bioreactor.

**5.2. Preparation of Large-Scale Bioreactors**

A Wheaton 500 mL suspension bioreactor (**Fig. 6.2**) (Wheaton Science Products, Cat#356922) with a 4-blade impeller (Wheaton Science Products, Cat#356902) and measuring probes (DO sensor, Broadley James Corporation; pH sensor, Mettler-Toledo; and temperature sensor, MINCO) is a common type of computer-controlled bioreactor, which has been successfully used to scale-up the expansion of mammary epithelial stem cells (27), murine NPCs (17), and human NPCs (11). The 500 mL bioreactor has two side-arms, which are provided for taking samples or adding growth medium to the bioreactor and give the option of adding filtered aeration and venting tubes. The protocol to prepare one 500 mL bioreactor for cell inoculation is described as follows.

1. Siliconize the inner surface of the 500 mL bioreactor and outer surface of the impeller using a 1:9 ratio of Sigmacote in hexane and allow the vessels to dry at room temperature.

2. Wash the inner surface of the 500 mL bioreactor and outer surface of the impeller with double distilled water.

3. Add 500 mL double distilled water into the bioreactor.

4. Insert the measuring probes into the bioreactor and calibrate the pH, temperature, and dissolved oxygen probes. The instructions to carry this out are typically provided by the manufacturer of the equipment.

5. Cover the probes, side-arms, and gas-inlet tubes with aluminum foil.

6. Sterilize the vessel in an autoclave using the wet-cycle.

7. Carefully transfer the vessel (contains hot water) to a biosafety cabinet and allow the bioreactor to cool down to room temperature.

8. Replace the water with 490 mL of fresh growth medium.

9. Place the bioreactor containing fresh growth medium on a stir plate, and connect the measuring probes and gas-inlet tubes to the control unit and gas pump, respectively.

10. Set the agitation rate on the stir plate. Note that the addition of probes to a culture fluid impacts the hydrodynamics and must be taken into consideration when choosing an appropriate agitation rate. An agitation rate that is too high will be detrimental as the excessive shear can decrease the viability of the culture. A low agitation rate can fail to provide adequate mixing. We recommend agitation rates of

60 rpm and 85 rpm for expansion of murine NPCs (17) and human NPCs (11) in 500 mL bioreactors, respectively.

11. Prior to inoculating the cells into the bioreactor, start the process control and allow the culture conditions to reach the set-point values. This allows the medium to equilibrate to the appropriate temperature, DO, and pH for expansion of the cells.

***5.3. Inoculation of Large-Scale Bioreactors and Cell Sampling Protocols***

Protocols for preparation of cell inoculum and cell sampling for a 500 mL bioreactor are conducted according to the protocols described for inoculation and sampling of standard 125 mL spinner flask bioreactors in **Sections 4.3** and **4.4**. In order to perform cell sampling or to feed the 500 mL bioreactor, the measuring probes must be disconnected and then the bioreactor must be moved into a biosafety cabinet. The probes need to be properly reconnected after cell sampling or feeding. Moreover, two 5 mL samples must be taken from each 500 mL bioreactor to perform cell counts.

# 6. Critical Steps and Troubleshooting

***6.1. Aggregate Dissociation***

1. The dissociation protocols described here can be used for handling small quantities of cells (i.e., cells derived from one T-25 or T-75) or samples ($\leq$10 mL) taken from suspension bioreactors. In order to dissociate large numbers of cell aggregates (i.e., harvested from 125 mL or 500 mL bioreactors), the aggregate suspension must be first divided into multiple 15 mL centrifuge tubes (each tube containing 10 mL of aggregate suspension). After harvesting the cell aggregates via centrifugation at $140g$ for 10 min, aggregate dissociation techniques described in **Section 3.1.1** can be used to produce a single cell suspension in each tube.

2. To dissociate human NPCs aggregates into single cells, enzymatic dissociation, which results in high cell viability (greater than 90%), is recommended.

3. Although mechanical dissociation has been widely used in the literature for dissociation of mammalian cell aggregates, it is not a favorable technique due to high cell trauma and significantly low viability (about 60–70%) when compared to other techniques.

4. A chemical dissociation technique has been recently invented for dissociation of murine NPC aggregates (13), but is not discussed here.

**6.2. Serial
Subculturing of NPCs
in Culture**

1. Prior to serially subculturing the cells, growth characteristics of each mammalian NPC cell line must be evaluated to determine the maximum growth rate and the culture day on which the maximum cell density is reached. Once this is known, cells should be subcultured 1–2 days before the maximum viable cell density is reached. This will ensure that the cells are passaged while in the exponential growth phase, and may reduce the lag-phase in the subsequent culture.

2. Cell morphology and culture conditions, including the medium color and amount of cell debris in the culture, must be frequently monitored to observe possible culture contamination or changes in cell characteristics. Morphology change may be a sign of cell transformation or change in the intrinsic growth characteristics. It should be noted that we do not use antibiotics when culturing these cells.

3. To perform future experiments or transplantation studies in animal models of neurodegenerative disorders, it is important to generate a cell bank both at early passages and following the production of large populations of cells in culture. Aggregates of mammalian NPCs grown in stationary culture or suspension bioreactors can be cryopreserved long term in the vapor headspace above liquid nitrogen at a temperature below −190°C. The cells can be cryopreserved in a freezing medium comprised of the serum-free growth medium supplemented with 10% dimethyl sulfoxide (DMSO), a cryoprotectant known to protect the cells from disruption during the freezing and thawing process.

4. After thawing the cryopreserved mammalian NPCs at 37°C in a water bath, the freezing medium must be replaced with fresh growth medium, and the cells need to be incubated (37°C and 5% $CO_2$) for 1–2 days in order to recover from the freeze-thaw process. Then, mechanical dissociation may be used to dissociate the cell aggregates into single cells for inoculation into fresh medium. Considering that the cell aggregates have been cryopreserved in a freezing medium containing DMSO, over-trituration of the cell aggregates must be avoided to lower detrimental effects of mechanical dissociation on the viability of the cells. In comparison with the number of triturations used for serial subculturing of the cells grown in culture (**Section 3.1.1.1**), it is recommended to use fewer triturations for dissociation of the aggregates obtained from cryopreserved vials. Obviously, using a lower number of triturations may cause some of the aggregates to remain intact or partially dissociated, which may affect the

results of cell counts performed for serial subculturing of the cells. However, this approach can improve the viability and encourage faster recovery of the cells after cryopreservation and dissociation procedure.

**A    Growth Characteristics in 125 mL and 500 mL Bioreactors**

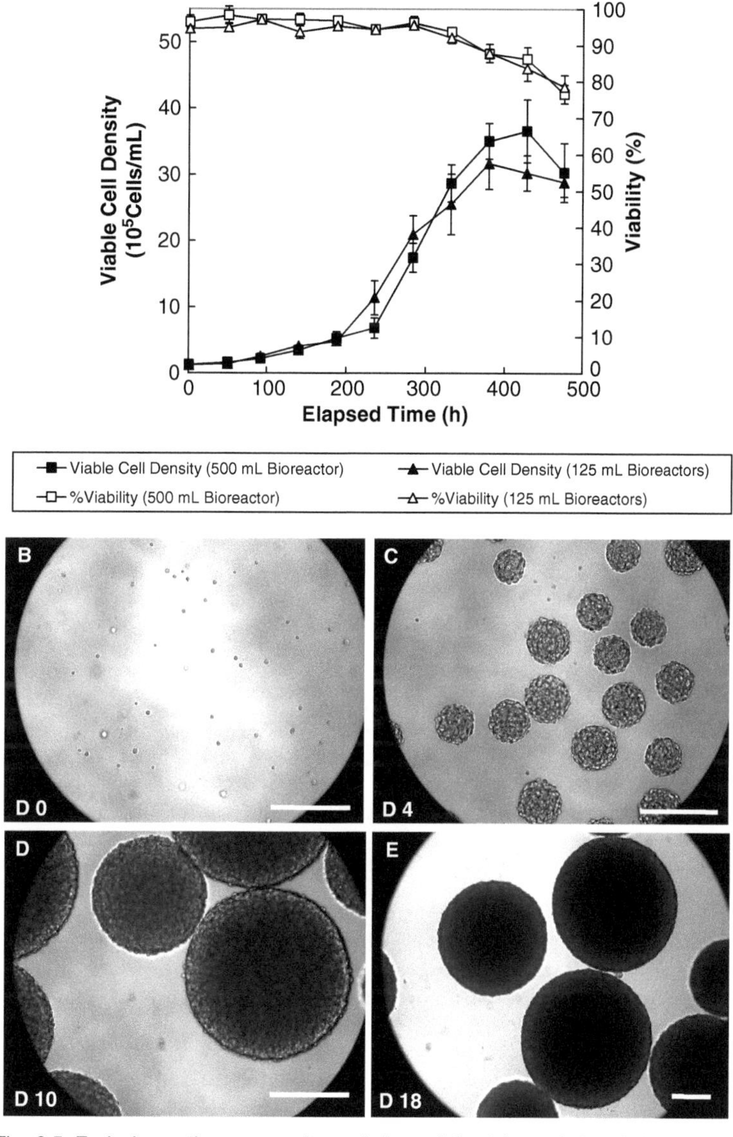



Fig. 6.5. Typical growth curves and morphology of the telencephalon-derived human NPCs in suspension cultures. Shown are growth curves of human NPCs grown in standard 125 mL suspension bioreactors and 500 mL computer-controlled bioreactors (**A**). Photomicrographs demonstrate the cells grown in 500 mL bioreactors immediately following inoculation (**B**), 4 days post-inoculation (**C**), 10 days post-inoculation (**D**), and 18 days post-inoculation (**E**). Scale bars represent 200 μm. The figures are adapted and modified from Baghbaderani et al. (11).

## 7. Typical Protocol Results

**Figure 6.5** demonstrates typical growth curves for human NPCs grown in the 125 and 500 mL suspension bioreactors. The cells reach a comparable maximum viable cell density of about $3.5 \times 10^6$ cells/mL after 16–18 days at both scales, exhibiting a doubling time of 84 h. The viability of the cells remains about 90% over the same period in culture. Moreover, human NPCs grow in the form of cell aggregates in suspension bioreactors, where the aggregate size increases over time in culture. Aggregates with diameters of approximately 500 μm can be observed by the end of culture day 18, which coincides with when the maximum viable cell density is observed.

## Acknowledgments

This study was supported generously by research grants from the Natural Sciences and Engineering Research Council of Canada and the Canadian Stem Cell Network.

## References

1. Carpenter MK, Cui X, Hu Z-y, et al. In vitro expansion of a multipotent population of human neural progenitor cells. Exp Neurol 1999;158(2):265–78.
2. Mukhida K, Baghbaderani BA, Hong M, et al. Survival, differentiation, and migration of bioreactor-expanded human neural precursor cells in a model of Parkinson disease in rats. Neurosurg Focus 2008; 24(3–4):E8.
3. Ostenfeld T, Joly E, Tai YT, et al. Regional specification of rodent and human neurospheres. Brain Res 2002;134(1–2):43–55.
4. Piao JH, Odeberg J, Samuelsson EB, et al. Cellular composition of long-term human spinal cord- and forebrain-derived neurosphere cultures. J Neurosci Res 2006;84(3):471–82.
5. Storch A, Sabolek M, Milosevic J, Schwarz SC, Schwarz J. Midbrain-derived neural stem cells: from basic science to therapeutic approaches. Cell Tissue Res 2004;318(1): 15–22.
6. Svendsen CN, ter Borg MG, Armstrong RJE, et al. A new method for the rapid and long term growth of human neural precursor cells. J Neurosci Methods 1998;85(2):141–52.
7. Butler M. Animal Cell Culture & Technology. 2nd Edition. London: BIOS Scientific Publishers; 2004.
8. Kallos MS, Sen A, Behie LA. Large-scale expansion of mammalian neural stem cells: a review. Med Biol Eng Comput 2003;41(3):271–82.
9. Sen A, Kallos MS, Behie LA. Effects of hydrodynamic on extended cultures of mammalian neural stem cell aggregates in suspension culture. Ind Eng Chem Res 2001;40:5350–7.
10. Sen A, Kallos MS, Behie LA. Expansion of mammalian neural stem cells in bioreactors: effect of power input and medium viscosity. Brain Res Dev Brain Res 2002; 134(1–2):103–13.
11. Baghbaderani BA, Mukhida K, Sen A, Hong M, Mendez I, Behie LA. Expansion of human neural precursor cells in large-scale bioreactors for the treatment of neurodegenerative disorders. Biotechnol Progr 2008;24(4):859–70.
12. Suzuki M, Wright LS, Marwah P, Lardy HA, Svendsen CN. Mitotic and neurogenic effects of dehydroepiandrosterone (DHEA) on human neural stem cell cultures derived

from the fetal cortex. PNAS 2004;101(9): 3202–7.

13. Sen A, Kallos MS, Behie LA. New tissue dissociation protocol for scaled-up production of neural stem cells in suspension bioreactors. Tissue Eng 2004;10(5–6):904–13.

14. Storch A, Paul G, Csete M, et al. Long-term proliferation and dopaminergic differentiation of human mesencephalic neural precursor cells. Exp Neurol 2001;170(2): 317–25.

15. Shuler ML, Kargi F. Bioprocess Engineering Basic Concepts. 2nd edition. New Jersey: Prentice Hall PTR; 2002.

16. Cherry RS, Kwon K-Y. Transient shear stresses on a suspension cell in turbulence. Biotechnol Bioeng 1990;36(6):563–71.

17. Gilbertson JA, Sen A, Behie LA, Kallos MS. Scaled-up production of mammalian neural precursor cell aggregates in computer-controlled suspension bioreactors. Biotechnol Bioeng 2006;94(4):783–92.

18. Moreira JL, Alves PM, Aunins JG, Carrondo MJ. Changes in animal cell natural aggregates in suspended batch cultures. Appl Microbiol Biot 1994;41(2):203–9.

19. Moreira JL, Alves PM, Rodrigues JM, Cruz PE, Aunins JG, Carrondo MJ. Studies of baby hamster kidney natural cell aggregation in suspended batch cultures. Ann N Y Acad Sci 1994;745:122–33.

20. Moreira JL, Aunins JG, Carrondo MJ. Hydrodynamics effects on BHK cells grown as suspended natural aggregates. Biotechnol Bioeng 1995;46:351–60.

21. Moreira JL, Feliciano AS, Santana PC, Cruz PE, Aunins JG, Carrondo MJ. Repeated-batch cultures of baby hamster kidney cell aggregates in stirred vessels. Cytotechnology 1994;15(1–3):337–49.

22. Chawla M, Bodnar CA, Sen A, Kallos MS, Behie LA. Production of islet-like structures from neonatal porcine pancreatic tissue in suspension bioreactors. Biotechnol Prog 2006;22:561–7.

23. Kallos MS, Behie LA. Inoculation and growth conditions for high-cell-density expansion of mammalian neural stem cells in suspension bioreactors. Biotechnol Bioeng 1999;63(4):473–83.

24. Sen A. Bioreactor Protocols for the Long Term Expansion of Mammalian Neural Stem Cells in Suspension Culture [PhD]. Calgary: University of Calgary; 2003.

25. Sen A, Kallos MS, Behie LA. Passaging protocols for mammalian neural stem cells in suspension bioreactors. Biotechnol Prog 2002;18(2):337–45.

26. Kallos MS, Behie LA, Vescovi AL. Extended serial passaging of mammalian neural stem cells in suspension bioreactors. Biotechnol Bioeng 1999;65(5):589–99.

27. Youn BS, Sen A, Kallos MS, et al. Large-scale expansion of mammary epithelial stem cell aggregates in suspension bioreactors. Biotechnol Prog 2005;21(3):984–93.

28. Youn BS, Sen A, Behie LA, Girgis-Gabardo A, Hassell JA. Scale-up of breast cancer stem cell aggregate cultures to suspension bioreactors. Biotechnol Prog 2006;22(3):801–10.

29. Svrcek WY, Mahoney DP, Young BR. A Real-Time Approach to Process Control. England: John Wiley & Sons Ltd.; 2000.

# Chapter 7

# Intracellular Calcium Assays in Dissociated Primary Cortical Neurons

Navjot Kaur, David V. Thompson, David Judd, David R. Piper, and Richard G. Del Mastro

## Abstract

In this chapter, the basic tools and methods to isolate neurons from the embryonic rat cortex are provided. We outline the isolation of fresh neuronal cells and their storage, post-thaw maintenance, and the application of techniques to measure intracellular calcium changes in response to the addition of neurotransmitters. Many of these techniques come in kit formats supplied by companies like Invitrogen and thus enhance the ease of use, time required, and reproducibility.

**Key words:** Primary embryonic neurons, cryopreservation, immunocytochemistry, rat, intracellular calcium, neurotransmitter.

## 1. Introduction

The architecture of the brain is complex in nature (1). This is observed by the intricate layout of the cells that work in concert with each other to perform multifunctional roles. Perturbation at the cellular or molecular level can lead to an alteration in the status quo and mark the beginnings of neurodegeneration. This can become augmented over time and turn into a loss in function, which is first generally observed as a phenotypic change. The advent of a neurological disease is slow and progressive and can lead to the eventual death of the individual.

The study of neurological disease is often refractory using traditional molecular biology and cell culture techniques. Research

L.C. Doering (ed.), *Protocols for Neural Cell Culture*, Springer Protocols Handbooks,
DOI 10.1007/978-1-60761-292-6_7, © Humana Press, a part of Springer Science+Business Media, LLC 1992, 1997, 2001, 2010

using animals has paved the path forward and provided a wealth of data through the use of these model systems. However, the technological breakthroughs witnessed in molecular biology (sequencing of the human genome, microarrays, and RNAi) and cell biology (improved media and broader availability of neuronal cell types) over the past decade have enabled scientists to embark on discoveries that were originally thought intractable. These powerful technology platforms have enabled scientists to dissect molecular pathways underlying neuronal dysfunction leading to identifying potential therapeutic treatments for neurological disorders (2).

The sequencing of the human genome has led to the discovery of millions of nucleotide variants that can be used to discover disease-associated loci (3). As neuronal diseases are polygenic in nature, the use of single-nucleotide polymorphisms and association studies have revealed the location of elusive genes that underlie these complex disorders (4). These discoveries are just the beginning to understanding the functional role of the genes in the context of disease. Well-trodden paths that interrogate the role of the gene and its protein product in the cell are taken. These typically represent the use of RNAi to downregulate the gene (5) and conversely its overexpression (6) to determine if these overt perturbations have any effect on cell health through pathways involved in neurodegenerative diseases, the generation of antibodies directed toward the protein to determine cell location (7), yeast-two hybrid to identify protein–protein interactions (8), and quantitative assessment by Northern blots, Western blots, and quantitative RT-PCR (9).

Other studies that employ a combination of statistical and biological methods to measure interactions of disease-associated genes and their SNPs have revealed that a complex interplay exists between them, which introduces a broad range of disruptive influences (10). The genomic variants in particular can reside within introns or exons of a gene and can disrupt pre-mRNA splicing by introducing exon skipping, enhance the use of cryptic splice sites, and alter the ratio of alternatively spliced isoforms. Such perturbations have been shown to be the cause of various human disease phenotypes (11–16).

This chapter provides the methodologies to isolate fresh cells from the cortex of rat E18 embryos, generate cryopreserved stocks, thaw and maintain the cells in tissue culture. In addition, protocols have been provided to measure intracellular calcium responses. These protocols represent progress toward understanding the basis of neuronal disease from which standard molecular and cell biology tools can be employed.

## 2. Isolation of Cells from Embryonic Cortical Tissue

### 2.1. Reagents and Equipment

1. Hibernate-E medium (BrainBits, LLC., Cat. no. HE)
2. B27 (Invitrogen, Cat. no. 17504-044),
3. Glutamax-I (Invitrogen, Cat. no. 35050-061)
4. Hibernate-E, without $Ca^{2+}$ medium (BrainBits LLC, Cat. no. HE-Ca)
5. Papain (Worthington, Cat. no. LS003119)
6. Neurobasal medium (Invitrogen, Cat. no. 21103-049)
7. Trypan blue (Invitrogen, Cat. No. 15250-061)
8. Pasteur pipettes
9. Hemocytometer
10. Conical tubes (15 mL, BD Biosciences, Cat. no. 352097) 50 mL, BD Biosciences, Cat. no. 352098)

### 2.2. Protocol

1. Dissect cortex pairs from ten E-18 rat embryo brains. Remove all the meninges thoroughly.
2. Collect all the tissue in a conical tube containing Hibernate-E medium supplemented with 2% B27, 0.5 mM Glutamax-I at 4°C.
3. Let the tissue settle to the bottom of the tubes and then carefully remove supernatant leaving only the tissue covered by medium.
4. Enzymatically digest the tissue in 4 mL Hibernate-E, without $Ca^{2+}$ medium containing 2 mg/mL filter-sterilized papain at 30°C for 30 min with gentle shaking the tube every 5 min.
5. Add 6 mL supplemented Hibernate-E medium.
6. Centrifuge the tube for 5 min at 150$g$.
7. Remove supernatant and suspend the tissue in 5 mL complete Hibernate-E medium by triturating with fire-polished glass Pasteur pipette.
8. Let the tube stand undisturbed for 2 min so that big debris settles down (if any).
9. Transfer cells to a new tube leaving behind all the debris.
10. Count the cells using a hemocytometer.
11. Centrifuge the tube for 4 min at 200$g$.
12. Remove supernatant and suspend the cell pellet either in neurobasal medium supplemented with 2% B27 and 0.5 mM Glutamax-I for culturing or in cryopreservation medium to freeze the cells as described in **Section 3** (17).

## 3. Cryopreservation of Neuronal Cells

### 3.1. Reagents and Equipment

1. Synth-a-freeze, cryopreservation medium, (Invitrogen Corp., Cat. no. R-005-50)
2. Cryogenic vials
3. Isopropanol chamber
4. Freezer, –80°C.
5. Liquid nitrogen freezer

### 3.2. Protocol

1. Add cold cryopreservation medium to the cell pellet obtained in Step12 in **Section 2.2** (12) to make cell suspension at a concentration range of $2.0 \times 10^6$–$1.0 \times 10^7$ cells/mL.
2. Dispense 1 mL suspension to prelabeled, prechilled cryovials.
3. Put the tubes in isopropanol chamber and leave at 4°C for 10 min.
4. Leave the chamber at –80°C overnight.
5. Transfer frozen vials to vapor phase of liquid nitrogen till further use.

## 4. Recovery of Frozen Neuronal Cells

### 4.1. Reagents and Equipment

1. Conical tubes (15 mL, BD Biosciences, Cat. no. 352097)
2. Neurobasal medium (Invitrogen, Cat. no. 21103-049)
3. B27 (Invitrogen, Cat. no. 17504-044),
4. Glutamax-I (Invitrogen, Cat. no. 35050-061)
5. Water bath (set at 37°C)
6. Trypan blue (Invitrogen, Cat. No. 15250-061)
7. Hemocytometer
8. A vial of home-made cryopreserved cells (as described above) or purchased cryopreserved primary rat cortex neurons (Invitrogen, RCN 1 M Cat. No. A10840-01, RCN 4 M Cat. no. A10840-02).

### 4.2. Protocol

1. Remove one vial of frozen cells from liquid nitrogen.
2. Thaw the vial in 37°C water bath by gentle swirling.
3. Wipe the vial with ethanol and tap in gently on the surface so that all the medium settles down.

4. Open the vial in the hood.

5. Rinse the pipette tip with medium and very gently transfer the cells to a pre-rinsed 15 mL tube.

6. Rinse the vial with 1 mL pre-warmed complete neurobasal/B27 medium, and add to the cells in the 15 mL tube extremely slowly at the rate of one drop per second. Mix by gentle swirling after each addition.

7. Slowly add 2 mL of complete neurobasal/B27 medium to tube (for a total suspension volume of 4 mL).

8. Mix the suspension very gently with P-1000 pipette without creating any air bubbles.

9. To a microcentrifuge tube containing 10 $\mu$L of 0.4% trypan blue add 10 $\mu$L cell suspension using a prerinsed tip. Mix only by gently tapping the tube. Determine the viable cell density using a manual (i.e., hemocytometer) counting method.

**4.3. Critical Steps and Troubleshooting**

Do not centrifuge the cells as they are extremely fragile upon recovery from cryopreservation. It is important to rinse every pipette tip and vial with complete neurobasal/B27 medium before using for cell suspension to avoid the cells sticking to the plastic.

# 5. Substrate Coating

**5.1. Reagents and Equipment**

1. D-PBS with $Ca^{2+}$ and $Mg^{2+}$ (Invitrogen, Cat. no. 14040).

2. Poly-D-lysine (BD Biosciences, 354210)

3. Distilled water (Invitrogen, 15230)

4. Multiwell plates/ multichamber slides

**5.2. Protocol**

1. Prepare Poly-D-lysine solution of 15 $\mu$g/mL in PBS with $Ca^{2+}$ and $Mg^{2+}$.

2. Coat 300 $\mu$L/cm$^2$ in plates or in multichambered slides to obtain 4.5 $\mu$g/cm$^2$ coating.

3. Incubate the plates at room temperature for 1 h.

4. Remove the solution, and rinse three times with sterile $H_2O$.

5. Leave the plates uncovered in the hood till the wells are completely dry, approximately for 30 min.

6. Use immediately, or store at 4°C.

## 6. Culturing and Maintenance of Neurons

### 6.1. Reagents and Equipment

1. Poly-D-lysine-coated plates

2. Neurobasal medium (Invitrogen, Cat. no. 21103-049)

3. B27 (Invitrogen, Cat. no. 17504-044)

4. Glutamax-I (Invitrogen, Cat. no. 35050-061)

5. A vial of home-made cryopreserved cells (as described above) or cryopreserved primary rat cortex neurons (Invitrogen, RCN 1 M Cat. No. A10840-01, RCN 4 M Cat no. A10840-02).

### 6.2. Protocol

1. Thaw primary rat cortex neurons as described above.

2. Plate ~1 × $10^5$ cells per well in poly-D-lysine (4.5 µg/cm$^2$) coated 48-well plate or 8-chambered slide. Dilute cell suspension to 500 µL per well by adding complete neurobasal/B27 medium.

3. Incubate at 36–38°C in a humidified atmosphere of 5% $CO_2$ in air.

4. After 4–24 h of incubation, aspirate half of the medium from each well and replace with fresh medium. Return to incubator.

5. Feed cells every third day by aspirating half of the medium from each well and replacing with fresh medium.

## 7. Post Thaw Evaluation: Immunocytochemistry

### 7.1. Reagents and Equipment

1. Poly-D-lysine coated multichambered slides (as prepared above)

2. D-PBS with $Ca^{2+}$ and $Mg^{2+}$ (Invitrogen, Cat. No. 14040)\

3. Goat serum (Invitrogen, Cat. No. 16210-064)

4. Paraformaldehyde (4%)

5. Mouse anti-MAP2 antibody (Invitrogen, Cat. no. 13-1500)

6. Rabbit anti-GFAP antibody (Invitrogen, Cat. no. 08-0063)

7. Alexa Fluor 488 goat anti-mouse IgG (Invitrogen, Cat. no. A-11029)

8. Alexa Fluor 594 goat anti-rabbit IgG (Invitrogen, Cat. no. A-11037)

9. 4′, 6-diamidino-2-phenylindole, dihydrochloride (DAPI) (Invitrogen, Cat. No. D-1306)

    10. ProLong Gold antifade reagent (Invitrogen, Cat. No. P36930)

    11. Fluorescence Microscope

**7.2. Protocol**

1. Plate cells on to a poly-D-lysine (4.5 $\mu g/cm^2$) coated eight-chambered slide by seeding $1 \times 10^5$ cells per chamber in 500 $\mu L$ of medium.

2. Incubate at 36–38°C in a humidified atmosphere of 5% $CO_2$ in air.

3. After 24 h of incubation, aspirate half of the medium from each well and replace with fresh medium.

4. Return to incubator.

5. Feed cells every third day by aspirating half of the medium from each well and replace with fresh medium.

6. When ready to perform immunocytochemistry procedure, aspirate supernatant and rinse cells twice with D-PBS with $Ca^{2+}$ and $Mg^{2+}$.

7. Fix the cells with 4% paraformaldehyde for 20 min.

8. Rinse cells three times with D-PBS with $Ca^{2+}$ and $Mg^{2+}$.

9. Permeabilize cells with 0.3% Triton-X (diluted in D-PBS with $Ca^{2+}$ and $Mg^{2+}$) for 5 min at room temperature.

10. Rinse cells three times with D-PBS with $Ca^{2+}$ and $Mg^{2+}$.

11. Incubate cells coated with 5% goat serum solution diluted in D-PBS with $Ca^{2+}$ and $Mg^{2+}$ for 60 min at room temperature.

12. Incubate cells coated with primary antibody (mouse anti-MAP2; 10 $\mu g/mL$; and/or rabbit anti-GFAP, 4 $\mu g/mL$) diluted in 5% goat serum solution at 2–8°C overnight.

13. Rinse cells three times with DPBS with $Ca^{2+}$ and $Mg^{2+}$.

14. Incubate with secondary antibody (Alexa Fluor® 488 goat-anti mouse (H+L), 10 $\mu g/mL$, and/or Alexa Fluor® 594 goat-anti rabbit (H+L), 10 $\mu g/mL$) diluted in 5% goat serum solution for 60 min at room temperature.

15. Rinse three times with DPBS with $Ca^{2+}$ and $Mg^{2+}$.

16. Stain with DAPI solution (3 ng/mL) for 10 min.

17. Mount with ProLong Gold antifade reagent and observe under the microscope using filters for FITC, Cy5, and DAPI.

**7.3. Typical Protocol Results**

The thawed cortical neurons cultured in neurobasal medium, supplemented with B27 and Glutamax-I, show >90% neuron population stained with MAP2 antibody with minimum or absence of astrocytes. Within 3–4 days in culture, the neurons display extensive neurite outgrowth, which keeps on increasing as long as they are kept healthy in culture (**Fig. 7.1**). Results vary if neurons are cultured in the presence of serum.

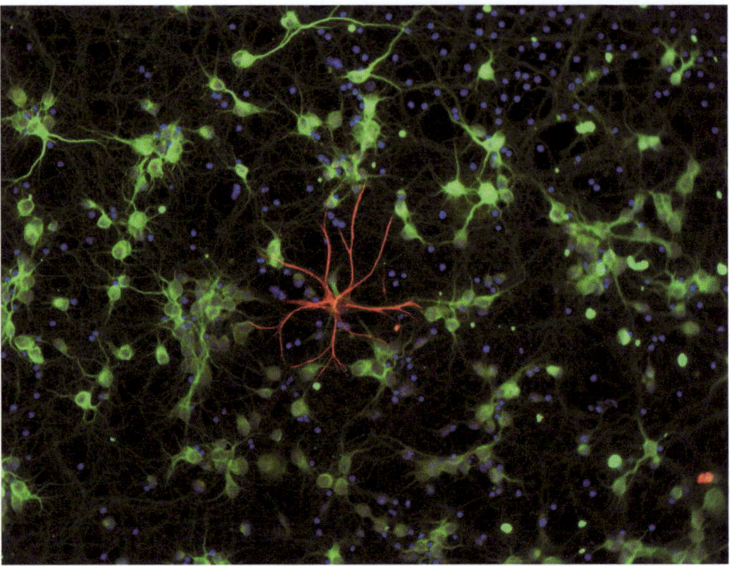

Fig. 7.1. Immunofluorescence detection of primary rat cortex neurons (Invitrogen, Cat. no. A10840-01): Neurons are stained with mouse anti-MAP2 antibody (Cat. no. 13-1500) and detected using Alexa Fluor 488 goat anti-mouse IgG antibody (*green*; Cat. no. A-11029), and astrocytes are stained with rabbit anti-GFAP antibody (Cat. no. 08-0063) and detected using Alexa Fluor 594 goat anti-rabbit IgG antibody (*red*; A-11037). Nuclei are stained with DAPI (*blue*).

## 8. Fluo-4-Based Measurements of Cytosolic Calcium Changes in Response to Neurotransmitter Applications

### 8.1. Reagents and Equipment

1. Hanks' balanced salt solution (HBSS, Invitrogen #14025)
2. Fluo-4 AM + 9% (w/v) Pluronic® F-127 (Invitrogen #F-14201, #P-6867)
3. T2000 inverted microscope (Nikon)
4. ORCA-ER digital camera (Hamamatsu)
5. Lamda DG-4 illumination system (Sutter instruments)

### 8.2. Protocol

1. Wash neuronal cells with 100 μL Hanks' balanced salt solution (HBSS, Invitrogen #14025)
2. Load neuronal cells by adding 100 μL solution containing 3 μM fluo-4 AM + 9% (w/v) Pluronic® F-127 (Invitrogen #F-14201, #P-6867) in HBSS for ∼60 min at RT.

3. Wash the neuronal cells once with 100 μL of HBSS and leave the plate at RT in the dark for ~60 min.

4. Wash neuronal cells with 100 μL HBSS and maintain at RT in 100 μL HBSS.

5. Place the 96-well microtiter plate on an inverted microscope (e.g., Nikon T2000) for visual inspection and fluorescent imaging.

6. Use an excitation light of 488 nm (Lambda DG-4) and collect images of emitted light at 520 nm (ORCA-ER).

7. Challenge the cells in one well with 20 μl 3 mM acetylcholine to achieve 500 μM acetylcholine final concentration.

8. Collect the data using appropriate software (MetaFluor, MDS Analytical Technologies).

9. Repeat for each neurotransmitter (separate wells, final concentrations): 500 μM glutamate, 500 μM dopamine (add 500 μM ascorbic acid with dopamine to prevent dopamine oxidation), 500 μM γ-aminobutyric acid, and 500 μM ATP.

***8.3. Critical Steps and Troubleshooting***

1. Collect data and perform analysis using MetaFluor imaging software.

2. Regions of interest (ROI) are defined around a random series of cells using an automated algorithm provided by the MetaFluor software.

3. The integrated fluo-4 signal for each ROI is normalized to the first ten data points ($F/F_0$) and then plotted against time.

4. A neuronal cell is considered responsive to a given neurotransmitter if the resulting normalized signal rises more than 10% compared to the baseline signal and within 60 s following neurotransmitter addition.

***8.4. Typical Protocol Results***

When performing these types of experiments we normally measure the response of 50–100 neurons in each well that is being assessed. Cells are tested after two weeks of culture following the recovery from thaw. Visually, the cells typically display a neuronal morphology with both dendritic and axonal processes clearly recognizable by cellular polarity and proportionate size. Examples of responses to each of the neurotransmitters that we test are shown in **Fig. 7.2**. The frequency of cells that respond to each neurotransmitter is collected and plotted in the bar graph (**Fig. 7.3**). We have found that the neuronal cells studied exhibit clear changes in intracellular $Ca^{2+}$ ($[Ca^{2+}]_i$) with varying frequency, which depends on the neurotransmitter. A high frequency

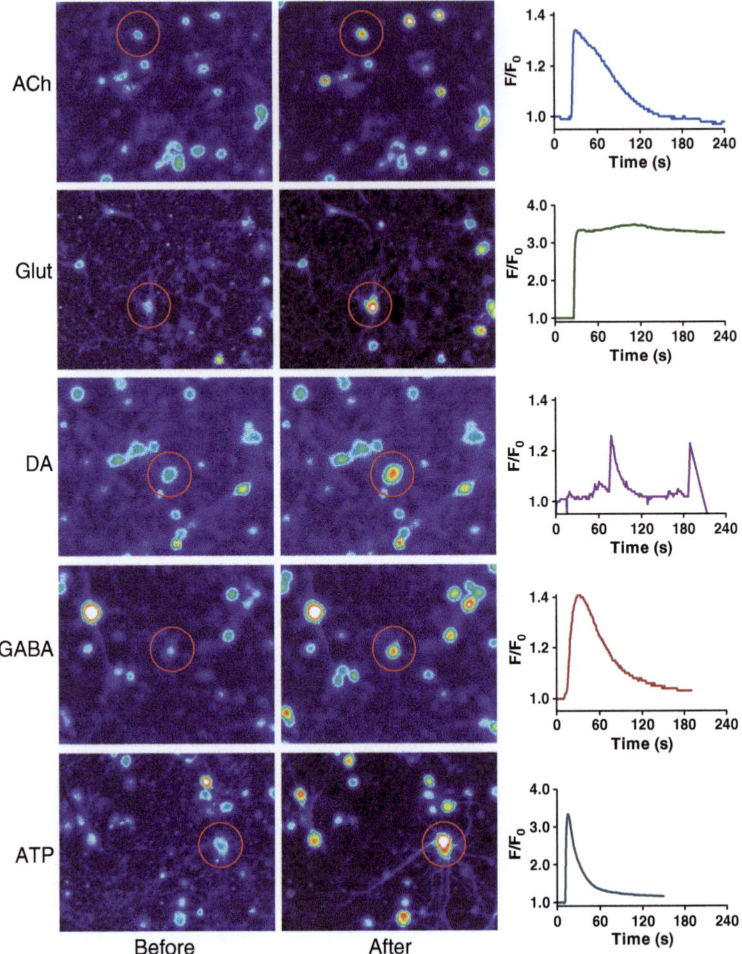



Fig. 7.2. Neurotransmitter-evoked changes in intracellular calcium: Cells are depicted in pseudo-color. Cool colors (*purple-blue*) represent low intracellular calcium and hot colors (*yellow–red–white*) represent higher intracellular calcium as measured by relative fluorescence from fluo-4. Neurotransmitters acetylcholine (ACh), glutamate (Glut), dopamine (DA), GABA, and ATP were added separately to individual wells. The left column represents the cells before addition, the right column after addition. The rightmost column indicates the change in fluorescence plotted against time for the cell circled in *red* for each row.

($\sim$50%) of cells respond to GABA with an increase in intracellular calcium, which suggests that these neuronal cells are at an early developmental stage. This response is due to the chloride equilibrium potential tending to depolarize instead of hyperpolarize the cell membrane and lead to activation of voltage-gated calcium channels and $Ca^{2+}$-influx. This observation of an inverted chloride gradient is the inverse to a terminally differentiated neuron (18, 19, 20).

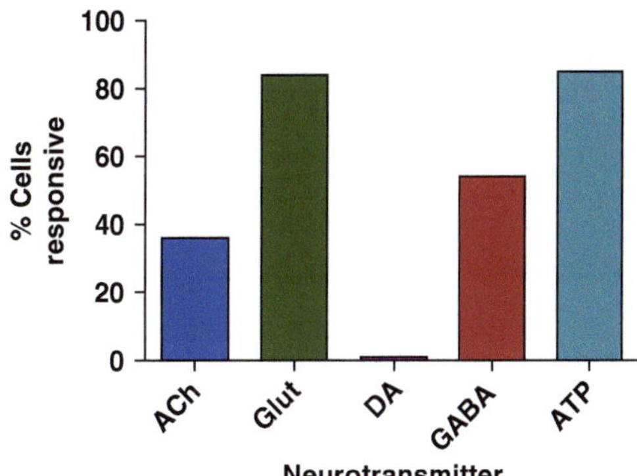


Fig. 7.3. Frequency of cells responding to neurotransmitters: Cells responded with varying frequency to a number of neurotransmitters following 2 weeks of recovery in culture. While responses to glutamate and ATP were particularly robust, many cells also responded to GABA, suggesting the population represents neurons at varying stages of maturity.

## References


1. Rowe DL, Cooper NJ, Liddell BJ, Clark CR, Gordon E, Williams LM. Brain structure and function correlates of general and social cognition. J Integr Neurosci 2007;6(1): 35–74.
2. Shaughnessy L, Chamblin B, McMahon L, et al. Novel approaches to models of Alzheimer's disease pathology for drug screening and development. J Mol Neurosci 2004;24(1):23–32.
3. Ropers HH. New perspectives for the elucidation of genetic disorders. Am J Hum Genet 2007;81(2):199–207.
4. Oksenberg JR, Baranzini SE, Sawcer S, Hauser SL. The genetics of multiple sclerosis: SNPs to pathways to pathogenesis. Nat Rev Genet 2008;9(7):516–26.
5. Paradis S, Harrar DB, Lin Y, et al. An RNAi-based approach identifies molecules required for glutamatergic and GABAergic synapse development. Neuron 2007;53(2):217–32.
6. Falk T, Xie JY, Zhang S, et al. Over-expression of the potassium channel Kir2.3 using the dopamine-1 receptor promoter selectively inhibits striatal neurons. Neuroscience 2008;155(1):114–27.
7. Saper CB. A Guide to the perplexed on the specificity of antibodies. J Histochem Cytochem 2008.
8. Navaratnam DS. Yeast two-hybrid screening to test for protein-protein interactions in the auditory system. Methods Mol Biol 2009;493:257–68.
9. Hyman BT, Augustinack JC, Ingelsson M. Transcriptional and conformational changes of the tau molecule in Alzheimer's disease. Biochim Biophys Acta 2005;1739 (2–3):150–7.
10. Del Mastro RG, Turenne L, Giese H, et al. Mechanistic role of a disease-associated genetic variant within the ADAM33 asthma susceptibility gene. BMC Med Genet 2007;8:46.
11. Woodley L, Valcarcel J. Regulation of alternative pre-mRNA splicing. Brief Funct Genomic Proteomic 2002;1(3):266–77.
12. Faustino NA, Cooper TA. Pre-mRNA splicing and human disease. Genes Dev 2003;17(4):419–37.
13. Pagani F, Baralle FE. Genomic variants in exons and introns: identifying the splicing spoilers. Nat Rev Genet 2004;5(5):389–96.
14. Schwerk C, Schulze-Osthoff K. Regulation of apoptosis by alternative pre-mRNA splicing. Mol Cell 2005;19(1):1–13.
15. Lewandowska MA, Stuani C, Parvizpur A, Baralle FE, Pagani F. Functional studies on the ATM intronic splicing processing element. Nucleic Acids Res 2005;33(13): 4007–15.
16. Venables JP. Unbalanced alternative splicing and its significance in cancer. Bioassays 2006;28(4):378–86.
17. Brewer GJ, Torricelli JR, Evege EK, Price PJ. Optimized survival of hippocampal neurons in B27-supplemented Neurobasal, a new serum-free medium combination. J Neurosci Res 1993;35(5):567–76.
18. Ben-Ari Y, Cherubini E, Corradetti R, Gaiarsa JL. Giant synaptic potentials in

immature rat CA3 hippocampal neurones. J Physiol 1989;416:303–25.

19. Ben-Ari Y, Cherubini E. Zinc and GABA in developing brain. Nature 1991; 353(6341):220.

20. Zhang L, Spigelman I, Carlen PL. Whole-cell patch study of GABAergic inhibition in CA1 neurons of immature rat hippocampal slices. Brain Res Dev Brain Res 1990;56(1):127–30.

# Chapter 8

# Dissociated Hippocampal Cultures

Francine Nault and Paul De Koninck

## Abstract

A popular approach to study living neurons involves the preparation of dissociated cultures. If isolated from neonatal or embryonic animals, neurons survive their removal and subsequent dissociation procedures. They grow processes on appropriate substrates, acquire mature neuronal morphology and polarity, and form functional synapses that resemble those of neurons developing in vivo. On the other hand, dissociated neurons produce neural circuits which differ from native circuits.

While many protocols have been developed to grow multiple types of neurons, this chapter focuses on one major protocol: the preparation of neonatal rat hippocampal neurons. It involves several steps including the meticulous preparation of coverslips and stock solutions, dissection of the hippocampi, cell seeding, and the long-term maintenance of the cultures. While dissociated neuronal cultures are used by many and can be routinely maintained for several weeks, they deserve particular attention through daily observations and rigorous laboratory practice.

**Key words:** Dissociated CNS cultures, gene transfer, neurite outgrowth, synaptogenesis, co-cultures, growth substrate, neural circuits.

## 1. Introduction

The silver staining method discovered by Camillo Golgi rendered possible the study of individual cells within the brain. Santiago Ramón y Cajal exploited it beautifully, leading to the neuron doctrine and setting the stage to tremendous advancement in our understanding of neurons and brain circuits. One of the next challenges at hand was to be able to study *live* individual neurons. Given the extreme complexity and fragility of the nervous system, such challenge required the development of tissue culture methods. After Harrison's (1) first discovery of a method to culture neurons, it took several decades to develop reproducible

L.C. Doering (ed.), *Protocols for Neural Cell Culture*, Springer Protocols Handbooks,
DOI 10.1007/978-1-60761-292-6_8, © Humana Press, a part of Springer Science+Business Media, LLC 1992, 1997, 2001, 2010

techniques for culturing neurons. In the 1970s, peripheral neurons were commonly cultured, whereas robust cultures from central nervous system became routine only in the 1980s. Today, tissue culture is widely spread and is almost taken for granted. We tend to forget that it is a "miracle" if a cell taken out of its natural environment not only can survive but often resembles and reacts as it does in vivo. Reality strikes back when our neuronal cultures do not grow or survive well, and the subsequent tedious troubleshooting reminds us of the complexity of maintaining cells alive and healthy outside of the animal. In fact, growing neurons in culture with steady success is not a trivial task.

### 1.1. The Use of Dissociated Neuronal Cultures

Several regions of the central and peripheral nervous systems can be maintained in culture, including the cortex, cerebellum, hippocampus, spinal cord, dorsal root ganglion, superior cervical ganglion, retinal, and others. The success and yield of these cultures are usually inversely correlated with the age of the animal from which the tissue is dissected. Cultures from embryos generally yield far more neurons than from neonatal animals, while cultures from adult animals are extremely difficult and generally impractical. There are also several culture configurations that can be used: high- or low-density – or even single – dissociated cells, explants cultures, or organotypic slice cultures. While the latter is covered in a different chapter by Michal E. Dailey, this chapter is focussed only on dissociated hippocampal neuronal cultures.

What are the advantages of using *dissociated* neuronal cultures? For one, dissociated cells are more easily visualized under the microscope, whether it is through simple phase-contrast or epifluorescence microscopy. By contrast, more sophisticated equipment, such as a confocal microscope or even a two-photon imaging system, may be needed to image fluorescent cells in a slice culture. Cultured cells in a dissociated configuration are also more convenient for applications that require gene transfer of plasmid DNA. Because the process of cell dissociation breaks apart all connections and most processes, such preparation is ideally suited to study neuritic growth and synaptogenesis. Despite the absence of many environmental factors present in slices or in vivo, dissociated neurons still acquire mature neuronal morphology and polarity over time and make functional synapses in most cases. Dissociating cells allows one to choose the cell type to culture, such as the neuronal or glial subtypes, through various enrichment protocols that have been developed, as well as to combine chosen cell types under co-culturing conditions. Because it is possible to control tightly (to some extent) the environment of dissociated cultures, this preparation is very good for cellular and molecular investigations of brain cell signalling. They can also serve as an excellent model preparation to study basic electrophysiological properties of neurons.

In some instances, dissociated cultures have even provided key insights into synaptic plasticity mechanisms, despite the lack of an organized circuitry as seen in vivo. Indeed, dissociated neuronal cultures do not provide specific circuits, since the synaptogenesis that occurs in these cultures is not controlled by the same cues as in vivo and the neurons are spatially laid out in a random fashion in a two-dimensional configuration. At the cellular level, investigators have seen differences, for instance, in the number of primary processes, the number and shape of dendritic spines, and the dynamic features of spines and filopodia. Thus, the dissociation of brain cells before culturing them necessarily leads to "imperfect" neural circuits. Nevertheless, they have provided a wealth of valuable insights into the functioning of the nervous system.

There are, to date, a very wide range of dissociated neuronal culture protocols and configurations, depending on the desired application, but also because of historical, traditional, cultural, and even financial factors. This chapter cannot even begin to cover them all. It will focus mainly on rat hippocampal cultures, with a few comparisons with hippocampal cultures from mice, which are obviously useful because of the accessibility of genetically modified mice. In the history of hippocampal cultures, two scientists had a major impact. Gary Banker (2) developed a popular culture configuration, which consists of low-density, nearly pure embryonic neurons grown on coverslips apposed to a glial feeder layer (3, 4). This preparation has led to many discoveries on the cell biology of neuronal development and synaptogenesis, taking advantage of immunocytochemical and imaging approaches. Later, Brewer et al. (5) developed a growth media recipe for higher-density cultures, which led to a simplified culture technique and allowed for large-scale cultures to be prepared more easily for biochemical investigations.

In this chapter, we provide detailed procedures from making solutions to maintaining cells for long-term culture, as well as tips and tricks to ease the process of preparing dissociated rat hippocampal cultures. Other excellent protocols are also available (3, 4, 6).

## 2. Reagents and Solutions

### 2.1. Stocks Solutions

All solutions are made with tissue culture-grade distilled water.

1. **Ara-C** (*cytosine β-d-arabinofuranoside*) (1 mM) (Sigma cat. # C-1768), 4.8 mg/20 mL of distilled $H_2O$, 0.22 μm filtered, 1 mL aliquots

2. **B-27 supplement** (50×) (Gibco Invitrogen Corporation cat. # 17504-044)

3. **FBS** (*fetal bovine serum*) (Hyclone cat. # SH30071.03)

4. Boric acid (100 mM) (EM SCIENCE cat. # 2710), 3.1 g/ 500 mL of distilled $H_2O$, 0.22 μm filtered

5. **Sodium borate** (100 mM) (EM SCIENCE cat. SX0355-1), 9.54 g/250 mL of distilled $H_2O$, 0.22 μm filtered

6. **BSA** (*bovine serum albumin*) (300 mg/mL) (Sigma cat. # A-9576) (*BSA fraction V, 30% solution*), 1.7 mL aliquots

7. L-**Cysteine** (100×) (Sigma cat. # C-2529), 42 mg/mL of distilled $H_2O$, 0.22 μm filtered, 320 μL aliquots

8. **Dissection solution w/o 40× solution**, 56.4 mL HBSS 10× w/o $Ca^{2+}/Mg^{2+}$/NaHCO3, 5.1 mL HEPES (1 M at pH 7.6), 443.6 mL distilled H2O, 1 mL phenol red (0.5%), *final solution will give a pH of 7.4*

9. **Distilled water** (tissue culture grade) (Gibco Invitrogen Corporation cat. # 15230-162)

10. **DNAse 1 solution** (Sigma cat # D-5025), 150,000 units/ 7.5 mL of distilled $H_2O$ (*to get 20 K units/mL*), 0.22 μm filtered, 1.25 mL aliquots.
**Note**: Depending on the lot number, it may not dissolve properly. According to Sigma's technical service; centrifuge the undissolved material should be centrifuged before filtration, the enzyme activity should not be affected.

11. **EDTA** (0.5 M) (Gibco Invitrogen Corporation cat. # 15575-038)

12. **Ethanol** (70%), 2800 mL/4 L of distilled $H_2O$

13. **Glucose** (200 mg/mL) (Gibco Invitrogen Corporation cat. # 15023-021), 40 g/200 mL of distilled $H_2O$, 0.22 μm filtered, 10 mL aliquots

14. **Glutamax-1** (*L-glutamine as a stabilized dipeptide*) (200 mM) (Gibco Invitrogen Corporation cat. # 35050-061), 1.4 mL aliquots

15. **HBSS w/o $Ca^{2+}/Mg^{2+}/NaHCO_3$,/phenol red** (10×) (Gibco Invitrogen Corporation cat # 14185-052)

16. **HBSS with $Ca^{2+}/Mg^{2+}$ w/o $NaHCO_3$/phenol red** (10×) (Gibco Invitrogen Corporation cat.# 14065-056)

17. **HEPES** (1 M) (Gibco Invitrogen Corporation cat. # 11344-041), 238 g/L pH 7.6 with NaOH, 0.22 μm filtered, 50 mL aliquots

18. **$NaHCO_3$** (7.5% w/v) (Gibco Invitrogen Corporation cat. # 25080-094)

19. **NaOH** (0.1 N) (EM Science cat. # B10252-34), 4 g/L distilled $H_2O$, 0.22 μm filtered, 1 mL aliquots

20. **Neurobasal media** (Gibco Invitrogen Corporation cat. # 21103-049)

21. **Nitric acid solution** (Sigma cat # 438073) (*stock is 70%*). Make a 10% solution for Aclar coverslips or 50–70% solution for glass coverslips using distilled $H_2O$.

22. **Papain** (Worthington cat. # LS003126) (*Enzyme activity varies with lots.*)

23. **Pen/Strep** (Gibco Invitrogen Corporation cat # 15140-031), penicillin (10,000 units/mL)/streptomycin (10,000 μg/mL), 10 mL aliquots

24. **Phenol red** (0.5%) (Gibco Invitrogen Corporation cat # 15100-43)

25. **Sodium pyruvate** (100 mM) (Gibco Invitrogen Corporation cat. # 11360-070)

26. **Sylgard 184** (Paysley products cat. # 184), Follow the manufacturer's recommendations.

27. **6.67×solution**, 80 mL HBSS 10×w/o $Ca^{2+}/Mg^{2+}/$ $NaHCO_3,$/phenol red, 23.5 mL $NaHCO_3$ (7.5% w/v), 0.8 mL EDTA (0.5 M), 16.5 mL distilled $H_2O$

28. **40×solution**, 10 mL pen/strep, 10 mL glucose (200 mg/mL), 20 mL sodium pyruvate (100 mM), 10 mL distilled $H_2O$, make 3.2 mL aliquots.

**2.2. Working Solutions**

1. **Borate buffer pH 8.5** (100 mM), Mix about 1 part of sodium borate 100 mM with about 3 parts of boric acid 100 mM until you reach pH 8.5.

2. **BSA columns**, *Solution for 4 columns, enough for dissection of 10 animals.* 20 mL trituration solution (*see below*), 1.2 mL BSA 300 mg/mL, 600 μL NaOH 0.1 N. Split into 4 tubes of 5 ml (use 15 mL centrifugation tubes).

3. **Complete growth media**, Neurobasal media, B-27 supplement 50×, use 1/50, pen/strep, use 1/200, Glutamax-1 (200 mM), use 1/400

4. **Cutting solution**, 5 mL dissection solution (*see below*), 5 mL papain solution (*see below*)

5. **Dissection solution**, 40 mL dissection solution w/o 40× solution, 1 mL 40× solution

6. **Papain solution**, 4.5 mL 6.67× solution, 300 μL L-cysteine 100×, 750 μL 40× solution, 400 μL DNAse 1 solution, 24 mL distilled $H_2O$. Right before dissection, add enough papain to get a concentration between 10 and 12 units/mL. Warm up the solution for 10 min at 37°C to allow the papain to dissolve. Since the papain stock is not sterile, filter the dissolved papain solution through a 0.22 μm syringe filter, previously rinsed with 10 mL of sterile distilled $H_2O$.

7. **Poly- D -lysine solution** (100 µg/mL for Aclar and 1 mg/mL for glass coverslips) (Becton Dickinson cat. # 354210), 20 mg/200 mL in distilled water or borate buffer pH 8.5 (100 mM)

8. **Trituration solution**, 47.8 mL neurobasal media, 332.5 µL BSA stock (300 mg/mL), 1.25 mL 40× solution, 660 µL DNAse I solution

## 3. Material and Equipment

### 3.1. Coverslips and Accessories

Two types of coverslips can be used for neuronal cultures: glass or Aclar (plastic).

1. **Aclar 22C sheet** (Honeywell cat # P5000HS). Copolymer film, 5 mil (approximately 0.125 mm) thick, consisting primarily of polychlorotrifluoroethylene. (Typically, we use 13 mm circles, but also 18 mm works when using 12-well plates).

2. **Glass coverslips no 1D** (Fisherbrand cat # 12-545-82 [*12 mm diameter*], #12-545-84 [*18 mm diameter*], these coverslips are *0.13–0.17 mm thick*)

3. **Multiwell plates** (Becton Dickinson cat # 353047 [24 wells], #353043 [12 wells])

4. **Petri dishes** (Becton Dickinson cat # 353001 [35 mm], #353004 [60 mm], #353003 [100 mm]) (Sarstedt cat # 83.1803 [150 mm])

5. **Transfer pipettes** (Becton Dickinson cat # 357575)

6. **9-inch Pasteur pipette** (Fisherbrand cat # 13-678-20D)
   **Note:** The tip diameter of the pipettes will vary within a box; use only the pipettes with largest opening to prevent damage of the cells while triturating.

### 3.2. Equipment

Biological safety cabinet, centrifuge with swing-out buckets, hemocytometer, heat block, incubator (37°C, 5% $CO_2$, 90% humidity) connected to $CO_2$ gas cylinder, inverted microscope (for the culture room), $O_2/CO_2$ (95%/5%) gas cylinder.

Other equipment (some of these are either optional, harder to find, or necessary only for culturing cells on Aclar coverslips):

1. **Arch punch to cut Aclar** (Duncan Instruments cat # 1271E) [*13 or 18 mm*]

2. **Teflon cutting board** (*3/8" × 15" × 18"*) (can be found through a local company that specializes in plastic materials)

3. **Dissection binocular,** Although the hippocampus can be removed from postnatal rat brain without a binocular, it is necessary for proper dissection from embryonic brain.

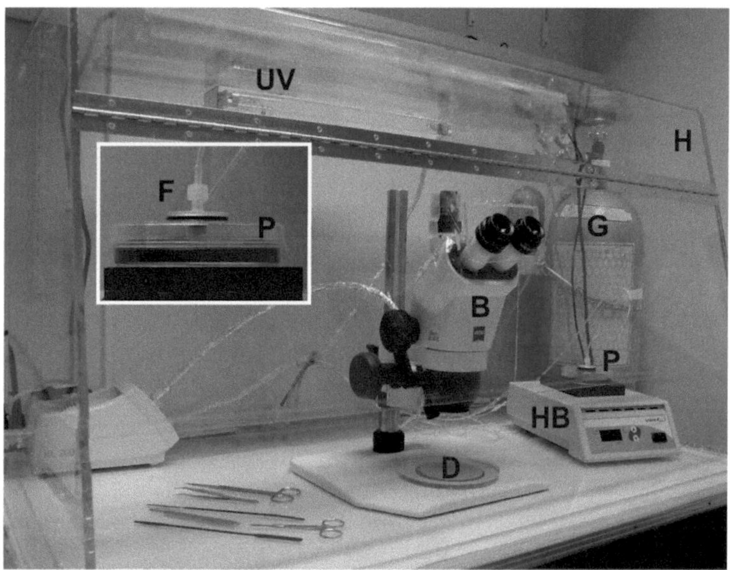



Fig. 8.1. Example of a dissection hood (H), equipped with a binocular (B), heat block (HB), Petri dish (P) for papain incubation, connected to the $O_2/CO_2$ gas cylinder (G) through a syringe filter (F). A UV light (UV) is used to sterilize the inside, for a 30–60 min period, before the dissection (hence the protecting foil around the electrical wires). A dish (D) with Sylgard, Whatman paper, and dissection solution is placed under the binocular.

4. **Dissection hood** (**Fig. 8.1**, *H*), Can be custom-designed to include the dissection binocular (*B*) and the heat block (*HB*) on which the papain-containing Petri dish (*P*) is placed; the hood helps keeping cultures sterile; a UV light (*UV*) can be installed and turned on for a 30–60 min period, before dissection.
   **Note**: place the Sylgard plates inside the hood during that time.

5. **Dissection instruments**, You will need scissors (big and small), a scalpel, 2 Teflon-coated spatulas (Fisher Scientific cat. # 57951-002), 1 stainless steel spatula (Fisher Scientific cat. # 57950-000), a few tweezers of different shapes and one specifically used for the handling of the coverslips, a 5/45 tweezer (Roboz cat. # RS-4918).

6. **Baffled Erlenmeyer flask 2 L** (VWR cat. # 89000-992)

## 4. Protocol for Coverslips Preparation

Neurons can be grown on glass or plastic, coated with poly-D-lysine (PDL) (some investigators also add laminin in addition to the PDL) or poly-L-lysine. For microscopy applications, the only plastic that we know of which is compatible with fluorescence is Aclar (22C, 5 mil). For electron microscopy, this plastic is very convenient because it can be cut. Our experience is that neurons tend to adhere more reliably on Aclar coverslips, but many laboratories have good success with glass. The latter is slightly more transparent, thus has a lower background for fluorescence applications, which can be critical for low-level signals or non-confocal applications.

For their preparation, the three main differences are that i) the glass cannot be shaken as hard during washes or it will shatter, ii) the Aclar can only withstand incubation in 10% nitric acid while glass will resist 50–70% nitric acid iii) the coating of the glass with Poly-D-lysine has to be done at a much higher concentration for the cells to adhere properly. Wear rinsed powder-free gloves and use dishes exclusively reserved for this purpose. Never wash the dishes using detergent. Use tissue culture grade, sterile distilled water to wash coverslips and to prepare solutions. Handle the coverslips with forceps or tweezers, not with hands or gloves.

### 4.1. Cutting the Aclar Coverslips

While glass coverslips can be purchased pre-cut, Aclar (22C, 5 mil), on the other hand, comes as a roll and coverslips have to be cut to desired size using a punch. Use a 13- or 18-mm arch punch and a hammer to cut the circles on a Teflon cutting board (inert, non-toxic surface). Collect the circles in a 50 mL conical tube (*a full tube will hold close to 1200 coverslips*).

### 4.2. Rinsing the Aclar Coverslips

A lot of debris (dust on the Aclar and small chips from the cutting board) can stick to the coverslips, but can be washed away easily.

#### 4.2.1. Rinsing Method 1

1. Use a clean tip box (200 µL will do) with the tip rack as a lid and use it as a strainer.

2. Put in 200–300 coverslips with distilled water, shake by hand and strain through the tip holder rack. Repeat until you see no debris.

#### 4.2.2. Rinsing Method 2

1. Use a bath sonicator with a pan used only for that purpose.

2. Put in 200–300 coverslips with distilled water, sonicate 2–3 min and decant the water. Repeat until you see no debris.

### 4.3. Washes with Water

From this step on, use sterilized dishes with foil as a lid.
1. Put 200–300 glass coverslips in a 2 L size Erlenmeyer flask or the same amount of rinsed Aclar coverslips in a 2 L size *Baffled* Erlenmeyer flask (for stringent washing).

2. Wash the coverslips with distilled water (~150 mL) 3–4 times for 15 min on an orbital shaker (running vigorously for Aclar, moderately for glass), and carefully pour off the solution.

   **Note:** Use a glass funnel to catch any coverslips falling off the Erlenmeyer. For this step and the following ones, make sure the sides of the Erlenmeyer are rinsed as you empty it.

For this step and the following ones, make sure the sides of the Erlenmeyer are rinsed as you empty it.

***4.4. Washes with 70% Ethanol***

1. Wash the coverslips with 70% ethanol (~150 mL), 3–4 times for 10 min on the shaker. Each time the Erlenmeyer is emptied, make sure the sides are rinsed as well.

2. Wash away the ethanol 3–4 times with distilled water, as above. Each time the Erlenmeyer is emptied, make sure the sides are properly rinsed so there is no carry over of the ethanol in the subsequent washes. This washing procedure with distilled water should be followed for all subsequent washes below.

***4.5. Washes with Nitric Acid***

1. Wash the coverslips at room temperature with 150 ml of nitric acid (10% for Aclar and 50–70% for glass coverslips) on the shaker overnight.

2. Wash away the nitric acid 4–5 times with distilled water, as above.

3. After the last wash, drain the water as much as possible.

***4.6. Coating the Coverslips with Poly-D-Lysine***

1. Add 30–40 ml of PDL solution to cover all coverslips in each flask. Incubate on orbital shaker 1–3 h at room temperature. **Note:** The shaking has to be very gentle, just enough to see the coverslips moving. Note that the coverslips will be sticking together at the beginning but they will eventually come apart, allowing the PDL to coat both sides.

2. Store the flasks in the refrigerator without shaking overnight.

3. Wash away the PDL 4–5 times with distilled water, as above.

4. Transfer the coverslips from 1 or 2 flask(s) into a 150 mm Petri dish using distilled water to help recover all of them. Aspirate most of the water, leaving just a little at the bottom of the dish.

***4.7. Drying and Sterilizing the Coverslips***

Some investigators prefer not to dry coverslips before their use and thus only prepare small batches before every culture preparation. We find that good coverslips (especially Aclar) with good batches of PDL can be dried and stored for months. This has two main advantages: (i) large batches can be made at once; and (ii) it also allows for a radial gradient of cell density to be seeded using small drops of cells of different concentrations (*see* below, **Fig. 8.3A**).

1. In the tissue culture hood, separate the coverslips with the help of two *5/45* tweezers and lay them side by side on large Kimwipes affixed with tape inside the hood. *Be careful not to scratch the coverslips with the tweezers; handle them on the edges.*

2. Allow the coverslips to dry for 30 min.

3. Expose the coverslips to the UV light inside the biological safety cabinet for 30 min.

4. Using sterile tweezers, transfer each of them to a sterile 24-well plate putting the UV-treated surface facing the bottom of the plate.

5. Expose the second side of the coverslips to the UV light for 30 min (lid off the plate).

6. Put the lids (still sterile) back on the plates and store them at 4°C in clean plastic bags.

   **Note**: Recycle the bags containing the 50 mL conical tubes.

---

## 5. Protocol for Hippocampal Neuron Preparation

This protocol mainly describes a preparation for rat neonatal hippocampal neurons, grown at high density (1500–2000 cells/mm$^2$) in serum-free media (Gibco Invitrogen Corporation), developed by Brewer et al. (5). A few alternative configurations are also presented.

Prepare the working solutions (section 2.2) from the stocks solutions just before dissection. Keep solutions on ice until ready for dissection.

### 5.1. Dissection

Before starting the dissection, you can prepare a 100 mm Petri dish (**Fig. 8.1**, *P*) in which a small hole is punctured on the cover, in order to fit the tip of a 0.22 μm syringe filter (**Fig. 8.1**, *F*), connected to a gas ($O_2/CO_2$; 95%/5%) tank (**Fig. 8.1**, *G*).

**Note:** The cover can be reused for subsequent dissections if cleaned with 70% ethanol.

Pour the papain solution into the dish a few minutes before starting the dissection, place the dish on the hot plate (~35°C) and start the flow of gas over the solution, in order to maximize oxygen concentration in the solution (without bubbling it, which is not advisable). The $CO_2$ will ensure that the papain solution is at a physiological pH (7.2–7.4), which can be verified via the coloration of the phenol red; if the coloration is appropriate (orange-red), this suggests that gas exchanged is good and thus that oxygenation should be maximal.

**Note:** Trypsin is often used for the digestion of embryonic tissue because neurons at that stage are less fragile and trypsin works much faster than papain. With postnatal tissue, we prefer to use papain.

1. Anaesthetize P0–P3 rats (up to 10 sequentially, as you dissect them) using a method approved by your institutional animal care organization.

   *All subsequent steps are carried out aseptically.*

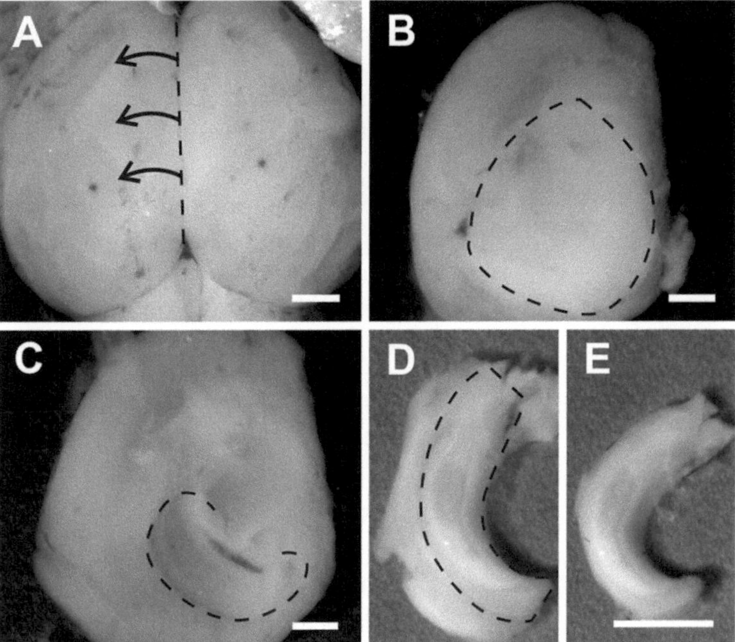

Fig. 8.2. Dissection of hippocampus from postnatal rat brain. (**A**) Rat brain; separate the two hemispheres with a scalpel along the *dotted lines*. Place one side in the ice-cold dissection solution while working with the other. Flip the hemisphere to expose the midbrain. (**B**) Left brain hemisphere with the inside facing up; remove midbrain region inside dotted lines. (**C**) Hippocampus is visible once the midbrain and meninges are removed; scoop up with a spatula. (**D**) Hippocampus with some extra tissue to remove (outside of *dotted lines*). (**E**) Hippocampus ready to be chopped in smaller chunks and transferred into papain. Scale bars: 1 mm.

2. Remove each brain (**Fig. 8.2A**) and transfer them in a 60 mm Petri dish placed on an ice pack containing dissection solution. Cut each brain in half with a scalpel on the longitudinal cerebral fissure (**Fig. 8.2A**, dotted line) and leave them in the ice-cold dissection solution.

   *Subsequent steps are all carried out at room temperature until plating.*

3. Dissect hippocampi (**Fig. 8.2B–D**), one at a time under the dissecting binocular, in a 150 mm Petri dish cover containing polymerized Sylgard covered with a # 1 Whatman paper (100 mm) on which there is enough dissection solution for the tissue to be covered (too little solution will reduce the yield of healthy neurons; **Fig. 8.1**, *D*).

   **Note:** Sylgard is not absolutely necessary, but it provides a good support to retain the dissection solution in place.

4. Once meninges are removed and the hippocampus is free from the brain (**Fig. 8.2D**), cut out the extra tissue (**Fig. 8.2E**) and transfer it with a spatula into a drop of cutting solution, standing on a 35 mm Petri dish cover containing polymerized Sylgard.

**5.2. Digestion with Papain**

1. Immediately after transferring one hippocampus into the drop of cutting solution, cut it in small pieces with a scalpel and transfer the pieces with the liquid (using a 3 mL transfer pipette with tip cut wider using a scalpel) to the 100 mm Petri dish containing the papain solution kept warm and gassed on the heat block.

2. After dissection of all the hippocampi, allow them to incubate on the heat block for another 15 min. Stop the digestion by transferring the hippocampi pieces to a 50 mL conical tube (already containing 15 mL of trituration solution) using the 3 mL transfer pipette with a wide opening.
   *All subsequent steps are carried out in a biological safety cabinet.*

3. Let the pieces settle down to the bottom of the tube for about 2 min.

4. Remove most of the solution but leave 3–5 ml (be careful when aspirating solution, DNA from dead cells increases viscosity and may cause the clumps to be drawn into the pipette). Use the leftover solution to resuspend the pieces and transfer the whole suspension to a 15 mL conical tube, complete the volume to ~10 mL using the trituration solution. Close the tube and invert it gently a few times; let the pieces settle down for about 2 min.

**5.3. Trituration**

1. Meanwhile, fire-polish a 9-inch Pasteur pipette (with cotton) to smoothen the tip, wet the pipette with trituration solution to coat the glass (keep the pipette in an empty 15 or 50 mL tube in a rack inside the cabinet).

2. Remove most of the supernatant from the tissue chunks using the suction
   (**Note:** Put a plastic pipette tip at the end of the suction pipette to improve the control of the suction.), leave 1–2 ml and finish by hand using the Pasteur pipette to prevent aspirating inadvertently the tissue chunks.

3. Add about 2 volumes of trituration solution to the tissue pellet and triturate with the Pasteur pipette 40–100 times (*Try not to introduce air bubbles during that step, do not fill the Pasteur pipette above the level where the glass widens*). Let visible chunks settle to the bottom for about 3 min and transfer most of the dissociated cells (supernatant) to a 15 mL centrifugation tube kept aside.

4. Fire-polish the Pasteur pipette further to reduce the tip size a little bit and cool it down. Add about 2 volumes of trituration solution to the undissociated material and triturate again. Let the visible chunks settle to the bottom for about

3 min and transfer most of the dissociated cells to the same centrifugation tube kept aside.

5. Repeat trituration steps once more. (*Usually, some small chunks remain in the trituration tube and they can be discarded; if they are still pretty big, it means that the digestion did not work optimally or that the trituration was not thorough enough*). Discard undissociated material.

**5.4. Cell Enrichment**

1. Bring the volume of the dissociated cell solution to 4 mL using the trituration solution; resuspend and gently lay 1 mL of it on top of the solution in each of the BSA columns. (*The BSA solution has a higher density allowing the cell suspension to float on top. Upon centrifugation, dissociated cells will pellet, while many cell debris will remain in the column. The BSA also helps inactivating the papain.*) We recommend splitting the cell suspension into 4 columns if you dissect 10 animals; you can scale down if you dissect less. The purpose of splitting the cell suspension is to prevent excess viscosity from DNA of dead cells to alter cell separation during centrifugation.

2. Centrifuge with an IEC clinical centrifuge at speed # 4 for 5 min. (*Time and speed will vary in different centrifuges, use 350–400 g for 5 min in a swing out bucket rotor.*)

**5.5. Cell Resuspension**

Remove supernatant (to prevent aspirating the pellet, remove the last milliliter with the trituration Pasteur pipette). Using a small volume of complete growth media, resuspend the pellets very gently (they are very small) and combine all the cells into a single tube. Bring the final volume to about 200 μL per animal and count the cells using a hemocytometer. Depending on the cell-seeding configuration (*see* below), you may adjust the cell suspension volume to obtain a specific cell concentration.

**5.6. Cell Seeding**

Cells can be plated either on wet or dry coverslips depending on the desired cell density and/or configuration on the coverslips. At the beginning, it is a good idea to test different cell densities and assess survival over time. Neurons plated at very low density (i.e., <10,000 cells/well or 2 cm$^2$) will not survive more than a few days in a neurobasal/B-27-based culture media, unless you use a glial feeder layer. The following description of cell seeding is to obtain a radial gradient of cells, with high density of neurons in the center and lower density on the edges (**Fig. 8.3A**). This configuration, which requires pre-dried coverslips, is useful to obtain longlasting cultures, where the high-density neurons help sustaining the low-density neurons on the edges (**Fig. 8.6A,B**), without the need of a glial feeding layer. Other configurations are proposed in **Section 5.8**.

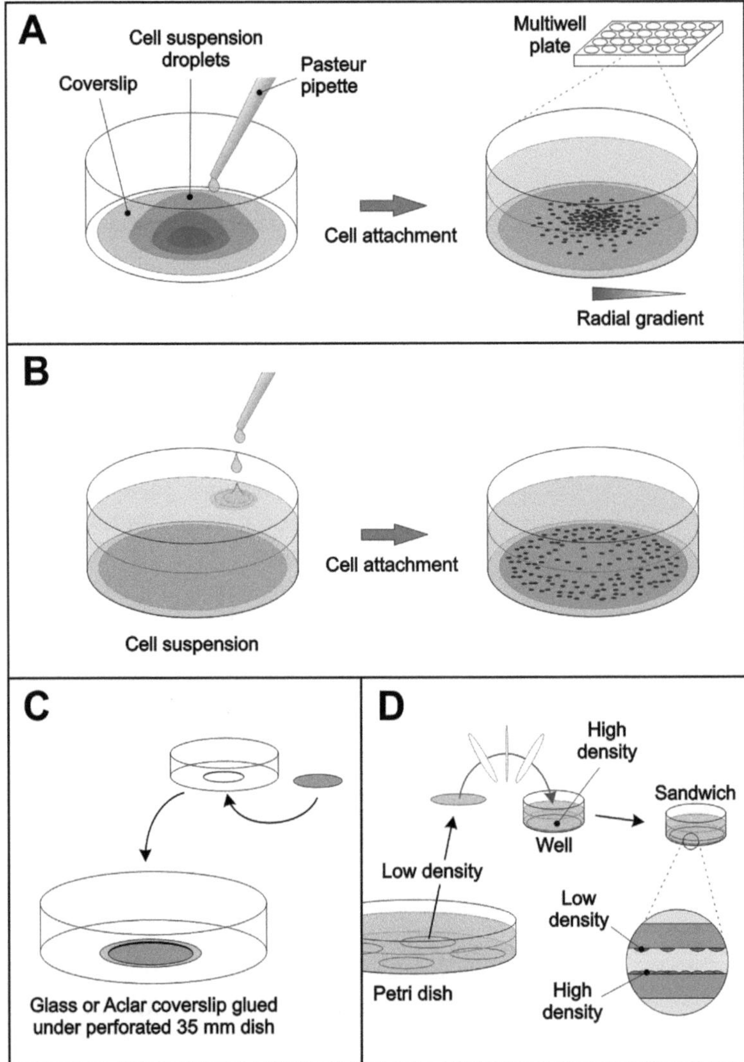



Fig. 8.3. Different culture configurations. (**A**) To obtain a radial gradient of density, three droplets of cell suspension (at decreasing density) are deposited on dried PDL-coated coverslips (*left*) with 10 min intervals between droplets to allow cell attachment. One to two hours later, the wells are rinsed and filled with growth media (*right*). (**B**) To obtain a relatively even distribution of cells across the coverslip, they are pre-mixed with growth media and 1 mL is added to each well. (**C**) Coverslip glued under perforated Petri dish. Cells can be plated in the center by (i) adding droplets of cells on the dried coverslip only (filled with media 1–2 h later), or (ii) by temporarily placing a sterile Teflon ring around the center hole in a pre-filled Petri dish (removed the next morning) to prevent cells from floating away from the center. (**D**) Neurons plated at low density on coverslips kept in a large Petri dish can be transferred into wells containing high-density neurons (plated previously), by flipping the coverslip to create a "sandwich" configuration. The high-density layer seems to provide factors that sustain the low-density neurons for several weeks.

1. Using a fire-polished Pasteur pipette, add a first drop (~30–35 µl) of cell suspension (1–2 million cells/mL) to the center of each coverslip already in the 24-well plates (**Fig. 8.3A**); put the plates into a single row in the incubator (*37 C, 5% CO2, 90% humidity*) to rapidly warm up the bottom of each plate, incubate for 10 min to allow most cells to settle down and attach to the coverslip.

2. Add a second drop of cell suspension 10 times diluted on top of the first drop and return to the incubator for another 10 min.

3. Add a last drop of cell suspension 100 times diluted on the top of the second drop and return to the incubator, this time for at least 90 min.
   **Note:** For the last drop, be gentle to prevent the liquid to run to the sides.

4. Rinse cells to remove remaining debris with 0.5 mL of pre-warmed neurobasal media without additive (*added dropwise directly over the cells*) and sequentially replace it (*right away*) with 1 mL of pre-warmed complete growth media supplemented with 2% of FBS (we find that the cells grow much better if we add 2% FBS at plating; we do not add FBS from then on). Aclar coverslips can occasionally trap an air bubble under and float. To remove them, push down on the edge of the coverslips (using a 1 mL pipette tip). This step is not necessary for glass coverslips.

***5.7. Cell Maintenance***

Feed cells on day 5, adding Ara-C (to kill dividing cells) by replacing 0.5 mL of cell media with 0.5 mL of complete growth media containing 10 µM Ara-C (this will give a final Ara-C concentration of 5 µM). Feed the cells again on day 9 by replacing half the media with fresh complete growth media (without Ara-C). From then on, feed cells twice a week by replacing half the media. We find that this feeding scheme, including the use of FBS during the first few days and the delayed addition of Ara-C on day 5, allows for a moderate amount of glial cells to proliferate and survive. Then, because the neurobasal media is not very permissive to glial cell proliferation, we see little increase in glial content over time. Indeed, the glial cells that no longer divide on day 5 will survive the Ara-C treatment, whereas those that continue dividing should die.

Try to use freshly prepared growth media every time you feed the cells. If you grow neurons at extremely high density, the cell media might turn yellow between feedings. In that case, feed 3 times a week instead of 2, or use a vessel in which you can put more media.

### 5.8. Alternatives

*5.8.1. Seeding
Configurations*

The radial density configuration described above is an excellent compromise for many applications. For example, the high cell density in the center will be useful for transfections (which tend to be low efficiency), while the low density on the periphery will often be useful for immunocytochemistry and imaging and will be appreciated when doing electrophysiology.

On the other hand, cells can also be seeded at a homogenous density throughout the coverslip. This can be achieved by adding 1 mL of complete growth media (+2% FBS) containing, for instance, 25,000–500,000 cells/mL directly into each well with a coverslip at the bottom (**Fig. 8.3B**). In our experience, the surface tension in the small wells tends to increase the cell density around the edges of the coverslips compared to the center.

In general, this seeding method yields more debris and dead cells compared to the previous method, but they can be washed off the next day by aspirating the media from the well and immediately washing the coverslip by flushing 0.5–1.0 mL of neurobasal media directly on the cells and then replacing it with complete growth media containing 2% FBS. The long-term viability of such cultures varies in our hands from 2 to 4 weeks, while the radial density cultures survive well up to 6 weeks. We find that at very high density, neurons tend to migrate into groups, but this usually does not affect the health of the culture.

When seeding the cells, you also have to consider the choice of vessel for your application (time lapse imaging, biochemistry, immunocytochemistry, transfections). For example, if you are planning to collect the cells for protein extraction, it is a good idea to seed the cells directly on the bottom of a PDL-coated Petri dish. Another interesting vessel is the one where a 25 mm glass or Aclar coverslip is glued, using Sylgard, under a perforated 35 mm Petri dish (**Fig. 8.3C**). Such modified Petri dish is very much appreciated because it can serve as a chamber on the microscope (upright or inverted) for imaging or electrophysiology. This can minimize manipulation of coverslips from the incubator to the microscope. Teflon inserts can be machined to introduce into the Petri dish and guide the flow of solutions over the cells or the ground electrodes when recording electrophysiologically. It is particularly useful for high-resolution imaging on an inverted microscope, using a 63–100× oil immersion objective, since the coverslipped bottom replaces the thick plastic. When using Aclar to cover the bottom, the plastic coverslip can easily be peeled off after cell fixation at the end of an experiment for immunostaining and mounting on a slide.

For growing low-density culture, the Banker method is the best characterized approach and is described in details in Kaech and Banker (4). The essence of this method is to grow the neurons on a coverslip apposed against a glial feeder layer. A fairly easy substitute for this approach, which has not been thoroughly characterized, however, consists of making a "sandwich" of coverslips into a well (**Fig. 8.3D**). The bottom coverslip supports high-density neurons (which could be plated the week before), while the other coverslip, on which low-density neurons are plated, is flipped upside down. As such, the latter are fed by factors released from the high-density neurons. The sandwich approach is thought to be beneficial also by restricting oxygen to the cells (7).

Finally, investigators have used microcultures, where the PDL can be sprayed as droplets onto coverslips to restrict neurons from growing their processes out to reach other neurons. This configuration is interesting for the study of autaptic synapses, and can be combined with a feeder layer. Many other configurations can be designed, such as those to guide neurites along specific paths of substrates.

*5.8.2. Embryonic Versus Postnatal Cultures*

Our hippocampal neuron dissociation method described above using P0–P3 rats can be used with rat embryos around E18. Embryonic preparations yield more neurons and tend to have less glial cells. For many applications, both types of cultures are interchangeable. However, depending on the question pursued (which may involve a developmentally regulatory gene), it may be important to choose between embryonic and postnatal animals. The growth media neurobasal was initially developed for embryonic cultures, whereas neurobasal A (which has a higher osmolarity) has more recently been developed for postnatal cultures. However, we find that the original neurobasal works well with both embryonic and postnatal preparations.

*5.8.3. Mouse Versus Rat Cultures*

Using the same method described here for postnatal rat hippocampal neurons, we have grown mice neurons with good success. However, the number of surviving neurons is significantly lower. It may be advisable to use embryonic mice to improve the yield.

## 6. Critical Steps and Troubleshooting

It is important to observe your cultures on a daily basis to detect any problem, otherwise planned experiments may be compromised. However, your observations should be done quickly to reduce stress (light exposure, temperature and pH variations) on the cells. It is preferable to have an inverted microscope in the culture room for daily observations; if you plan to transfect the cells with fluorescent markers, having fluorescence capability on the microscope is very useful to check quickly for transfection efficiency.

### 6.1. Poly-D-Lysine and Coverslip Preparation

One of the most common problems with hippocampal neurons is the quality of the substrate, mainly the PDL or the glass coverslips. For this reason, we decided to put a lot of information in this section.

Over the years, we have tried several brands and molecular weights of PDL. Our best results were achieved using Becton Dickenson's PDL which has a very high molecular weight. Nevertheless, the quality of the poly-D-lysine still varies from lot to lot and even within a lot over time. In our experience, a PDL batch that resuspends easily in 10 mL of water will give a good substrate to work with. You may use water only or borate buffer (pH 8.5) to dissolve the PDL to the desired concentration, usually 50–250 μg/mL. We find that resuspending the PDL first in water and then adding borate buffer helps dissolving the PDL. When the PDL does not dissolve well in water (slime or clumps visible) or if it precipitates when adding borate buffer, there is something wrong with it, and in that case, it is preferable to use only water for resuspension. Incubating the solution at 37°C a few hours to overnight might help. To detect a bad batch of PDL, dilute the solution to its final concentration (50–250 μg/mL) into a tissue culture flask, swirl around the solution in the flask. If you see droplets or undissolved material on the walls of the flask, that is a bad sign. The solution should leave no trace on the walls of the flask. You may try to warm it up at 37°C overnight and transfer the PDL to a fresh flask and repeat the test. If you still see deposits sticking to the walls of the new flask, the PDL should be discarded.

A good PDL solution will keep for weeks to months at 4°C if not contaminated. We do not sterilize our PDL solution because the molecular weight (500–550 KD) is pretty high and it plugs filters. On the other hand, we use sterile water and good aseptic techniques to prepare it, and coated coverslips are subsequently UV-treated. It is also advisable to test for the adherence of the coated coverslips after seeding the cells. You can achieve that by checking the cell density under the microscope before and after washing with 0.5 mL of media (do only a few coverslips at this point). If the cells are detaching during the wash, leave the coverslips longer in the incubator so the cells will have time to adhere properly.

A good solution of PDL will not, however, support neuronal growth and survival for long if the coverslips are not properly cleaned. Be very, very meticulous while preparing the coverslips (glass or plastic), otherwise the PDL coating will be compromised and cells will either not adhere properly or die within a few days.

Another nightmare for neuron "growers" is the quality of the commercially available glass coverslips. Not all types and sources are suitable for neuronal adhesion. The most commonly recommended one is German glass specified no 1D (4).

**6.2. Sylgard 184**

To prevent bad curing of the Sylgard, contaminants such as detergent, phosphate, sulphur, and metal residues should be avoided. In our laboratory, we use a bran new disposable Petri dish and a wood stick (Q-tip handle) when mixing the solution with its curing agent.

**6.3. Long-Term Cell Maintenance**

It is generally fairly easy to grow neurons for a week, but not everyone is able to maintain these cultures for a month. A lot of factors, including osmolarity and pH fluctuations, contaminations, and microglia, will contribute to the stress in the cultures, which can inevitably cause cell death.

*6.3.1. Maintain Constant Osmolarity*

Watch for media evaporation over time in the incubator. Avoid too much traffic in the incubator as the humidity level in the chamber takes very long to recover from opening the door. Make sure your water pan contains water at all time. Be aware also that for live cell applications (imaging, electrophysiology, etc.), if you transfer the coverslips into another buffered saline, its osmolarity should be adjusted to that of the growth media. This is particularly important if you use neurobasal, which has a very low osmolarity ($\sim$220 mOSM depending on lots).

*6.3.2. Maintain Proper pH of the Media at All Time*

Never use a growth medium if its color has turned red/pink (indicating high pH) as it may give a shock to the cells upon media change. Never leave a plate of cells out of the incubator for more than a few minutes. In the ambient air, $CO_2$ level is low; thus the $CO_2$ in the media will rapidly escape and the bicarbonate will raise the pH. It is easier for beginners to work with only a few coverslips per plate, leaving the other plates in the incubator while working on one at a time. You can aliquot the 500 mL media bottle to avoid warming up the entire bottle or leaving it open for too long when feeding lots of plates.

*6.3.3. Maintain Culture Sterility*

It is extremely important to use very good sterile methods (aseptic techniques) every step of the way. A contamination can ruin the cultures of an entire laboratory, and since they are sometimes kept for weeks it has a lot of consequences. It can be advisable to have one person in charge of the cultures for the whole group (including the preparation of all stock solutions and materials, as well as the cultures). This not only ensures more reproducibility in the results but also helps pinpointing the origin of any problem.

*6.3.4. Microglia*

While the presence of microglia is unavoidable, a sign of concern is the presence of *activated* microglia in the cultures which usually correlates with unhealthy cultures. The difficulty for the inexperienced person is to actually recognize them under the microscope (examples are shown in **Fig. 8.4A,B**).

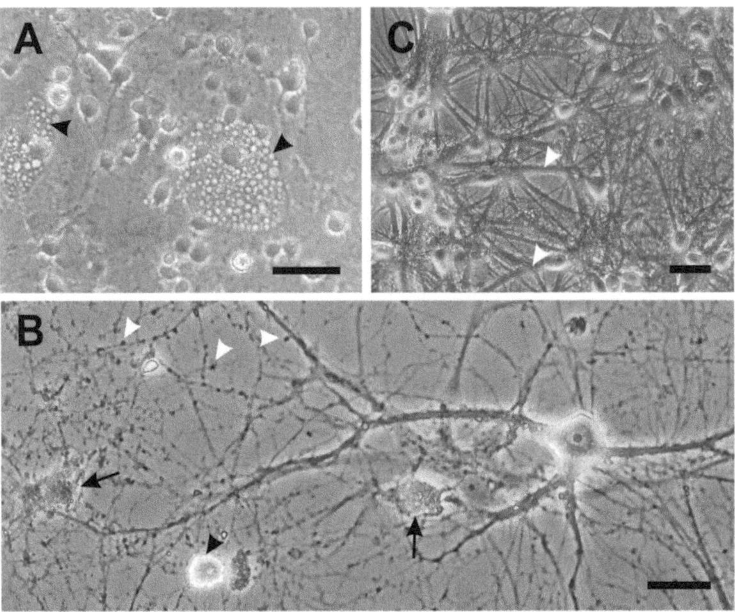

Fig. 8.4. Signs of trouble. (**A**) Presence of attached microglia (*arrowheads*). Too much microglia can be a sign of unhealthy cultures. (**B**) Low-density culture with processes showing signs of blebbing (*white arrowheads*), free floating microglia (black arrowhead), and dead neurons (*arrows*), (**C**). High-density culture showing a high degree of process fasciculation (*arrowheads*). Scale bars: 50 μm

*6.3.5. Other Stresses*

To minimize the stress to the cells, do not aspirate the media in the well completely, always leave a bit at the bottom to prevent direct exposure of the cells to air. Never leave a well almost empty for any period of time, replenish with media or buffer right away, otherwise the cells will die.

Another unpredictable problem that may occur is the quality of the B-27 batch that you receive. Indeed, issues with B-27 have been raised in the past (which is one of the reasons why we adopted the 2% FBS supplement for plating), so keep good track of your lot numbers. If your cultures suddenly go bad when you change lot, call the manufacturer to check if others have filed complaints with the same lot. Such advise is good for any new stock or solution used for your cultures. Keep track of when you change anything, this will help you pinpoint any problem rapidly.

There are several other recipes of growth media, supplemented or not with serum, that are used worldwide and which were not described in this chapter. Each recipe has its pros and cons and it would be difficult to compare them fairly.

*6.3.6. Cell Death*

When you examine your cells under the microscope, there are a number of signs that will let you know that your culture is dying or already dead. The sight of a lot of floating cells in the media after a few days in culture is a clear one. If you see that the neurites are segmented and/or blebbing (**Fig. 8.4B**), those cells should be discarded. As mentioned above, the presence of microglia, especially the round floating form, is usually a warning sign.

# 7. Typical Protocol Results

The hippocampal culture preparation described above will yield half to one full 24-well plate of viable cells (equals ~1–2 million cells) at high density per animal. This includes all type of cells. **Figure 8.5** shows examples of cells plated at relatively high density (~100,000/well). At the time of plating (**Fig.8.5A**), the cells are mostly spherical and devoid of processes, although some pyramidal neurons still have part of their apical dendrite. Small debris of broken processes and dead cells are also visible. Rinsing the cells 1–2 hours after plating can remove most of the debris. You will see very short processes extending from many of the neurons by then. On the following day, their processes have developed to various lengths (**Fig. 8.5B**). By 1 week, the neurons

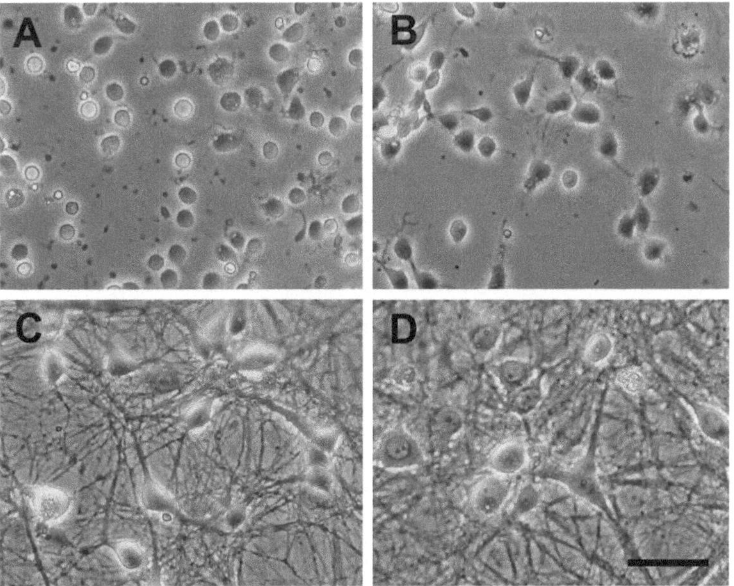

Fig. 8.5. Hippocampal neurons growing in culture. Time after plating: (**A**) 2 hours, (**B**) 1 day, (**C**) 1 week, (**D**) 2 weeks. Scale bar: 50 μm

have formed an extensive network of processes and have begun intense synaptogenesis (**Fig. 8.5C**). The mesh of processes densifies even further by 2 weeks (**Fig. 8.5D**). When plating the cells at a "radial gradient" density, we obtain high-density neurons in the center of the coverslips (**Fig. 8.6A**) and low-density neurons on the edges (**Fig. 8.6B**). However, to grow low-density neurons (**Fig. 8.6D**) throughout the coverslip, it is necessary to place it apposed to a feeder layer, made of either glial cells, as described by Banker and colleagues (3, 4), or high-density neurons (**Fig. 8.6C**).

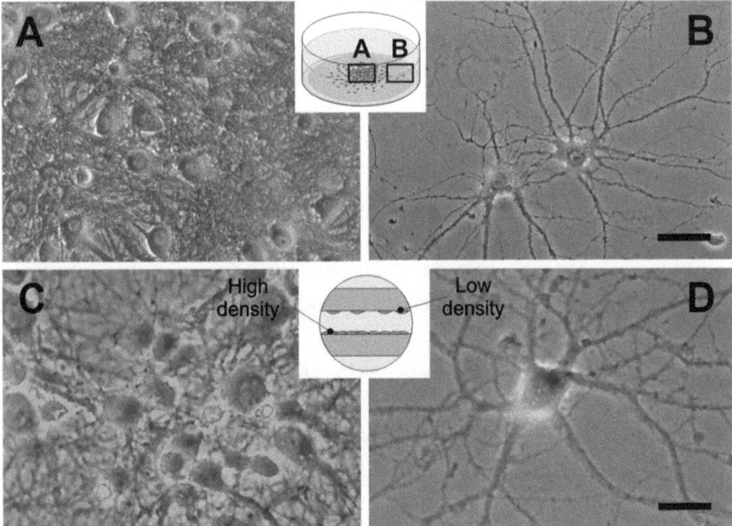


Fig. 8.6. Possible plating configurations: (**A**, **B**) Radial gradient density of neurons (21 DIV); photographs taken from the same coverslip (**A**, *center*; **B**, *edge*). Scale bar: 50 μm. **C**, **D**) Sandwich of two coverslips, one at high density (**C**), the other at low density (**D**). Scale bar: 30 μm.

These rat hippocampal cultures can be maintained for several weeks; however, for many applications where synaptically connected neurons are desired, 10–18 days in culture is quite a useful range. Older neurons can be more difficult to patch clamp (more difficult to obtain a gigaohm seal) and to transfect with plasmid DNA using lipid- or $Ca^{2+}$/phosphate-based methods. Mature spines can appear during the second week in culture, but they are more prominent >14 days in culture (**Fig. 8.7B**). Many laboratories prefer to do their experiments on young (6–9 DIV) neurons because they are easy to transfect or because their cultures do not survive for longer periods. Once you have cultures going routinely in your laboratory, do not expect that they will have the same properties weeks after weeks. The speed at which

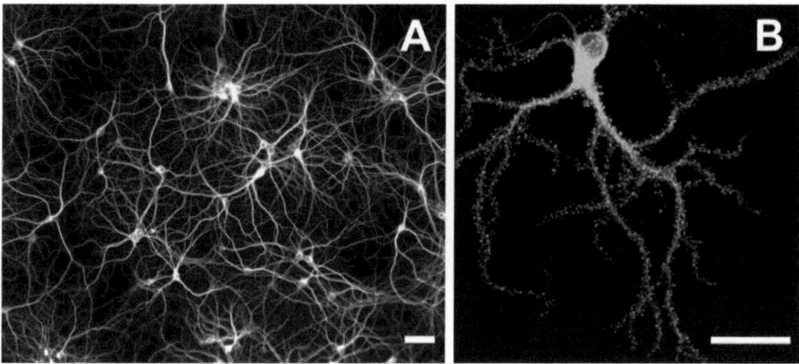

Fig. 8.7. (**A**) Field of hippocampal neurons (18 DIV), immunostained for MAP-2, showing somatodendritic domains. (**B**) Hippocampal neuron (21 DIV) transfected with GFP-CaMKII; note the numerous dendritic spines and spine-free axon. Scale bars: 100 μm.

they develop and form synapses as well as their longevity can be quite variable.

## Acknowledgments

We thank most particularly Mr. Sylvain L. Côté for his tremendous help with illustrations. We also thank Mado Lemieux for revising the chapter. PDK's laboratory is funded primarily by the Canadian Institutes of Health Research of Canada and the Natural Science and Engineering Research Council of Canada

## References

1. Harrison RG (1907) Observations on the living developing nerve fiber, Anat Rec, 1: 116–8.
2. Banker GA (1980) Astroglial cells in primary culture release factors into the medium that promote the growth and prolong the survival of rat hippocampal neurons in vitro, Science, 209: 809–10.
3. Banker G and Goslin K (1998) Culturing nerve cells, 2 nd edition, The MIT Press, Cambridge MA.
4. Kaech S, Banker G (2006) Culturing hippocampal neurons, Nature protocols, 1: 2406–15.
5. Brewer GJ, Torricelli EK, Evege EK, and Price PJ.(1993) Optimized survival of hippocampal neurons in B27-Supplemented Neurobasal, a new serum-free medium combination. J Neurosci Res, 35: 567–76.
6. Vicario-Abejón C (2004) Long term culture of hippocampal neurons, Current Protocols in Neuroscience, supplement 26, John Wiley & Sons, New York.
7. Brewer GJ and Cotman CW. (1989) Survival and growth of hippocampal neurons in defined medium at low density: advantages of a sandwich culture technique or low oxygen, Brain res, 494: 65–74.

# Chapter 9

# Primary Sensory and Motor Neuron Cultures

## Andrea M. Vincent and Eva L. Feldman

## Abstract

The cell culture study of neuronal activity in health and disease requires careful consideration of the system used. All neurons are definitely not created equal, and isolation of mature neurons from the tissue of interest is crucial to the understanding of that particular population of neurons. We are particularly interested in neurodegenerative diseases and so altering cell death for the purposes of cell culture would defeat the object of the investigations. This chapter describes our principal sensory and motor neuron cell culture techniques.

**Key words:** Adult neurons, defined media, dorsal root ganglia, embryonic motor neurons.

## 1. Introduction

The ability to culture primary neurons is critical to the study of neuronal biochemistry and molecular biology. Because committed neurons do not proliferate, neuronal cell lines are fundamentally altered in all areas of activity. Transformation of neuronal lines necessarily alters their survival, signaling, and metabolism, so for these and many other studies primary cells are essential. The issue of cell identity is also an important advantage of primary neuron cultures. An investigator tends to have high confidence in the identity of neurons reproducibly harvested from a specific site from an animal. Issues of senescence and enrichment of more robust population subtypes are eliminated in primary cell culture. We have particularly focused on enriched neuronal cultures so that we can study neuronal biology in isolation. These cultures are amenable to extraction of protein, lipid, and nucleic acid for assessment of expression and post-translational modifications

L.C. Doering (ed.), *Protocols for Neural Cell Culture*, Springer Protocols Handbooks,
DOI 10.1007/978-1-60761-292-6_9, © Humana Press, a part of Springer Science+Business Media, LLC 1992, 1997, 2001, 2010

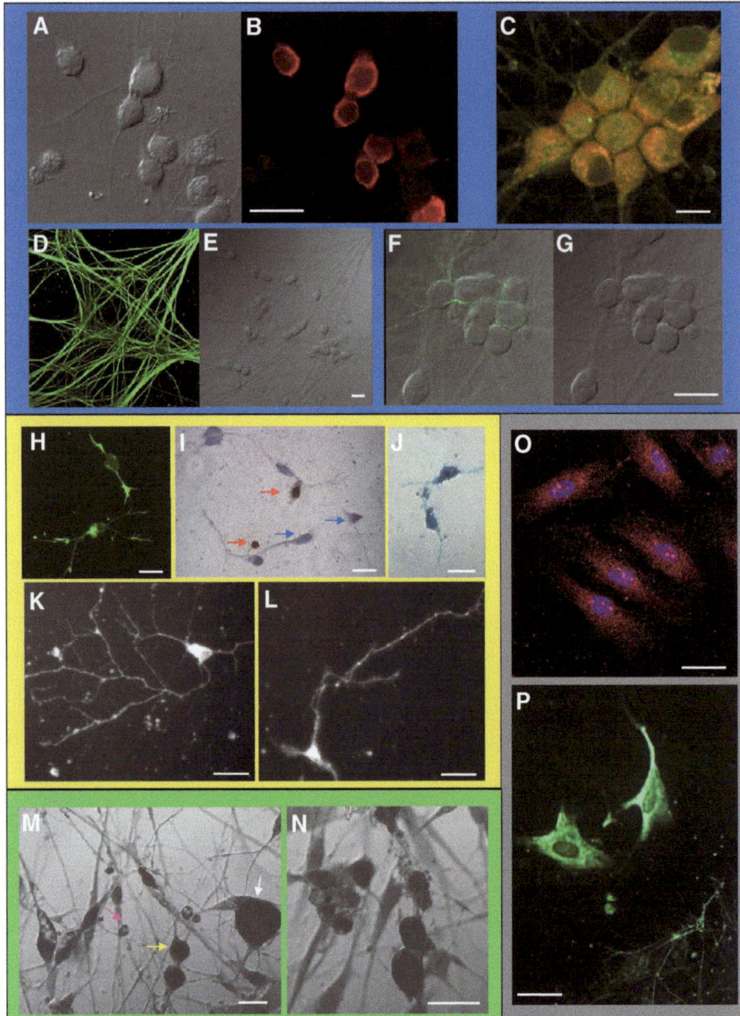

Fig. 9.1. Primary neuronal and Schwann cell cultures. The *blue* panel contains images of embryonic DRG neurons, the *yellow* panel contains images of embryonic motor neurons, the *green* panel shows adult DRG neurons, and the *gray* panel contains Schwann cells. (**A**) A phase-contrast image of well-dissociated DRG neurons that corresponds to the fluorescent image in (**B**), labeled with anti-nitrotyrosine and a red secondary antibody. (**C**) A fluorescent image of several neurons labeled with 50 nM green Mitotracker and 3 μM MitoSOX (11). In (**D–G**), neurons are stained for a membrane protein NQO1. (**D**) is a low-magnification green fluorescent micrograph of the phase image in (**E**). (**F**) is a higher magnification overlay of green fluorescent and phase-contrast images and (**G**) is the phase-contrast only image. (**H**) shows an F-actin stain in motor neurons after just 3 h in culture. The lower neuron was also exposed to 100 μM VEGF. (**I**) Motor neurons are TUNEL stained after exposure to 100 μM glutamate. Dead, TUNEL-positive neurons are *brown* and examples are indicated with *red* arrows. Live, TUNEL-negative neuronal nuclei are *blue* and examples are indicated with *blue* arrows. (**J**) Motor neurons are stained *blue* for acetylcholinesterase activity. (**K and L**) are fluorescent micrographs of different neurons on the same coverslip that were loaded with 50 nM Mitotracker. The two images underline the differences in the degree of neurite length and branching. (**M and N**) Phase-contrast images of TUNEL-stained neurons. In (**M**) the range of neuronal sizes found in adult, but not embryonic, cultures is highlighted with very large

with high confidence that the measurements are neuronal and not occurring in accessory cells.

We have established protocols for many embryonic neuronal cultures and also a limited number of adult neurons. In some studies, co-culture with glial cells is important, so we also culture various glial cells. These cultures are used for immunohistochemistry (1), biochemical enzyme assays (2), gene expression at mRNA and protein level (3, 4), live cell fluorescence-based metabolic assays (5, 6), survival (5, 6), and signaling measurements (1, 7). We have found that the most effective method for gene transfer to primary neurons is viral infection. Using adeno- and adeno-associated viruses, we achieve high expression levels and greater than 90% infection rates within 24–48 h (3, 8). This fits well within our 3-day culture period. These gene transfer experiments prove powerful for the rescue of a knockout phenotype or the study of gene overexpression (2, 8). We also can use lower multiplicities of infection to achieve 10% or 50% infection. With a bicistronic vector that expresses GFP as well as the gene of interest, we can microscopically compare neurons in the same dish with and without the transgene and separate them by green fluorescence.

We also have devised serum-free defined media for our neuronal cultures. This development confers many advantages over other protocols. The media permits the culture of neurons at very low density in the absence of a glial feeder layer. This is important for our studies of neurite growth on different extracellular matrix proteins and aligned nanofibers (9) and for immunohistochemistry (6, 7). The absence of serum also is important for our studies into the effects of growth factors on signaling and survival (1, 3).

We and others have found that embryonic and neonatal neurons require basal 25 mM glucose in order to remain viable in culture, and this is contained in the formulation of neurobasal media (Gibco, #21103). This is consistent with many neuronal cells, but contrasts with most other primary cell types (5, 10). Adult DRG neurons may be grown in either low- or high-glucose media, although very large neurons are decreased in high glucose and fluctuations in glucose are detrimental to the neurons (11).

Representative examples of the cells isolated in the following protocols are shown in **Fig. 9.1**.

◄————————————————————————————————————

Fig. 9.1. (continued) (*white arrow*), medium (*yellow arrow*), and small (*pink arrow*) cell bodies. (**N**) is a higher-magnification image of medium neurons. (**O**) Schwann cells are labeled with a red fluorescent stain for a transcription factor normally sequestered in the cytoplasm (Nrf-2) and blue nuclear DAPI. (**P**) Schwann cells are stained *green* with the same antibody against the membrane protein NQO1 shown in (**D**) and (**F**). In each image the *white* bar is 50 μm.

## 2. Protocols

### 2.1. Dorsal Root Ganglia Neurons

#### 2.1.1. Dishes

Coat with collagen the day before and allow to dry in a sterile environment overnight. We prefer type 1 collagen from rat tail (Sigma #C7661), which we dissolve at 1 mg/mL in 1 M acetic acid.

#### 2.1.2. Media

The culture medium is made in neurobasal medium (Gibco #21103). Add 1 mL B27 (Gibco #17504.044), 50 ng/mL NGF (2.5S from mouse, Harlan Bioproducts for Science, Inc. #BT-5017), 7 μM aphidicolin (Sigma #A0781), 1× peni-cillin/streptomycin/neomycin (Invitrogen 15140-122), 2 μM L-glutamine (Sigma G-3126).

##### 2.1.2.1. Plating Medium

##### 2.1.2.2. Feed Medium

Neurobasal plus 1 mL B27, 50 ng/mL NGF, 7 μM aphidicolin, 1× pen/strep/neo.

##### 2.1.2.3. Treatment Medium

Neurobasal plus 4 ng/mL selenium 4 ng/mL hydrocortisone, 0.01 mg/mL transferrin, 3 ng/mL $\beta$-estradiol, 50 ng/ml NGF, 7 μM aphidicolin. Additives are from Sigma, cell culture grade.

##### 2.1.2.4. Dissection Setup

1. In the laminar flow hood
2. Instruments for microdissection:
   (a) 1 forceps Roboz Inst RS-5242
   (b) 1 scissors Roboz Inst RS-5910
   (c) 2 forceps Roboz Inst RS-5045
   (d) 1 scissors Roboz Inst RS-6010
3. Instrument holder
4. Leibowitz L-15 media (Invitrogen 11415)
5. Dissecting microscope
6. 100 mm plastic tissue culture dishes (Falcon)
7. In the animal room
8. 15 day pregnant rat (Sprague Dawley)
9. Ketamine:rompun 10:3
10. 0.3 ml Beuthanasia-D
11. Instruments for large dissection – forceps and scissors
12. 1 sterile 150 mm tissue culture dish
13. Bag for animal disposal
14. 75% alcohol in a squirt bottle

### 2.2. The Dissection

#### 2.2.1. Large Dissection

1. Anesthetic – 0.5 mL of ketamine:rompun IM
2. When rat is no longer conscious, lay animal on her back with belly exposed

3. Inject Beuthanasia-D (0.3 mL) into the chest (in the heart if possible)

4. Soak belly with alcohol and wipe with sterile gauze

5. Open the belly and remove the uterus. Be careful not to cut the bowel.

6. Place uterus in the 150 mm dish

7. Take immediately to the hood

8. Remove the embryos and put four each in 10 mL of L-15 in a 100 mm dish.

*2.2.2. Microdissection*

1. Place embryo on its side and remove the head (save heads for cortical cultures).

2. Turn the embryo (facing the side) and hold it still with the microforceps by spearing the abdomen where the liver is.

3. Remove all the organs.

4. Flip the spinal column area on its back so the inside is facing up.

5. Remove any residual organs or large blood vessels.

6. Turn the column so that it runs horizontal.

7. Hold the ribs on either side of the column with the microforceps.

8. Make an incision along the top of the spinal column to expose the cord. Make an upward motion with the incision so the cord is not damaged.

9. Once the spinal cord is exposed, grab the cord at the top and pull it out with DRG attached.

10. DRG that remain in the column can be collected by plucking out.

11. Place all of the cords with DRG into another clean 100 mm dish containing 10 ml L-15.

**2.3. Cell Dissociation**

1. Continuing under the microscope, pull off the DRG and discard the spinal cords (or place in a clean 100 mm dish containing 10 mL L-15 for spinal motor neuron or glia cultures below).

2. Transfer the DRG to a 50 mL centrifuge tube, remove most of the L-15 and add 3 mL trypsin/EDTA solution (Gibco #25200).

3. Incubate at 37°C for 15 min.

4. Aspirate most of the trypsin, then add 3 drops of serum.

5. Add 2 mL L-15.

6. Triturate gently using a serum-coated glass pipette.

7. Centrifuge to pellet cells: approx $1000g$ for 5 min.

8. Resuspend cells in plating media and transfer to dishes. Approx $1 \times 10^6$ cells/mL, and 100 μL per well of 96-well, 500 μL in 12-well.

9. Feed medium is added 6–12 h after plating.

10. Treatment medium is applied after a further 48 h and experiments performed on day 3 in culture.

## 3. Critical Steps and Troubleshooting

The timing of experiments in these cells is important. Aphidicolin is an antiproliferative agent that kills the majority of Schwann cells in these cultures. These Schwann cells die over the first 48 h in culture. A second wave of Schwann cell death occurs around day 5 in culture. The Schwann cell death may interfere with experimental assessments. After the second wave of Schwann cell death, the survival of neurons is markedly decreased and many detach from the collagen-coated surface. For co-culture experiments with DRG neurons and Schwann cells, the aphidicolin may be omitted (12). In earlier studies, we used 2-fluorodeoxyuridine (FUDR; 40 μM) rather than aphidicolin (5, 8). Generally, the results are similar, but their mechanisms of action are different. Aphidicolin inhibits DNA polymerase-α that does not alter mitochondrial DNA synthesis but blocks nuclear DNA synthesis (13). Since our recent studies have focused on mitochondria, we now prefer aphidicolin.

## 4. Spinal Motor Neurons

### 4.1. Dishes

Coat with poly-L-lysine (50 μg/mL dissolved in sterile water) the day before. On the day of the dissection, plates are rinsed twice with sterile water and then plating media added. The poly-L-lysine is not allowed to dry.

### 4.2. Media

#### 4.2.1. Plating Medium

The culture medium is made in neurobasal medium (Gibco #21103). Add 2.5 mg/mL albumin, 2.5 μg/mL catalase, biotin, 2.5 μg/mL SOD, 0.01 mg/mL transferrin, 15 μg/mL galactose, 6.3 ng/mL progesterone, 16 μg/mL putrescine, 4 ng/mL selenium, 3 ng/mL β-estradiol, 4 ng/mL hydrocortisone, × penicillin/streptomycin/neomycin (Invitrogen 15140-122), 1× B27 additives (Gibco #17504.044), and 2 μM L-glutamine. Additives are from Sigma, cell culture grade.

#### 4.2.2. Feed Medium

Neurobasal medium plus 2.5 mg/mL albumin, 2.5 μg/mL catalase, biotin 2.5 μg/mL SOD, 0.01 mg/mL transferrin, 15 μg/mL galactose, 6.3 ng/mL progesterone, 16 μg/mL putrescine, 4 ng/mL selenium, 3 ng/mL $\beta$-estradiol, 4 ng/mL hydrocortisone, × penicillin/streptomycin/neomycin, 1× B27.

*4.2.3. Treatment Medium*

Neurobasal medium plus 2.5 mg/mL albumin, 2.5 μg/mL catalase, biotin 2.5 μg/mL SOD, 0.01 mg/mL transferrin, 15 μg/mL galactose, 6.3 ng/mL progesterone, 16 μg/mL putrescine, 4 ng/mL selenium, 3 ng/mL $\beta$-estradiol, 4 ng/mL hydrocortisone, × penicillin/streptomycin/neomycin.

### 4.3. Procedure

1. Refer to DRG cultures to obtain E15 spinal cords.
2. Under the dissecting microscope, strip meningeal membranes from the spinal cords using microforceps.
3. Chop cords into 2–3 mm pieces using scissors.
4. Transfer to a 50 ml sterile centrifuge tube.
5. Remove most of the L-15 is removed and 3 mL trypsin/EDTA solution.
6. Incubate at 37°C for 15 min.
7. During the incubation, dilute Optiprep (Sigma #D1556) in L-15 and prepare 10, 15 mL tubes each containing 2 mL 5.4% Optiprep.
8. After incubation, triturate cord pieces gently using a serum-coated glass pipette.
9. Centrifuge to pellet cells: approx 1000*g* for 5 min.
10. Resuspend in 1 mL L-15, then layer 100 μL aliquots on each Optiprep tube.
11. Centrifuge at 1800*g* for 15 min, then collect the media layer at the top of the gradient containing motor neurons into a clean 50 mL tube.
12. Dilute with L-15 and pellet cells at 1000*g* for 5 min.
13. Resuspend motor neurons at $10^6$ cells/mL in plating media and place 100 μL per well of 96-well, 500 μL in 12-well, 1.5 mL in 6-well.
   The motor neurons survive greater than 2 weeks if fed every 3 days with feed media. The cultures are >90% motor neurons by staining for islet-1 and SMI-32 and are positive for acetylcholinesterase activity.

## 5.  Spinal Glia

These are designed for co-culture experiments with motor neurons. A glial feeder layer is prepared as follows:

Plates, media, dissection, and procedure are identical to spinal motor neurons above, until the centrifugation over Optiprep.

### 5.1. Procedure

1. Remove all of the Optiprep and suspended cells to leave the pellet.

2. Resuspend the pellet at $10^6$ cells/mL in plating media and place 100 µL per well of 96-well, 500 µL in 12-well.

3. Allow to grow to confluence (2–3 days).

4. Harvest the cells by applying trypsin/EDTA solution for 1 min at 37°C.

5. Inhibit trypsin with excess serum, then resuspend cells in feed media.

6. Split the cells 1:4 and repeat for 3 passages.

7. These cells can now be used as a feeder layer and enriched motor neurons are plated on top of the glia.
   This procedure is currently unpublished (manuscript submitted). We find that 80% of the glia are GFAP-positive, 10% are OX-42-positive, indicating oligodendrocytes, and less than 2% are positive for neurofilament. The cultures may be harvested and cryopreserved for future use without any change in the cellular composition of the culture. To cryopreserve the cells, we use feed media containing 10% DMSO and 20% serum.

## 6. Cortical Neurons

### 6.1. Plates

Coat with poly-L-lysine (50 µg/mL) the day before. On the day of the dissection, plates are rinsed twice with sterile water and then plating media added. The poly-L-lysine is not allowed to dry.

### 6.2. Media

#### 6.2.1. Plating Media

The culture medium is made in neurobasal medium (Gibco BRL). Add 2.5 mg/mL albumin, 2.5 µg/mL catalase, biotin 2.5 µg/mL SOD, 0.01 mg/mL transferrin, 15 µg/mL galactose, 6.3 ng/mL progesterone, 16 µg/mL putrescine, 4 ng/mL selenium, 3 ng/mL $\beta$-estradiol, 4 ng/mL hydrocortisone, × penicillin/streptomycin/neomycin, 1× B27 additives, and 2 µM L-glutamine (same as spinal motor neurons).

#### 6.2.2. Feeding Media

Neurobasal medium plus 2.5 mg/mL albumin, 2.5 µg/mL catalase, 2.5 µg/mL SOD, 0.01 mg/mL transferrin, 15 µg/mL galactose, 6.3 ng/mL progesterone, 16 µg/mL putrescine, 4 ng/mL selenium, 3 ng/mL $\beta$-estradiol, 4 ng/mL hydrocortisone, × penicillin/streptomycin/neomycin, 1× B27 additives (same as spinal motor neurons).

**6.3. Procedure**

1. Proceed per **Section 2.2** above until heads are removed.
2. Place heads in 100 mm dish containing 10 mL L15 media.
3. Remove the skull.
4. Remove brain meningeal membrane using a sterile, dry cotton swab.
5. Separate the brain hemispheres.
6. Remove the cerebellum, brainstem, and white matter by teasing apart with iridectomy scissors.
7. Place cortices and L-15 into a 15 mL conical tube.
8. Allow tissue to settle to the bottom of the conical tube.
9. Carefully pipette off the media without disturbing the tissue.
10. Add 3 mL trypsin/EDTA (pre-warmed) and incubate at 37°C for 6 min.
11. Remove most of the trypsin without disturbing the cortices.
12. Add serum to neutralize trypsin.
13. Add 3 mL L-15 and triturate gently using a serum-coated glass pipette.
14. Bring sample up to 10 mL using L15 media and continue to triturate if needed.
15. Centrifuge to pellet cells: approx $1000g$ for 5 min.
16. Resuspend pellet in 15 mL of plating media and count cells, then plate at $10^6$/mL.
17. Feed the cells next day with feed media.
    Cells are fed every 2–3 days. Generally, these cells are used after 1 week in culture. We do not use an inhibitor of proliferation to remove non-neuronal cells. While these cultures survive long term (>3 weeks), we use at 1 week to prevent non-neuronal overgrowth.

## 7. Schwann Cells

Schwann cell isolation is based on the method of Wrabetz and Feltri (14).

**7.1. Dishes**

Coat with poly-L-lysine (50 µg/mL dissolved in sterile water) the day before. On the day of the dissection, plates are rinsed twice with sterile water and then plating media added. The poly-L-lysine is not allowed to dry.

### 7.2. Media

#### 7.2.1. Plating Medium

High-glucose DMEM (Gibco 10-013-CV) containing 10% calf serum (Hyclone #SH300723.03), 1× penicillin/streptomycin/neomycin (Invitrogen 15140-122), and 6 μM L-glutamine

#### 7.2.2. Feed Medium

High-glucose DMEM containing 10% fetal bovine serum (Hyclone #SH30070.03), 2 μL/mL bovine pituitary extract (Becton Dickinson #354123), 0.04 μL/mL forskolin (Sigma #F6886), and 10 μM cytosine-B-arabino furanoside hydrochloride (Ara-c).

#### 7.2.3. Treatment Medium

For glucose toxicity experiments, or other experiments requiring defined media: Low-glucose DMEM (Gibco #10-014-CV)/Ham's F-12 (Gibco #10-025-CV) 1:1 containing transferrin (10 μg/mL), putrescine (10 μM), progesterone (20 nM), and sodium selenite (30 nM) (Sigma).

#### 7.2.4. Dissection Setup

We have used C57 BL/6 J mice pups (3–4 days old) and 3-day-old Sprague Dawley rat pups.

#### 7.2.5. Prepare Beforehand

1. 1% collagenase (Sigma C9722)
2. 10 mL of 2.5% trypsin in water (Gibco 610-5095)
3. Two 100 mm Petri dishes containing PBS without calcium and magnesium with 1× penicillin/streptomycin/neomycin (Invitrogen 15140-122). Place on ice.

### 7.3. The Dissection

1. Sterilize dissecting board and instruments.
2. Sterilize pups with 75% ethanol and decapitate.
3. Pin pup dorsal side up with legs spread and use forceps to tear away the skin.
4. Using the dissecting microscope, go to L4 and dissect down to the sciatic nerve.
5. Remove entire sciatic nerve and place in PBS plate.
6. Remove the perineurium with fine forceps and place the remaining nerve in the second dish.
7. Using the dissecting microscope, cut the cleaned nerve into small sections.
8. Using a serum-coated glass pipette, transfer the nerve pieces to a conical tube, wash dish with PBS and add to tube up to a volume of 8 mL.
9. Add 1 mL 0.1 % collagenase (pre-warmed to 37C) and 1 mL of 2.5% trypsin to the tube for a total volume of 10 mL.

10. Incubate at 37°C for 30 min in a water bath, mixing every 10 min.

11. Centrifuge at 1800$g$ at room temperature for 5 min.

12. Remove the supernatant, wash the pellet with DMEM containing 10% calf serum, and centrifuge again.

13. Resuspend the pellet with 2 mL DMEM, 10% fetal calf serum and triturate 20 times.

14. Plate $10^6$ cells/mL
    DAY 2

15. 18–24 h after plating the cells, plates are washed with HBSS, then re-fed with feed media containing Ara-c
    DAY 4–5

16. Passage cells using trypsin/EDTA solution (Gibco #25200) 1:4 when they reach confluency.

17. Schwann cells can be used for up to 4 passages.
    The identity of the Schwann cells can be confirmed by immunoreactivity for S100 proteins. If contaminating fibroblasts are observed, they can be removed by complement-mediated lysis using antibody to Thy1.1 per published protocols (15).

    *Cultures from adult rodents*
    Clearly, the use of adult rodents presents significant advantages. In particular, it is more physiologically relevant to use mature neurons for the study of neurodegenerative diseases that affect adult animals and patients.

## 8. Dorsal Root Ganglia

Dorsal root ganglia contain the cell bodies of sensory neurons. These can be cultured and used for the study of signaling, metabolism, survival, and gene expression. The procedure for rats and mice is identical.

### 8.1. Dishes

Tissue culture plates are coated with rat tail collagen and air-dried overnight in a sterile environment.

### 8.2. Media

#### 8.2.1. Plating Media

The media is a 50:50 mix of DMEM (Cellgro 10 013-CV) and F10 (Cellgro 10 025-CV) plus 4 ng/mL selenium, 4 ng/mL hydrocortisone, 0.01 mg/mL transferrin, 3 ng/mL $\beta$-estradiol, 1 mL B-27, 1 mL pen/strep/neo, 7 μM aphidicolin (Sigma A0781), and 2 μM glutamine.

#### 8.2.2. Feed Media

The media is a 50:50 mix of DMEM and F10 (Gibco BRL) plus 4 ng/mL selenium, 4 ng/mL hydrocortisone, 0.01 mg/mL transferrin, 3 ng/mL $\beta$-estradiol, 1 mL B-27, 1 mL penicillin/streptomycin/neomycin.

**8.3. Procedure**

1. Rodent is euthanized using ketamine:rompun 10:3 followed by 0.3 mL Beuthanasia-D as described above.

2. Spray the dorsal area with 75% ethanol.

3. Open skin along the back from tail to neck using scissors.

4. Transect the spinal column at the base and dissect out with approximately 1 cm tissue either side up to ribs 7–8.

5. In the laminar flow hood, pin the spinal column to the dissecting board.

6. Open the dorsal side of the spinal column to expose the spinal cord using bone rongeurs.

7. Remove the spinal cord by pulling gently and cutting roots.

8. Dissect out the DRG from between the vertebrate and cut off all nerve roots.

9. Place the DRG in L-15 media until dissection is complete.

10. Aspirate the L-15 media and replace with 1 mg/mL collagenase (Sigma C9722) in L-15.

11. Incubate at 37°C for 30 min.

12. Aspirate the collagenase and replace with 3 mL trypsin/EDTA solution.

13. Incubate at 37°C for 15 min.

14. Aspirate trypsin/EDTA and add three drops serum to neutralize remaining trypsin.

15. Add 2 mL L-15 and triturate DRG gently with a serum-coated glass pipette.

16. Pellet cells by centrifugation at 1200$g$ for 5 min.

17. Resuspend cells in plating media and plated at $10^6$/mL.

18. Replace media with feed media after 24 h.

19. Replace media with treatment media after a further 24 h and perform experiments on day 3.

## 9. General Tips for Successful Cultures

We always dissociate tissue using trituration with a serum-coated glass pipette. The serum prevents loss of tissue through sticking to the pipette surface. It is important that the glass is not broken or cracked, as this leads to undue cell stress and a significant loss of neuron yield. To decrease the risk of neuronal damage, glass pipettes may be flame-polished. The viability of DRG neurons is greater, and neurite outgrowth is increased on laminin. If longer-term cultures or extensive neurite growth is warranted, replace collagen coating with 50 μg/mL laminin.

## Acknowledgments

This work was supported by, Juvenile Diabetes Research Foundation, the Amyotrophic Lateral Sclerosis Association, NIH-NIDDK, and the Program for Neurology Research and Discovery.

## References


1. Vincent AM, Mobley BC, Hiller A, Feldman EL. IGF-I prevents glutamate-induced motor neuron programmed cell death. Neurobiology of disease 2004;16(2): 407–16.
2. Vincent AM, Russell JW, Sullivan KA, et al. SOD2 protects neurons from injury in cell culture and animal models of diabetic neuropathy. Experimental neurology 2007;208(2):216–27.
3. Vincent AM, Feldman EL, Song DK, et al. Adeno-associated viral-mediated insulin-like growth factor delivery protects motor neurons in vitro. Neuromolecular medicine 2004;6(2–3):79–85.
4. Wiggin TD, Kretzler M, Pennathur S, Sullivan KA, Brosius FC, Feldman EL. Rosiglitazone Treatment Reduces Diabetic Neuropathy in STZ Treated DBA/2 J Mice. Endocrinology 2008;149(10): 4928–37.
5. Russell JW, Golovoy D, Vincent AM, et al. High glucose-induced oxidative stress and mitochondrial dysfunction in neurons. FASEB Journal 2002;16(13):1738–48.
6. Vincent AM, McLean LL, Backus C, Feldman EL. Short-term hyperglycemia produces oxidative damage and apoptosis in neurons. FASEB Journal 2005;19(6):638–40.
7. Vincent AM, Perrone L, Sullivan KA, et al. Receptor for advanced glycation end products activation injures primary sensory neurons via oxidative stress. Endocrinology 2007;148(2):548–58.
8. Vincent AM, Olzmann JA, Brownlee M, Sivitz WI, Russell JW. Uncoupling proteins prevent glucose-induced neuronal oxidative stress and programmed cell death. Diabetes 2004;53(3):726–34.
9. Corey JM, Gertz CC, Wang BS, et al. The design of electrospun PLLA nanofiber scaffolds compatible with serum-free growth of primary motor and sensory neurons. Acta biomaterialia 2008;4(4):863–75.
10. Joseph-Bravo P, Perez-Martinez L, Lezama L, Morales-Chapa C, Charli JL. An improved method for the expression of TRH in serum-supplemented primary cultures of fetal hypothalamic cells. Brain research 2002;9(2):93–104.
11. Vincent AM, Feldman EL. Can drug screening lead to candidate therapies for testing in diabetic neuropathy? Antioxidants & redox signaling 2008;10(2):387–93.
12. Berent-Spillson A, Russell JW. Metabotropic glutamate receptor 3 protects neurons from glucose-induced oxidative injury by increasing intracellular glutathione concentration. Journal of neurochemistry 2007;101(2):342–54.
13. Zimmermann W, Chen SM, Bolden A, Weissbach A. Mitochondrial DNA replication does not involve DNA polymerase alpha. The Journal of biological chemistry 1980;255(24):11847–52.
14. Wrabetz L, Feltri ML, Kim H, et al. Regulation of neurofibromin expression in rat sciatic nerve and cultured Schwann cells. Glia 1995;15(1):22–32.
15. Brockes JP, Fields KL, Raff MC. Studies on cultured rat Schwann cells. I. Establishment of purified populations from cultures of peripheral nerve. Brain Research 1979;165(1):105–18.

# Chapter 10

# Retinal Cell and Tissue Culture

## Francisco L.A.F. Gomes and Michel Cayouette

## Abstract


A method for culturing retinal progenitor cells and their progeny is described. The dissociated cell culture is useful to study isolated cells and the influence of cell intrinsic and environmental signals on various cellular processes. The retinal explant assay reproduces the three-dimensional environment observed in vivo and is useful in the study of retinas from mutant mice that are, for example, embryonic lethal. In addition, such retinal explants can be used to easily manipulate gene expression in embryonic retinas using viral vectors or electroporation.

**Key words:** Rat, progenitor cell, explant, dissociated retina.


## 1. Introduction

The retina contains six different neuronal cell types (rods, cones, horizontals, bipolars, amacrines, and ganglion cells) and one type of glia (Müller cells) that are generated from multipotent retinal progenitor cells (RPCs). Astrocytes also populate the retina, but they are not produced by RPCs and migrate into the retina from the brain through the optic nerve. All these different cell types are organized into three distinct layers. The ganglion cell layer (GCL) contains the cell bodies of ganglion cells, some displaced amacrine cells, and astrocytes. The inner nuclear layer (INL) contains the cell bodies of amacrines, horizontals, bipolars, and Müller cells, whereas the outer nuclear layer (ONL) contains the cell bodies of rods and cones. The small number of cell types and simple anatomy of the retina, combined with its accessibility, makes it a model of choice for the study of cell physiology, neuronal


L.C. Doering (ed.), *Protocols for Neural Cell Culture*, Springer Protocols Handbooks,
DOI 10.1007/978-1-60761-292-6_10, © Humana Press, a part of Springer Science+Business Media, LLC 1992, 1997, 2001, 2010

circuit formation, and mechanisms of cell fate specification and differentiation. Although the protocol described here could easily be adapted to study postmitotic retinal neurons, it was originally developed to study cell fate decision mechanisms (1). For this purpose, RPCs are obtained from embryonic or early postnatal retina and cultured such that they will divide and differentiate to generate retinal neurons and Müller glia.

Mechanisms of cell fate choice depends on both cell intrinsic and environmental factors, but the relative contribution of these signals to specific cell fate decisions remains unclear. Retinal cell culture provides a major advantage to study the contribution of intrinsic and environmental signals on cell fate decisions. Dissociated retinal cell cultures in serum-free and extract-free medium provide the investigator with a control over the environment, providing an assay to address questions about the effects of a particular molecule on the development of retinal neurons (1, 2). In addition, dissociated cell cultures can be used to study the importance of cell intrinsic mechanisms by isolating RPCs from their normal environment. The method described below is suitable for the clonal-density culture of RPCs. In such cultures, RPCs are isolated from each other and from the postmitotic neurons. They divide and differentiate into different retinal cell types to form small colonies, or "clones". In a recent study, we found that these clones are indistinguishable from the clones that develop in situ in the retina, in terms of both cell number and cell type composition (1). Based on these results, we suggested that RPCs are pre-programmed to generate a particular combination of cell types and that this program is played out largely cell intrinsically.

Another widely used retinal cell culture method is the explant assay, which consists in culturing a small piece of retina on an organotypic filter (3–5). Retinal explants can be prepared any time between embryonic day 13 (E13) and postnatal day 4 (P4). Although retinal ganglion cells tend to degenerate within 48 h after they are generated in such explants, and photoreceptor cells do not grow extended outer segments, the explants will develop very similarly to a retina in vivo and generate all the different retinal cell types that end up in the appropriate layer. The retinal explant culture assay is particularly useful in cases where a particular mouse mutant is embryonic lethal and its retinal development cannot be studied in vivo. Because retinal explants can be prepared from embryonic animals and electroporated or infected with viral vectors (6–9), it is also a useful approach for the study of gene function at embryonic stages, as manipulation of gene expression in vivo at embryonic stages is difficult. Here, we present a retinal explant culture method that includes serum, but serum-free culture conditions can also be achieved (10).

## 2. Reagents and Equipment

1. DPBS 1× – (Ca$^+$, Mg$^+$), Invitrogen, cat.# 14040
2. Neurobasal medium – Invitrogen, cat.# 21103-049
3. DMEM/F12 – Invitrogen, cat. # 10565-018
4. RGM- retinal growth medium, *see* Appendix
5. Paraformaldehyde 16%–EMS, cat.# 15710
6. Distilled water – Invitrogen, cat.# 15230
7. Penicillin/streptomycin – Invitrogen, cat.# 15070
8. L-Cysteine – Sigma, cat. #C-7352
9. LoOvo 10×, *see* Appendix
10. Goat anti-mouse IgG$_{2b}$ Alexa Fluor 488 (GAM 488-2b) Molecular Probes, cat.# A21141
11. Goat anti-rabbit IgG Alexa Fluor 594 (GAR 594), Molecular Probes, cat.# A11037
12. Rabbit anti-Pax6 – Chemicon cat.# AB5409
13. Mouse anti-Islet1 (clone 4D1), Developmental Study Hybridoma Bank
14. Poly-L-lysine –Sigma cat. #P4707
15. Trypsin inhibitor – Boerhinger Mannheim cat. # 109878
16. BSA –Sigma, cat.# A4161
17. DNAse – Worthington, 100 mg, cat.# LS002007
18. Laminin – Invitrogen, cat.# 23017-015
19. NaOH 1 N; ethanol 70%
20. Goat serum – Sigma, cat.# G6767
21. Triton X-100 – Sigma, cat.# T8787
22. Scissors (1), surgical, 14.5 cm, cat.# FST14001-14
23. Scissors (1), spring-type, straight, cat.# FST 15000-00
24. Forceps (3), straight, fine (Dumont-type, no. 5) cat.# FST 11252-23
25. Forceps (1), straight, 14.5 cm (FST cat.# 11002-14)
26. Snap-off blade, X-acto cat.# HUNX32444
27. Forceps (1), curved, 11.5 cm, Dumostar, FST cat.# 11297-10 (Dumont-type, no.7)
28. Hot bead sterilizer, FST 250 sterilizer, cat.# 18000-45
29. Petri dishes, 35 mm, plastic, sterile, Falcon cat.# 353001
30. Petri dishes, 100 mm, plastic, sterile, Fisherbrand, cat.#.08-757-13

31. Petri dishes, 150 mm, plastic, sterile, VWRbrand, cat.# CA73370-002

32. Pasteur pipette long (9"), VWR cat.# 14672-380

33. 0.22 μm filter (Millex-GP), cat.# SLGP 033RS

34. 20 mL syringe, sterile, BD cat.# 309604

35. Levy hemacytometer, VWR cat.# 15170-208

36. $CO_2$ incubator set at: 8% $CO_2$, 37°C

## 3. Protocols

### 3.1. Dissociated Retinal Progenitor Cell Culture

#### 3.1.1. Animals

Retinas from albino rat Sprague Dawley are used for preparation of dissociated cell culture. The best results in terms of survival and reproducibility are obtained with animals aged from embryonic day 17 (E17) to postnatal day 1 (P1). Mouse tissue can be used, but overall cell survival is drastically decreased.

#### 3.1.2. Preparation of Retinal Growth Medium

Preparation of medium should be done under sterile conditions. See Appendix for detailed protocol on the preparation of stock solutions.

##### 3.1.2.1. RGM Incomplete

On the day before the experiment, add in a 50 mL screw cap tube:

10 mL neurobasal

10 mL DMEM-F12

100 μL $N_2$ supplement

200 μL NAC

50 μL pen/strep 100×

40 μL cpt-cAMP

##### 3.1.2.2. RGM Complete

On the day of the experiment, add to the previous prepared medium (RGM incomplete):

200 μL insulin

200 μL B27

20 μL Forskolin

20 μL BDNF

5 μL NT-3

4 μL EGF

2 μL b-FGF

Filter solution through a 0.22 μm filter, previously rinsed with PBS. Keep RGM complete at 37°C until use.

<table>
<tr><td>

*3.1.3. Preparation of Other Solutions*

</td><td>

These solutions are prepared on the day of the experiment.

</td></tr>
<tr><td>

3.1.3.1. Papain

</td><td>

Add in a 15 mL tube:

10 mL DPBS (+ $CaCl_2$ and $MgCl_2$)

100 U papain

100 µL DNAse 0.4%

5 µL NaOH 1 N

2 mg L-cysteine crystal (to be added only prior to using the solution for dissociation)
Keep at 37°C.

</td></tr>
<tr><td>

3.1.3.2. LoOvo

</td><td>

Add in a 15 mL tube:

9 mL DPBS (+ $CaCl_2$ and $MgCl_2$)

1 mL LoOvo 10×

100 µL DNAse 0.4%
Keep solution at 37°C until use.

</td></tr>
<tr><td>

*3.1.4. Coating Dishes*

</td><td>

One or two days prior to the experiment coat the dishes as follow:

1. Make a 50% solution of Poly-L-lysine (PLL) by diluting PLL (Sigma P-4707) in equal amount of sterile water. Add 1.5 mL per 35 mm dish.

2. Leave at RT for at least 1 h. Aspirate PLL, open the lid, and let dry in the hood (approx. 1 h).

3. Rinse 3 times with sterile water; pour at least 2 mL water/dish for each rinsing.

4. Add 1.5 mL of laminin solution (5 µg/mL diluted in cold neurobasal medium) per dish.

5. Incubate at least overnight in incubator equilibrated at 37°C, 8%, $CO_2$. Longer incubation time (up to 2 days) has proven to improve quality of the preparation.

6. On the day of the experiment, aspirate laminin solution (DO NOT rinse or dry), and add 1 mL *RGM complete*. Return the dish to incubator.

</td></tr>
<tr><td>

*3.1.5. Dissection*

3.1.5.1. Eye Dissection

</td><td>

Eye extraction and dissection of the retina are carried out under sterile conditions (flame or laminar flow hood). Before starting the procedure, spray ethanol 70% thoroughly over the working area.

1. Prepare three (3) 100 mm Petri dishes with 10 mL DPBS ($Ca^+$, $Mg^+$).

2. Sterilize all surgical instruments by placing them in a hot bead sterilizer (set at 250°C) for at least one (1) min (if hot bead sterilizer is not available, wash instruments in 70%

</td></tr>
</table>

ethanol). Wait until the instruments reach room temperature (RT).

3. Put on gloves.

4. Euthanize the rat in a $CO_2$ chamber.

5. With the rat in dorsal decubitus, spray 70% ethanol solution thoroughly over the skin.

6. Pinch the skin with the straight (14.5 cm) forceps and with the utility scissor make a large cut across the lower abdomen exposing the uterus.

7. Transfer the uterus containing embryos to a 150 mm Petri dish. Remove the embryos from uterus/placenta and decapitate the embryos.

8. Transfer the heads to a 100 mm Petri dish filled with DPBS.

**3.1.5.2. Retina Dissection**

Work sterile under a dissection microscope.

1. Locate the eye (**Fig. 10.1A**) and remove the overlying skin, exposing the eye (**Fig. 10.1B**). Place the curved forceps on each side of the eye (**Fig. 10.1C**). Push it downward, closing the blades underneath the eye. Exerting little force, pull out the eye.

2. Transfer the eyes to another 100 mm dish with DPBS (**Fig. 10.1D**).

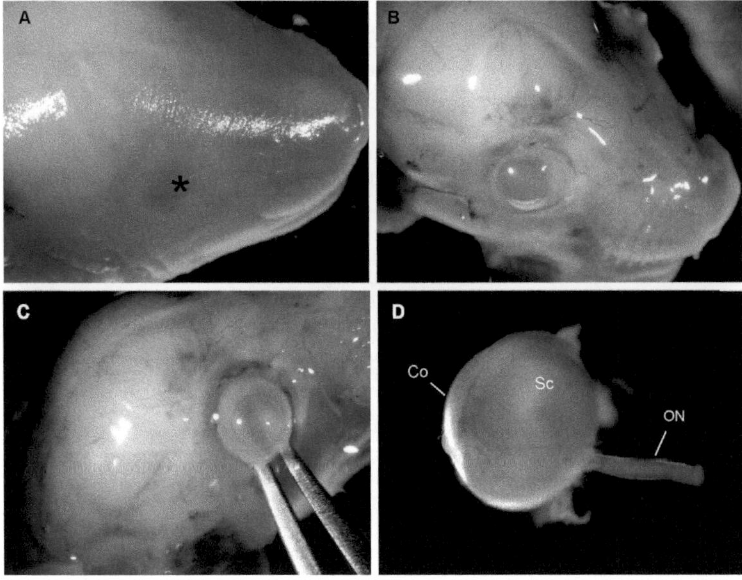


Fig. 10.1. Embryonic day 20 (E20) rat eye dissection procedure. (**A**) Rat head; note that the eyes are still covered with skin, * depicts the position of the right eye. (**B**) Rat head with skin peeled away, exposing the eye. (**C**) Right eye being extracted with the curved forceps. (**D**) Extracted eye. Co: cornea; Sc: sclera; ON: optic nerve.

3. With the lens facing down, secure the eyes in position using a fine straight forceps (no. 5). Strip away the optic nerve and any other extra-ocular tissue (**Fig. 10.2A, B**). Introduce one blade of the other forceps through the optic nerve opening between the choroids and the retina.

4. While gripping the eye through the optic nerve opening with one forceps, pinch the sclera with the other forceps. Pull them apart exposing the retina. Extract out the retina and lens (**Fig. 10.2C**).

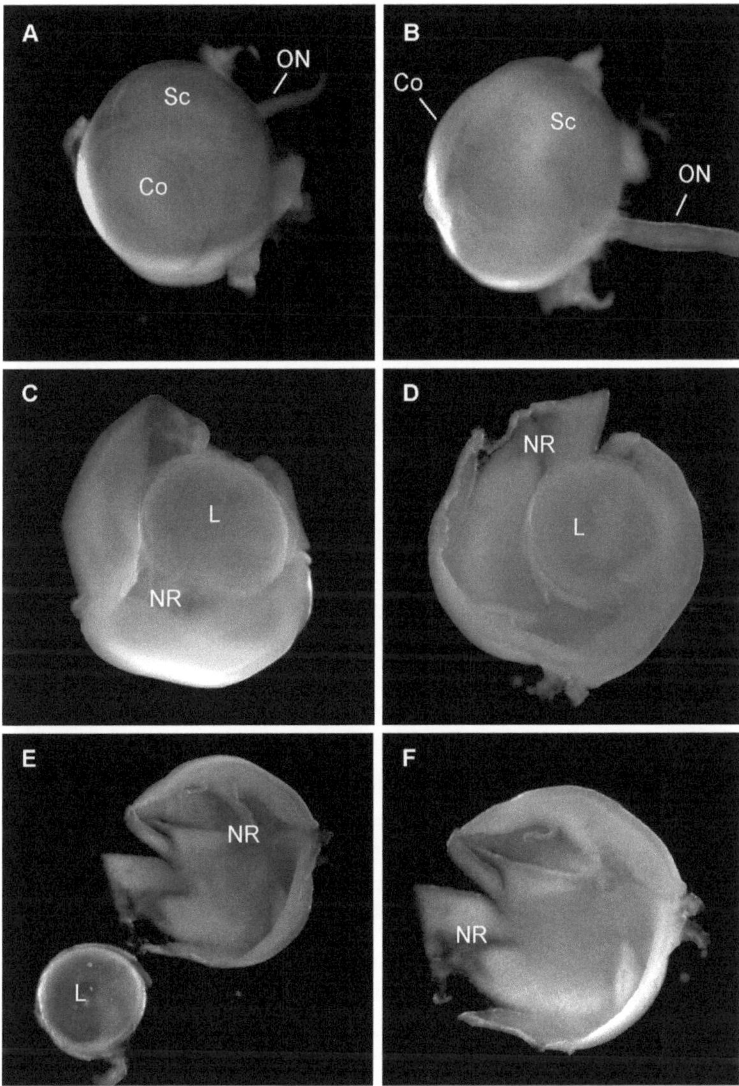


Fig. 10.2. Retinal dissection of an E20 rat eye. (**A**) Extracted eye front-view. (**B**) Extracted eye side-view. (**C**) Extracted neural retina with lens still attached. (**D**) Neural retina partially detached from the lens. (**E**) Neural retina completely separated from lens. (**F**) Entire neural retina. ON: Optic nerve; Sc: sclera; Co: cornea; NR: neural retina; L: lens.

5. Carefully detach the lens and isolate the retina (**Fig. 10.2D–F**).

6. From E18-19 onward it is required to remove the blood vessels covering the retina, which appear slightly reddish color and form a thin membrane overlying the inner face of retina. This can be removed by pinching and pulling with the forceps. Omitting this step will lead to endothelial cell contamination in the culture.

7. Transfer the dissected retina (**Fig. 10.2F**) to another 100 mm Petri dish using a P1000 pipette with a cut tip. Slice the retina in small (about 2 mm$^2$) pieces using the microscissors (spring-type).

The number of retinas to dissect depends on your need. Typically, the yield is about $0.5 \times 10^6$ cells for one E17 retina and $1.6 \times 10^6$ cells for one P0 retina. Although very few cells are needed for clonal-density cultures, it is best to dissect at least 6–8 retinas as it helps to improve cell recovery.

**3.1.5.3. Retina Dissociation**

Work sterile under a laminar flow hood.

1. Remove the Papain tube from the hot water bath and add 2 mg L-cysteine. Transfer the pieces of retina into this tube.

2. Return the Papain tube with the pieces of retina to the 37°C water bath. Incubate 5 min for E17 retinas and 8 min for P0 retinas (Note: This incubation time can be adjusted. The younger the retinas, the shorter the incubation in Papain.).

3. *Trituration.* With the pieces of retina at the bottom of the tube, aspirate the supernatant. (Be careful not to aspirate the retinas! If using an aspirator, switch to a pipette to remove the last few microliters.) Slowly add 4 mL of LoOvo over the retinas. Let the pieces of retina settle to the bottom of the tube, and then aspirate the LoOvo.

4. Add 1 mL of LoOvo. With the P1000 pipette flush gently up and down to break the pieces of tissue. After a few passages, press the P1000 tip against the bottom of the conical tube and flush up and down the cell suspension several times until the suspension appears cloudy.

5. Make sure that the cell suspension is well dissociated and that there are no clumps by observing a small aliquot (10 μL) on the microscope (you might have to dilute this aliquot to see individual cells.). If clumping is observed, continue pipetting up and down until unicellular suspension is obtained. (Note: If you still have problems to obtain a single cell suspension, the incubation time in Papain can be increased.)

6. Add the remaining 5 mL LoOvo to the solution and centrifuge 5 min at 900 rpm on a tabletop centrifuge (162 rcf) at RT.

7. Discard the supernatant. Add 1 mL RGM solution (37°C) and redissolve the pellet as described in Step 4.

8. Count the cells in a hemacytometer. For clonal-density cultures, add 5,000 cells per 35 mm dish, for low-density culture add 20,000 cells per 35 mm dish, and for high-density cultures add >100,000 cells per 35 mm dish. Return the dish to the incubator.

9. Wait 2 h for the cells to adhere to the bottom of the dish. Add 2 mL RGM and return the dish to the incubator.

10. Feed the cells every 3–4 days by removing 50% of the culture medium and adding back the same amount of fresh RGM.

In optimal conditions, the cells can be cultured for up to 10–14 days.

*3.1.6. Immunostaining*

**Note**: In these culture conditions, retinal progenitor cells will divide and differentiate into the different retinal cell types normally produced from E17 (amacrines, bipolars, rod photoreceptors, and Müller cells). Various cell-specific markers can be used to identify the different retinal cell types. Here, we present a protocol for the detection of Islet-1 and Pax-6. Islet-1 is a marker of bipolar cells and a subtype of amacrines and ganglion cells, whereas Pax-6 is a marker of amacrines and ganglion cells, as well as retinal progenitors. This protocol can be adapted to use with other markers.

1. Aspirate the culture medium and fix the cells by adding cold PFA 4% for 15 min on ice.

2. Rinse three times with PBS 1×.

3. Block/permeabilize for 1 h at RT with block/permeabilization solution.

4. *Primary Antibody*: in 1 mL antibody buffer (AB) add:
   (a) R-Pax6; add 10μL of 1:100 dilution in 1 mL AB (1:10,000)
   (b) M-Islet 1; add 0.5 μL in the same 1 mL AB (1: 2000)
   (c) Incubate overnight at 4°C.
   (d) Discard primary antibody.

5. Rinse three times with PBS 1×.

6. *Secondary Antibody*: in 1 mL AB add:
   (a) GAR 594; add 1 μL in 1 mL AB (1:1000).
   (b) GAM 488-2b; add 1 μL in the same 1 mL AB (1:1000).

  (c) Incubate 1 h at RT.

  (d) Discard secondary antibody.

7. Rinse three times with PBS 1×.

8. Observe under inverted microscope.

---

## 4. Retinal Explant Culture

### 4.1. Animals

Mouse or rat retinas can be used to prepare retinal explants. The age of the animals depend on the goal of your experiment, but explants can be prepared from animals ranging from embryonic day 12 to postnatal day 4.

### 4.2. Reagents and Material

1. 0.4 μm culture plated insert-millicell-CM(Millipore) cat.# PICMORG50

2. Petri dishes – 35 mm, plastic, sterile, Falcon cat.# 353001

3. FBS – Invitrogen, cat.# 12483-020 (heat inactivated)

4. Fungizone 1000× (FZ) – Invitrogen cat.# 15290-018

5. Penicillin, streptomycin (pen/strep), Invitrogen cat.# 15140-122

6. Glutamax 100×, Invitrogen, cat.# 35050

7. DMEM, Invitrogen, cat.# 10313

8. $CO_2$ incubator set at 5% $CO_2$, 37°C

### 4.3. Preparation of Explant Medium (EM)

In a 50 mL screw cap tube, add:

45 mL DMEM

5 mL FBS (10%)

500 μL pen/strep 100×

50 μL FZ 1000×
Keep the solution at 4°C for up to 1 month.

### 4.4. Preparation of the Culture Plate Inserts

At least 1 hour prior to the experiment, add 1.2 mL of EM in a 35 mm dish and place the insert into it using sterile forceps. Place the dish in the incubator.

### 4.5. Preparation of Explant Culture

1. Dissect out the retinas as in **Section 3.1.5**.

2. For retinas older than embryonic day 15, cut small pieces of 5–6 $mm^2$. For younger retinas (e.g., E13), place the whole retina on the insert. It is important to have a piece of retina that is as perfect as possible (i.e., no folds or cuts).

3. With sterile scissors cut the tip of a P1000 pipette and transfer the pieces of retina on the inserts (up to four explants per insert). Lay the piece of retina as flat as possible on the insert.

If the pieces of retina do not lie flat on the insert and folds (best seen by observing under an inverted microscope), use a fire-polished glass pipette or a pipette tip to gently flatten out the retina. (Using a dissecting microscope will help in this operation.) The final setup should look like the diagram in **Fig. 10.3**.

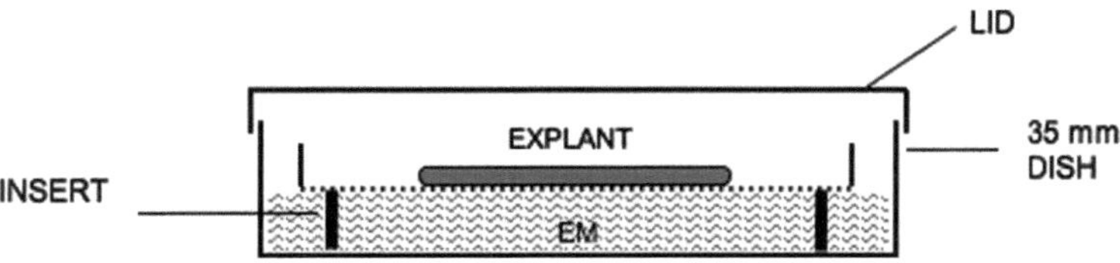


Fig. 10.3. Diagram showing the setup of the tissue culture insert with explant on top.

4. Return the dish to the incubator.

5. Feed the explants by replacing 80% of the medium with fresh medium every 3 days.

***4.6. Explant Fixation***

1. After the desired culture period, remove the insert from the dish and aspirate the culture medium. Add 1.5 mL of 4% cold PFA in the dish and replace the insert bearing the explants. Add more PFA directly on top of the explants to completely cover them.

2. Incubate 30 min at 4°C.

3. Rinse 3 times in PBS.

4. Proceed with cryosectionning and immunostaining.

## 5. Critical Steps and Troubleshooting

***5.1. Dissociated Cell Culture***

After laminin coating, debris are observed under the Microscope

1. Make sure you thaw laminin slowly. Remove from –80°C directly to 4°C then to RT. Never thaw laminin at 37°C; this might cause the laminin to agglomerate.

Too much cell death in optimal conditions, very little cell death should be observed (less than 20%). If cell death is a problem, pay attention to the following details:

1. Make sure that the procedure is done as quickly as possible. We normally take about 30 min to dissect 8 retinas and less than 45 min to dissociate the cells.

2. Use the RGM medium ingredients described here (the supplier matters for many components!). Once we obtain a good survival rate, we normally buy a large quantity of the same lot ingredient. Pay special attention to laminin and

B27. Keep track of the lot number, as variation from one lot to another has been observed for some components.

3. Overtrituration may damage the cells. If you overtriturate the cells, you will see debris when you count the cells with the hemacytometer chamber. Avoid making bubbles when pipetting up and down to dissociate the retina.

4. The entire procedure is done at room temperature.

5. Use *rat retinas*! The same procedure with mouse retinas will result in much more cell death. It is unclear why mouse cells do not survive as well as rat cells to this procedure.

### 5.2. Retinal Explants

Rosette formation and improper structure of the retina after the culture period.

1. Make sure that the pieces of retinas are flat on the insert.

2. Make sure there is serum in the culture medium, as it is required for normal lamination of the retina. If you want to use serum-free medium, add Sonic Hedgehog (Shh) to the culture medium.

## 6. Typical Protocol Results

### 6.1. Dissociated Cell Culture

Examples of typical results from a clonal-density culture are shown in **Fig. 10.4**. Over 10 days in culture, the retinal progenitor cells plated on day 1 will divide and differentiate to generate different retinal cell types. Immunostaining for various cell type-specific markers is used to identify the different cell types. At clonal density, however, some markers such as rhodopsin (rod photoreceptors) are not expressed. To identify photoreceptor cells, we use the typical chromatin condensation pattern in the nucleus (1). In addition, cell morphology is characteristic for each retinal cell type and helpful when identifying the different cell types obtained in culture. When preparing the culture from E17 retinas, only amacrines, bipolars, rod photoreceptors, and Müller cells are expected to be generated in this culture (the other three cell types are generated before E17).

### 6.2. Explant Culture

Examples of retinal explants after different period of time in culture are shown in **Fig. 10.5**. Note that the retinal layer develops normally. In addition, expression of the different cell type-specific markers can be detected by immunohistochemistry (not shown).

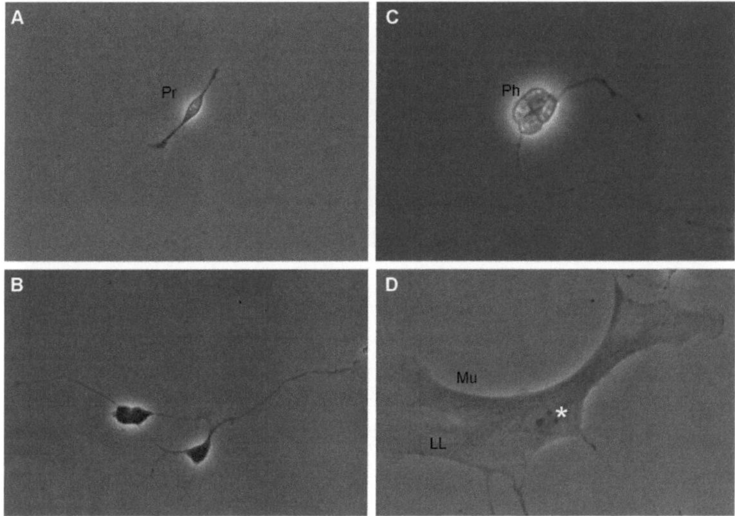



Fig. 10.4. Photomicrographs of retinal cells. (**A**) A retinal progenitor cell 4 h after plating. (**B**) A three-cell clone containing an amacrine and two photoreceptor cells. (**C**) A four-cell clone containing only rod photoreceptors; note the presence of white spots (heterochromatin condensation) inside the nucleus, a distinct characteristic of photoreceptor cells. (**D**) A Müller cell. The morphology of glial cells is very distinct from neurons. Note the absence of axons and dendrites and the presence of lamelipodia-like structures, * nucleus. Pr, progenitor cell; Ph, rod photoreceptor; Am, amacrine; Mu, Muller. LL; lamelipodia-like structure.

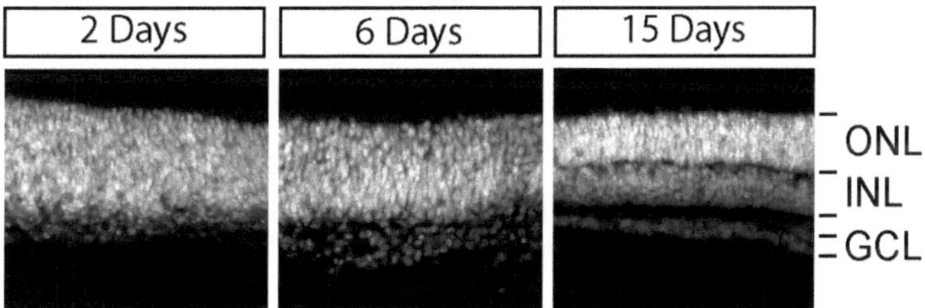



Fig. 10.5. Cross-sections of retinal explants after different period in culture. Embryonic day 13 retinal explants were prepared and cultured for 2, 6, or 15 days, as indicated. The explants were then sectioned in a cryostat and stained with a DNA dye (Hoechst) to reveal cell nuclei. Note that the development of the different retinal layers over time. ONL: outer nuclear layer; INL: inner nuclear layer; GCL: ganglion cell layer.

## Acknowledgments

We thank the Canadian Institutes of Health Research (CIHR) and the Foundation Fighting Blindness – Canada (FFB-C) for supporting our work. We would also like to thank Amel Kechad for providing the photographs used in **Fig. 10.5**.

## Appendix

*Retina Growth
Medium (RGM)*

| Reagent | To make 20 mL add: | Final | Stock | Keep stock at |
|---|---|---|---|---|
| DMEM/F12 | 10 mL | 1× | – | –20°C |
| Neurobasal | 10 mL | 1× | – | –20°C |
| N$_2$ | 100 μL | 0.5× | 100× | –20°C |
| B27 | 200 μL | 2× | 50× | –80°C |
| Pen/strep | 50–200 μL | 0.25–1× | 100× | 4°C |
| NAC | 200 μL | 60 μg/mL | 6 mg/mL | –20°C |
| cpt-cAMP | 40 μL | 0.1 mM | 50 mM | –20°C |
| Forskolin | 20 μL | 0.01 mM | 10 mM | –20°C |
| Insulin | 200 μL | 25 μg/mL | 2.5 mg/mL | –20°C |
| EGF[a] | 4 μL | 50 ng/mL | 250 μg/mL | –80°C |
| b-FGF[a] | 2 μL | 10 ng/mL | 100 μg/mL | –80°C |
| BDNF[a] | 20 μL | 50 ng/mL | 50 μg/mL | –80°C |
| NT-3[a] | 5 μL | 25 ng/mL | 100 μg/mL | –80°C |

[a]To be added just prior to use.

*Stock Solutions*

Neurobasal – Invitrogen cat.# 21103-049 – 1×, aliquot in10 mL store at –20°C.

DMEM/F12 – Invitrogen cat.# 10565-018 – 1×, aliquot in 10 mL store at –20°C.

Penicillin/streptomycin – Invitrogen 15140-122 – 100× stock, store at 4°C.

*Lo Ovomucoid –
LoOvo – (10×);
40 mL*

To 40 mL DPBS (without Ca and Mg, Invitrogen cat.# 14190):

Add 600 mg BSA (BSA, Sigma cat.# A-4161). Mix well.

Add 600 mg trypsin inhibitor (Boerhinger Mannheim, cat.# 109878) and mix to dissolve. Adjust pH to 7.4; requires the addition of approx. 400 μL of 1 N NaOH.

When completely dissolved, filter through 0.22 μm filter.

Make 1.0 mL aliquots and store at –20°C.

Papain:

Worthington Biochemicals cat.# LS003126

B27 (50×):

without vitamin A (Invitrogen, cat.#12587-010)

*DNAse 0.4%*

Dissolve 40 mg (Worthington, cat. # LS002007) in 10 mL of sterile water (do not vortex) mix by gentle inversion

Aliquot in 200 μL and store at –20°C.

**Insulin (2.5 mg/mL)**  To 10 mL sterile water add: 25 mg insulin (Sigma, cat.# I-6634), 50 μL HCl 1 N. Mix well.

Aliquot in 200 μL and store at –20°C or store at 4°C for no more than 4–6 weeks.

**cpt-cAMP (50 mM, 500×)**  cpt-cAMP (Sigma, cat.# C3912) in sterile water at 25 mg/mL. Aliquots in 40 μL and store at –20°C.

**NAC (6 mg/mL)**  Dissolve 6 mg *N*-acetyl cysteine (Sigma, cat.# A-9165) in 1 mL NB (will turn yellowish).

Make fresh every week. Aliquot in 200 μL and store at 4°C.

**Forskolin (10 mM)**  Add 2.4 mL DMSO (Sigma, cat.# D2438) to a 10 mg bottle (Sigma, cat.# F-6886) to make 4.2 mg/mL or 10 mM. Make 20 μL aliquots; store –20°C.

**$N_2$ Supplement**  Prepare $N_2$ stock solutions:

*Apo-transferrin*: (Sigma, cat.# T-1147). Dissolve in 5 mL of water to make 100 mg/mL stock solution (1000×). Filter (0.22 μm), aliquot 1 mL and store at –20°C.

*Progesterone*: (Sigma, cat.# P8783-1G). Dissolve 6 mg in 10 mL of ethanol to make 0.6 mg/mL stock solution (30,000×). Filter (0.22 μm), aliquot 40 μL and store at –20°C.

*Putrescine*: (Sigma, cat.# P5780). Dissolve 1.6 g in 10 mL of water to make 1 M (160 mg/mL) stock solution (10,000×) Filter (0.22 μm), aliquot 100 μL and store at –20°C.

*Sodium selenite*: (Sigma, cat.# S5261) Dissolve 2.59 mg in 5 mL of water to make 3 mM stock solution (100,000×). Filter (0.22 μm), aliquot 10 μL and store at –20°C.

*BSA*: (Sigma, cat.# A4161). Dissolve 500 mg in 10 mL of water to make 50 mg/mL stock solution (1000×). Filter (0.22 μm), aliquot 1 mL and store at –20°C.

*To make 10 mL of a 100× stock solution of $N_2$:*

| | |
|---|---|
| Apo-Transferrin | 1 mL |
| Progesterone | 33 μL |
| Putrescine | 100 μL |
| Sodium selenite | 10 μL |
| BSA | 1 mL |
| DMEM-F12 | 7857 μL |

Aliquot in 100 μL and store at –80°C.

**Growth Factors**

b-FGF (Peprotech, cat.# 100-18B). Reconstitute in 10 mM Tris pH8.5 to a concentration of 100 μg/mL. Aliquot in 5 μL and store at –80°C.

EGF (Peprotech, cat.# 315-09). Reconstitute in water to a concentration of 250 μg/mL. Aliquot in 5 μL and store at –80°C.

BDNF (Peprotech, cat.# 450-02). Dilute to a concentration of 50 μg/mL in water. Aliquot in 5 μL and store at –80°C.

NT3 (100 μg/mL): (Peprotech, cat.# 450-03) Dissolve 10 μg in 100 μL of sterile water. Aliquot in 5 μL and store at –80°C.

**Immunochemistry**

## Antibody Buffer

| | |
|---|---|
| Distilled water | 198 mL |
| NaCl, Fisher cat.# CAS 7647-14-5 | 2.25 g (150 mM final) |
| Tris Base, Fisher cat.# CAS 77-86-1 | 1.5 g (50 mM final) |
| BSA, Sigma, cat.# A2153 | 2.0 g (1% final) |
| L-Lysine, Sigma cat.# L5501 | 3.6 g (100 mM final) |
| Sodium azide, Sigma cat.# S2002 | 2 mL of a 4% stock in water (0.04% final) |

Adjust to pH 7.4.
Store at 4°C indefinitely.

**PFA 4%**

Prepare a 4% paraformaldehyde solution in 1× PBS.

Dilute a vial of 16% PFA (10 mL) (Electron Microscopy Science, cat# 15710) in 26 mL of water + 4 mL of 10× PBS. This solution can be kept for no longer than a month at 4°C.

**Block/Permeabilization Solution (40 mL) with 0.4% Triton**

| | |
|---|---|
| Goat serum | 8 mL (20% final) |
| Antibody buffer | 31.2 mL |
| Triton 20% | 800 μL (0.4% final) |

## References

1. Cayouette M, Barres BA, Raff M. Importance of intrinsic mechanisms in cell fate decisions in the developing rat retina. Neuron 2003;40:897–904.

2. Jensen AM, Raff MC. Continuous observation of multipotential retinal progenitor cells in clonal density culture. Dev Biol 1997;188:267–79.

3. Sheedlo HJ, Nelson TH, Lin N, Rogers TA, Roque RS, Turner JE. RPE secreted proteins and antibody influence photoreceptor cell survival and maturation.

Brain Res Dev Brain Res 1998;107: 57–69.
4. Sheedlo HJ, Turner JE. Influence of a retinal pigment epithelial cell factor(s) on rat retinal progenitor cells. Brain Res Dev Brain Res 1996;93:88–99.
5. Zhang SS, Fu XY, Barnstable CJ. Tissue culture studies of retinal development. Methods 2002;28:439–47.
6. Cayouette M, Raff M. The orientation of cell division influences cell-fate choice in the developing mammalian retina. Development 2003;130:2329–39.
7. Matsuda T, Cepko CL. Electroporation and RNA interference in the rodent retina in vivo and in vitro. Proc Natl Acad Sci U S A 2004;101:16–22.
8. Zigman M, Cayouette M, Charalambous C, et al. Mammalian inscuteable regulates spindle orientation and cell fate in the developing retina. Neuron 2005;48:539–45.
9. Donovan SL, Dyer MA. Preparation and square wave electroporation of retinal explant cultures. Nat Protoc 2006;1:2710–8.
10. Wang Y, Dakubo GD, Thurig S, Mazerolle CJ, Wallace VA. Retinal ganglion cell-derived sonic hedgehog locally controls proliferation and the timing of RGC development in the embryonic mouse retina. Development 2005;132:5103–13.

# Chapter 11

# Preparation of Normal and Reactive Astrocyte Cultures

Jean de Vellis, Cristina A. Ghiani, Ina B. Wanner, and Ruth Cole

## Abstract

Here, we describe methods to prepare primary cultures of astrocytes and reactive astrocytes. In 1980, McCarthy and de Vellis reported that highly purified cultures of astrocytes and oligodendrocytes can be obtained from neonatal brain tissue. The primary glial cultures prepared with this method have the great advantage of being devoid of neurons, enabling the analysis of astrocytes and oligodendrocytes separately. This culture model has been extensively used to advance our knowledge in the field of glial biology in normal conditions and after injury. The purification of astrocytes from primary mixed glial cultures, protocols that apply two types of mechanical injury (the scratch-wound model and the pressure-stretch model), as well as neuron-glial co-cultures are described in detail. These in vitro models, albeit limited by design, are invaluable tools to better understand the cellular and molecular mechanisms of astrocyte reactivity, scar formation and axonal growth inhibition in response to trauma.

**Key words:** Rat, astrocyte, reactive astrocyte, injury-induced gliosis.

## 1. Introduction

The study of glial cell development and function has been considerably enhanced by the development of methods to culture oligodendrocytes, astrocytes, and microglia from central nervous system tissue. A primary mixed glial culture, composed of astrocytes, oligodendrocytes, and microglia, is obtained when newborn disaggregated cerebral brain cells from rat are plated at high cell density ($2 \times 10^5/cm^2$) in serum-supplemented medium (1). Neurons fail to develop or survive in this culture model.

L.C. Doering (ed.), *Protocols for Neural Cell Culture*, Springer Protocols Handbooks,
DOI 10.1007/978-1-60761-292-6_11, © Humana Press, a part of Springer Science+Business Media, LLC 1992, 1997, 2001, 2010

At low cell density (e.g., $5 \times 10^4$ cells/cm$^2$), few oligodendrocytes develop, and the culture consists mostly of astrocytes. At high cell density, phase-dark, process-bearing spindle or spider-shaped cells appear by 4 days and stratify into clusters and individual cells above the bed-layer of cells. The bed layer consists of astrocytes rich in glial filaments. This observation led to the development of the shaking procedure, which results in selective removal of the process-bearing cells from the underlying astrocytes (1).Thus, highly purified cultures of astrocytes and oligodendrocytes, as wells as microglia, can be obtained from the same piece of brain tissue. Microglia cells (2) can be harvested from the stationary cultures by harvesting the medium on days 6 and 7, when they can be microscopically observed to be suspended in the medium. The remaining microglia and loosely adhering astrocytes are then removed from the mixed culture by a 6-h pre-shake, before the oligodendrocyte lineage cells are removed. The microglia cultures are about 95% pure, as characterized with immunocytochemistry using the microglia marker, ED 1 (3).

At the time of harvesting the process-bearing cells from the 7- to 9-day-old cultures, the cell population is mostly composed of oligodendrocyte progenitor cells and immature oligodendrocytes (4). If the process-bearing cells are placed in a chemically defined serum-free medium, <4% of the cells express astrocyte markers, such as glial fibrillary acidic protein (GFAP) (5). In fetal bovine serum (FBS)-supplemented medium, 25–35% of the cells express GFAP, originally called astrocyte type II. It has now been determined that astrocytes type II are an in vitro phenomenon with no in vivo counterpart (6). The bed-layer cells can be maintained as pure astrocyte cultures by keeping the flasks shaking slowly, to keep removing dividing oligodendrocyte progenitors and microglia. When cultured in the presence of horse serum, astrocytes grow flat processes and acquire a highly differentiated quiescent phenotype while continuing to form a lawn (7).

This primary culture of astrocytes can be used to set up pure secondary astrocyte cultures in serum or serum-free medium (8, 9). Astroglial and oligodendroglial cell lines have been developed from the cultures described above (10, 11). The usage of astrocyte cultures derived from neonatal brains has the advantage of using a population of astroglial cells with a purity of 99% (1).

Primary cultures of astrocytes offer an invaluable tool to characterize the changes that may occur in these cells following, for instance, a brain injury. Astrocytes respond to injury by becoming hypertrophic, hyperplastic, and increasing GFAP. This process, named reactive gliosis or astrogliosis, occurs in the brain and spinal cord in response to different types of injury including ischemia and inflammatory diseases. Its hallmark is an increase in GFAP (12, 13). Astrogliosis leads to the formation of a glial scar, a boundary necessary to contain the injury, but also an inhibitory obstacle to successful regeneration of damaged neuronal networks and axon tracts. The differences in the response to a harmful event, either chemical or physical, have been examined in astrocytes cultured from either the neonatal or the adult brain (12). It is important to point out that neonatal astrocytes in culture as well as in vivo display higher levels of GFAP compared to adult astrocytes. Yet, cultured astrocytes derived from adult brain or spinal cord were shown to arise from glial progenitor cells, similarly to those prepared from newborn brains (14–16). Hence, a protocol to obtain purified and highly differentiated astrocytes in culture is desirable to study mechanisms of astrogliosis and scar formation. Here, we provide a set of detailed protocols preparing and using purified astrocytes that were kept in culture for up to 8 weeks, as well as of reactive astrocytes (1, 7, 12, 17).

## 2. General Reagents, Equipment, and Media

1. $H_2O$, Invitrogen, cat #15230-162

2. DPBS, Invitrogen, cat #14200-075

3. Dulbecco's modified Eagle's medium/Ham's F-12 (DMEM/F12), Invitrogen cat #11320

4. Hank's balanced salt solution (HBSS), $Ca^{2+}$- and $Mg^{2+}$-free, Invitrogen cat #14170

5. Fetal bovine serum (FBS), Atlanta Biologicals cat # S11550 or Hyclone cat #SH30070

6. Versene (ethyhenediamine tetraacetic acid [EDTA]) solution 1:5000, Invitrogen cat # 15040

7. 2.5% Trypsin solution with EDTA 4Na, Invitrogen cat #15090

8. 0.25% Trypsin-1 mM EDTA, Invitrogen cat # 25200

9. 100-mm cell culture dishes

10. 60-mm cell culture dishes

11. 35-mm cell culture dishes

12. 15-mL and 50-mL sterile centrifuge tubes

13. Hemocytometer (Neubauer chamber), Hausser Scientific (VWR #15170-208)

14. IEC Clinical centrifuge (Rotor 221) Model #428-16106, International Equipment Company

---

## 3. Preparation of Newborn Rats

### 3.1. Reagents and Equipment

1. Newborn rats, postnatal day 1–2
2. Scissors (1), $5\frac{1}{2}$ in., straight, operating
3. Scissors (2), 4 in., straight, microdissecting
4. Forceps (2), microdissecting
5. Forceps (1), $5\frac{1}{2}$ in., tissue
6. Forceps (1), curved, fine
7. DMEM/F12, serum-free (10 mL)
8. Cell culture-treated dishes (5), 35-mm, tissue culture
9. Sponges (6), 4 × 4 in., sterile
10. Disposable towels (3), double-thickness, sterile
11. Ethanol (70%)
12. Containers (2), $4\frac{1}{2}$-oz (Falcon, cat no. 4014), containing 80 mL 70% ethanol
13. Plastic bag

### 3.2. Protocol for Dissection

1. Clean the area for dissection with 70% ethanol, and set up instruments in an array that logically follows the dissection procedure (**Fig. 11.2**):
   a. Place a $4\frac{1}{2}$-oz container filled with 70% ethanol next to container of pups.
   b. Place sterilized towels in an area next to a plastic bag, for disposal of beheaded pups.
   c. Place the three sterile scissors on sterile towels next to another $4\frac{1}{2}$-oz container with 70% ethanol.
   d. Also, near this container of ethanol, place tissue forceps and fine curved forceps on sterile toweling.
   e. Place fine microdissecting forceps on sterile gauze. Fill three 35-mm cell culture dishes with 2.5 mL DMEM/F12 (serum-free). One empty 35-mm cell culture dish is used to dissect both cerebra and one for disposal of the remaining brain tissue.
2. Removal of the brain:
   a. Immediately after euthanizing the newborn pup by $CO_2$ inhalation, dip it into 70% ethanol, allowing the thumb and forefinger also to be immersed.

b. Using a pair of sterile operating scissors, and holding the body of the rat, decapitate tile brad onto a sterile towel. The scissors are dipped in ethanol and returned to the sterile towel with the other scissors. The rat's body is disposed of in the plastic bag.

c. With the cut portion of the head facing you and with the head lying flat on the surface of the sterile towel, use a pair of microdissecting scissors to cut the skin at the midline of the head. Begin cutting from the base of the head, and cut to the mid-eye area. The scissors are dipped in ethanol and returned to the sterile towel with the other scissors. These scissors are used only for procedures b. and c. The skin is folded back by pulling the loose skin forward, using the index finger and thumb, and held open.

d. The second pair of microdissecting scissors is utilized to cut the skull at the midline fissure. The scissors are dipped in ethanol and returned to the sterile towel.

e. Using the thumb and forefinger, hold and apply slight pressure to the skull. Remove the raised portion of the skull with sterile tissue forceps.

f. Remove the brain by running the curved forceps along the bottom and sides of the brain calvarium, from the olfactory lobes to the posterior of the midbrain. Place the freed brain into a 35-mm cell culture dish containing 2.5 mL DMEM/F12 serum-free medium.

g. Repeat step 2, a–f, for each pup, placing all the brains into the same culture dish.

3. Removal of the cerebral hemispheres:

a. Remove one brain from the 35-mm culture dish to the inverted lid of a 35-mm culture dish, and remove the cerebral hemispheres as follows: (i) Using one pair of the microdissecting forceps steady the brain at the median fissure. Using the second pair of microdissecting forceps, gently cut the tissue along the median fissure, and pull one cerebral hemisphere away from the rest of the brain. (ii) Remove the cerebral hemisphere by pinching it against the culture dish surface. Remove the other hemisphere in a similar manner. Make sure that the olfactory lobes are no longer attached. (iii). Immediately immerse the cerebral hemispheres in serum-free medium contained in another sterile 35-mm cell culture dish. Repeat step a. i–iii. for each pup.

**Note:** Hemispheres from 15 to 20 pups are usually collected.

b. Use the same microdissecting forceps, which were used to remove cortices, to remove meninges. Place a hemisphere onto the inverted lid of a second cell culture dish. While using one pair of forceps to hold the hemisphere steady, pull off the meninges with the second pair of microdissecting forceps. The meninges can be easily teased away from the hemisphere by beginning with an exposed corner of the membrane on the outer surface, and gently pulling the meninges off. The removed meninges are blotted onto a piece of sterilized gauze.

c. Place the hemisphere piece into a fresh cell culture dish containing serum-free medium. Repeat this process (steps b. and c.) until all the cerebral hemisphere pieces are free of meninges and blood vessels.

d. Dip both forceps in ethanol, allow to drain and place on sterile towels.

---

## 4. Preparation of Glial Cell Cultures: Nitex™ Bag Method

The Nitex bag method was developed by Lu et al.(18). The procedure is done in a laminar flow hood.

### 4.1. Reagents and Equipment

1. Forceps (2)
2. DMEM/F12 + 10% FBS (75 mL)
3. HBSS (25 mL)
4. Cell culture dish, plastic, 100-mm
5. Flask, 75-cm$^2$
6. Flask, 25-cm$^2$
7. Nitex (No. 210) bag, 1 $\frac{1}{2}$ × 3 in., sterile
8. Glass rod, 7 × $\frac{5}{32}$ in., sterile
9. Beaker (1), covered with 130-μm Nitex mesh (or Cellector™ sieve with a 140-μm, mesh, (no.100) and a sterile 4 $\frac{1}{2}$ -oz container)
10. Beaker (1), covered with 230-μm Nitex mesh (or Cellector™ sieve with a 230-μm mesh (no. 60) and a sterile 4 $\frac{1}{2}$-oz. container)
11. Centrifuge tube, 50-mL, sterile
12. Hemocytometer
13. Pasteur pipette
14. Handheld digital counter
15. Inverted microscope

**4.2. Disaggregation of Brain Tissue**

1. Add 15 mL DMEM/F12-10% FBS to a 100-mm sterile culture dish. Place the Nitex bag in this dish and hold the mouth of the bag open with sterile forceps.

2. Pour or pipette the tissue pieces into the bag. Close the bag with sterile forceps and immerse it in the medium in the cell culture dish.

3. Use the sterile glass rod to tease the tissue through the mesh of the Nitex bag. Light strokes are used to prevent cell damage.

4. When teasing is completed, hold the closed bag upright with forceps, above the culture dish containing the screened cell suspension, and rinse 10 mL medium down the side of the bag. The combined cell suspension is then removed and filtered through the Nitex 230-$\mu$m covered beaker (or Cellector$^{TM}$ sieve, 230-$\mu$m, no. 60), which has first been wetted with 1 mL HBSS to allow the fluids to flow through properly.

5. Next, filter the cell suspension through the Nitex 130-$\mu$m covered beaker (or Cellector$^{TM}$ sieve, 140-$\mu$m, no. 100). Collect the cell suspension in a 50-mL centrifuge tube.

6. Centrifuge the resulting filtrate at 200$g$ for 5 min at room temperature. Resuspend the pellet in 10 mL culture medium.

7. Determination of cell number:
   a. Prepare a sample from the cell suspension for counting with a hemocytometer.
   b. Dilute 0.1 mL of the suspension with 1.9 mL HBSS. This is a cell dilution of 1:20. Because this is a total cell count, no trypan blue dye is required.
   c. Fill the hemocytometer with the diluted cell suspension, using a Pasteur pipette.
   d. Count the number of cells in at least one hemocytometer chamber (four squares).

8. Dilute the cell suspension in DMEM/F12-10% FBS, Plate $15 \times 10^6$ cells/75-cm$^2$ culture flask in a total volume of 10 mL or $5 \times 10^6$ cells/25-cm$^2$ culture flask in a total volume of 5.0 mL.

9. Gas the flask with 5% $CO_2$ and incubate at 37°C, with the lid tightly closed, or incubate in a 5% $CO_2$/95% air incubator, without gassing procedure and lid loosened.
   **Note:** Care should be taken not to disturb the flasks for 2 days. Movement of flasks during this initial 2-days period reduces cellular attachment. Culture medium is changed after 2–3 days, and every other day thereafter.

**4.3. Nitex
Mesh-Covered
Beakers**

1. Suppliers of Nitex mesh:
   a. B. & S. H. Thompson, 140 Midwest Road, Scarborough, ON, MIP 3B3 Can.
   b. B. & S. H. Thompson, 8148 Chemin Devonshire, Ville Mont Royal. PQ, H4P 2K3 Can.
2. Preparation of mesh: Nitex mesh is soaked in a large container of triple-distilled water. The water is changed $3\times$. The mesh is then coiled into a large beaker and dried in a drying oven at 65°C.
3. Preparation of Nitex mesh-covered beakers:
   a. Equipment
      i. Beaker (1), 50 mL.
      ii. Nitex, mesh (1), 7-cm square.
      iii. Aluminum foil (2) 8-cm squares.
      iv. Tape, masking and autoclave.

   b. Procedure:
      i. Place the square of Nitex mesh on the beaker, and tape all four corners down, using masking tape.
      ii. Cut autoclave tape to about 1-cm wide, and long enough to encircle the beaker.
      iii. Wrap the tape securely over the mesh, just under the lip of the beaker. The mesh should not be very taut. It is preferable to avoid taping the mesh to the beaker, since the tape is difficult to remove for washing.
      iv. Remove the masking tape from the corners of the mesh.
      v. Cover the beaker with two layers of aluminum foil.
      vi. Place a small piece of autoclave tape in the center of the foil cover. Autoclave for 30 min, using wrapped setting.

**4.4. Cellector Tissue
Sieves**

For the filtration of cell suspensions, Cellector tissue sieves maybe used in place of Nitex mesh-prepared beakers. The sieve is first sterilized with the selected screen in place, then is placed into a sterile cup (4 $\frac{1}{2}$-oz, Falcon, cat. #4014) and the tissue suspension is poured into it, allowing the liquid to flow through by gravity only. Forcing the suspension through the sieve can result in damage to the cell membranes. Any remaining cells are rinsed into the sterile container with medium or a balanced salt solution. The Cellector, complete including 85-mL pan, glass pestle, key, and nine screens (cat. #1985-85000) with various replacement screens, is available from Bellco Glass (Vineland, NJ).

## 5. Alternate Disaggregation Method: Stomacher Blender

1. DMEM/F12 + 10% FBS (60 mL)
2. Tubes, centrifuge, 50 mL
3. Lab-blender bag with 80-mL capacity
4. Cellector sieve with 230-μm screen (no. 60)
5. Cellector sieve with 140-μm screen (no. 100)
6. Containers (2), 4'/₂-oz, sterile (Falcon, cat, no. 4014)
7. Stomacher blender attached to a variable transfer set at 80 V (Stomacher Blender Tekmar, Available from VWR CanLab)

### 5.1. Reagents and Equipment

### 5.2. Disaggregation

1. Carefully pipette the tissue-medium suspension into a sterile blender bag, and add enough medium to bring the total amount in the bag to 50–60 mL.
2. Place the bag into the Stomacher blender, with about 2 cm of the bag visible above the closed door.
3. The blender should be attached to a variable transformer that is set at 80 V. This setting controls the speed of the blender paddles and prevents the disruption of the cell membranes.
4. Blend the tissue for 2.5 min and remove the bag with the cell suspension to the hood.
5. Pour the cell suspension into the no. 60 sieve, allowing it to filter by gravity only.
6. Pour this filtrate into the no. 100 sieve, again allowing the cells to flow by gravity only. Wash the adhering cells through the screen with 10 mL complete medium.
7. Centrifuge the cell suspension at $80g$ for 5 min.
8. Pour off the supernatants, resuspend the cells in DMEM/Fl2 (approx 100 mL/10 brains), and prepare a 1:100 dilution of cells for counting in a hemocytometer.
9. Dilute the cell suspension in DMEM/F12 + 10% FBS, and plate at $15 \times 10^6$ cells/75 cm$^2$ culture flask, in a total volume of 10–11 mL.
10. Incubate at 37°C in a humidified 5% $CO_2$ incubator, with the lid tightly closed.

## 6. Isolation of Astrocytes

### 6.1. Introduction

The initial cortical cell cultures should be cultured for 7–9 days. Periods shorter than this are not sufficient for stratification of astrocytes and oligodendrocytes. Periods longer than this can result in clustering of astrocytes above the bed layer, affecting the subsequent isolation of oligodendrocytes. In fact, the clustered astrocytes may break up and contaminate the oligodendrocyte preparation. We have found that the oligodendrocytes tend to adhere more strongly, the longer they remain in mixed culture (longer than 12 days), making the shaking process less efficient.

### 6.2. Reagents and Equipment

1. DMEM/F12
2. FBS
3. HBSS
4. Puck's BSS (15 mL)
5. Versene
6. 2.5% Trypsin solution (2 mL)
7. Cell culture-treated dishes (4), 35-mm
8. Beaker (1), covered with 30-μm Nitex mesh (or Cellector sieve, 25-μm, no. 500)
9. Centrifuge tubes (2), 15-mL
10. Lab-Line Junior orbit shaker in a dry 37°C incubator
    **Note:** A layer of packing foam will help insulate the flasks from the heat generated by the shaker.

### 6.3. Method for Isolation of Astrocytes

1. At the end of the 8–9 day culture period, microglia can be microscopically observed floating in the medium of the stationary cultures. Phase-dark, process-bearing cells are observed on top of a confluent phase-gray bed layer of astrocytes. Change the culture medium, close and tighten the plastic culture lids, and place the flasks on a rotary shaker to remove microglia and oligodendrocytes. Shake flasks at 200 rpm with a 1.5-in. stroke diameter at 37°C for 24 h. Remove flasks from the rotary shaker, change medium and return them to the rotary shaker for an additional 24 h.

2. After most of the oligodendroglial and microglial cells have been removed from the mixed-glial cultures by the two consecutive overnight shakes at 200 rpm, replenish the medium with 10 mL of fresh DMEM/F12 + 10% FBS, shake the flask at 100 rpm for an additional 48–72 h. Observe the cells daily to avoid overshaking. Replace the medium every 48 h. Shake the cells until fewer than 10 oligodendrocytes per microscope field ($\times$ 100) are observed. (These are the phase-dark cells, while the astrocytes are the phase-light cells.) The floating oligodendrocytes and microglia will be further removed every time the medium is changed.
   Replenish the medium every 2 days. The attached cells are 98% astrocytes.

3. Purified cultures can subsequently be subcultured by utilizing a versene–trypsin wash. Remove all medium from the flasks and wash the cells with 5 mL versene (EDTA) solution. Pour off and wash the cells with 2.0 mL/flask of a 2.5% trypsin solution, making sure that all the cells have been bathed. Pour off and incubate the flasks at 37°C, until the cells are completely disaggregated and run freely when the flask is inverted, usually 5–10 min.

4. Resuspend the cells in 8 mL DMEM/F12 + 10% FBS. Remove cells to a centrifuge tube and centrifuge for 5 min at 80$g$. Aspirate off old medium, and discard. Resuspend the pellet in 10 mL fresh culture medium. Determine the cell number by hemocytometry, and plate $4 \times 10^5$ cells/35-mm cell culture dish or as desired.

   Suggested optimal plating densities for astrocytes are as follows:

   a. $1 \times 10^6$ cells/75-cm$^2$ culture flask.

   b. $2 \times 10^4$ cells/2.0-cm$^2$, 24-well cell culture plate.

   c. $4 \times 10^5$ cells/35-mm cell culture-treated dish.

      **Note:** As cells are passaged in culture, they change in biochemical and immunological properties and we generally do not use astrocytes past the initial passage.

---

## 7. Reactive Astrocytes: Scratch-Wound Model

### 7.1. Introduction

The scratch-wound model, developed by Yu and colleagues (17), is a simple mechanical injury model in which a monolayer of confluent astrocytes is wounded by scratching it with the tip of a sterile pipette tip. The wounded astrocytes display a gliotic phenotype in the absence of injured neurons. This model (17), since described in 1993, has been used to investigate the role that GFAP upregulation plays in triggering reactive gliosis (17, 19), as well as other mechanisms likely involved in the astroglial response to an insult, such as cell proliferation and migration (20–23), and the contribution of specific signal transduction pathways or growth factors (22, 24). Furthermore, this model has been useful to test possible strategies to block or prevent astrogliosis (17, 19), since astrogliosis creates an environment unfavorable to neurite extension and regeneration.

Hence, the scratch-wound model offers the unique possibility of investigating astrogliosis in the absence of damaged neurons or other neural cells, which may release additional growth factors, cytokines, or other neuromodulators. Additionally, by using this model, it was shown that astrocytes can produce and release cytokines (25) in response to a trauma.

**7.2. Reagents and Equipment**

*General Items:* See **Section 2**

*Additional Items:*

1. Sterile plastic p20 pipette tips

2. Sterile washed round coverslips 18-mm

3. Lab-Line orbital shaker in a dry 37°C incubator

4. Poly-D-lysine, SIGMA cat # P6407

**7.3. Protocol: Purification of Astrocytes from Primary Rat Mixed Glial Cortical Cultures**

1. Primary cortical astrocytes are isolated from mixed glial cultures prepared following the McCarthy and de Vellis method (1), as described in the first two steps (a and b) in **Section 6.3**.

2. To passage the astrocytes, aspirate the medium and wash the cells with 5 mL DMEM/F12 without FBS/flask or with 3 mL trypsin–EDTA/flask to remove all the serum. Add 3 mL/flask of 0.25% trypsin–1 mM EDTA solution, and put back the flasks in the incubator until the cells are dissociated, about 3–5 min. Add 5–10 mL DMEM/F12 + 10% FBS and collect the medium with the cell in suspension into 15-mL conical sterile Falcon tubes. Centrifuge at 200$g$ for 5 min. Aspirate the supernatant and resuspend the cell pellets in 3–5 mL of DMEM/F12 + 10% FBS. Count the cells and plate at the desired density onto cell culture dishes or poly-D-lysine-coated glass coverslips. Suggested density:

   18-mm glass coverslips = $1.5 \times 10^5$ cells/mL

   35-mm cell culture plates = $4 \times 10^5$ cells/mL

**7.4. Protocol: Scratch-Wound Model**

1. One hour after plating, change the medium to remove possible debris and dead cells. Change the medium every 2–3 days and let the cells grow to 99% confluence in DMEM/F12 + 10% FBS, then change the medium to DMEM/F12 + 5% FBS and maintain the cells for at least two additional weeks. Change the medium every 3–4 days, at least twice/week. Change the cells to DMEM/F12 + 1% FBS 2–3 days before wounding.

2. Using a p20 sterile pipette tip make two parallel scratches and then two more scratches at a right angle to the previous two. Scratch the cells according to a simple grid to achieve about 37–40% injury (**Fig. 11.1A** and (17)).

3. Soon after the injury, wash the dishes once to remove all the debris and floating cells with DMEM-F12 + 10% FBS. Add the medium and place the culture dishes into the incubator. Replace the medium every 2 days.

**7.5. Critical Step and Trouble Shooting**

1. While shaking the flasks, it is important to check the cultures under the microscope daily to make sure not to overshake the mixed cultures as this can result in detaching the astrocytes from the flasks.

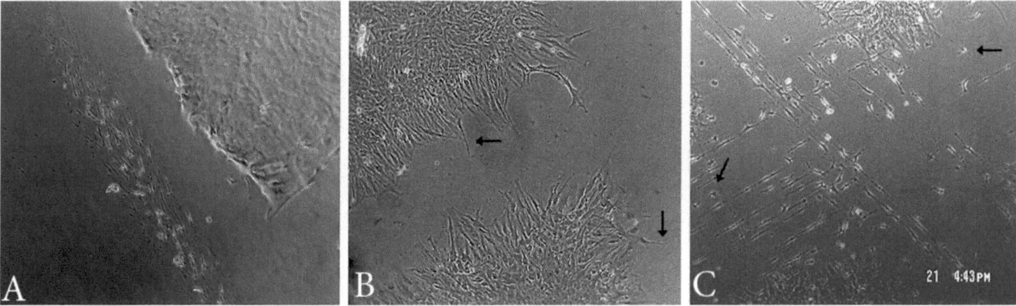

Fig. 11.1. Phase-contrast micrographs of live astrocytic cells after scratch-wound. Pictures were taken (**A**) soon after the cells were scratch-wounded with a sterile pipette tip, and the medium was changed to remove all the detached cells and debris. (**B**) Twenty-four hours and (**C**) forty-eight hours after the scratch.

2. Warm up the trypsin at 37°C before using it, since this is the optimal temperature for its enzymatic activity, the trypsinization will be faster. Wash the cells very well with DMEM-F12 without FBS to remove all the serum before adding the trypsin, because the serum neutralizes the enzymatic activity of the trypsin. Make sure that the trypsinization does not last more than 5 min. If necessary, the flasks can be gently tapped or knocked to detach the cells after adding some FBS to neutralize the trypsin.

### 7.6. Typical Protocol Results

During the first 24 h following the "wound-scratch," the astrocytes that surround the "wounded part" project their processes into the empty space (**Fig. 11.1B**). Some of the cell bodies may also move within the empty space, but without loosing contact with the astrocytic monolayer (**Fig. 11.1B**, see arrows). Occasionally, big stellate-shape reactive astrocytes may move into the middle of the empty part. Two days after the scratch (**Fig. 11.1C**), several bipolar cells with thick processes and smaller cells with short processes (**Fig. 11.1C**, arrows) migrate and colonize the middle of the "wounded cell culture." At 4 days post-scratch (**Fig. 11.2A**), some of the cells, in the middle of the "wound," appear flat and with a more resting-looking phenotype (**Fig. 11.2A**, arrowhead). Four and six days after the wound, long processes are crossing the "wounded area," projected toward the opposite intact monolayer of astrocytes (**Figs. 11.2A** and **B**, long arrow). After 6 days, more flat resting-looking astrocytes are seen in the middle of the scratch, which is about 80–85% closed (**Fig. 11.2B**, arrowhead), together with stellate-reactive cells (**Fig. 11.2B**, long arrow). The gap/wound is about 99% closed within 8 days after the scratch-wound (**Fig. 11.2C**). The regrowth can be easily recognized by the different orientation of the cells.

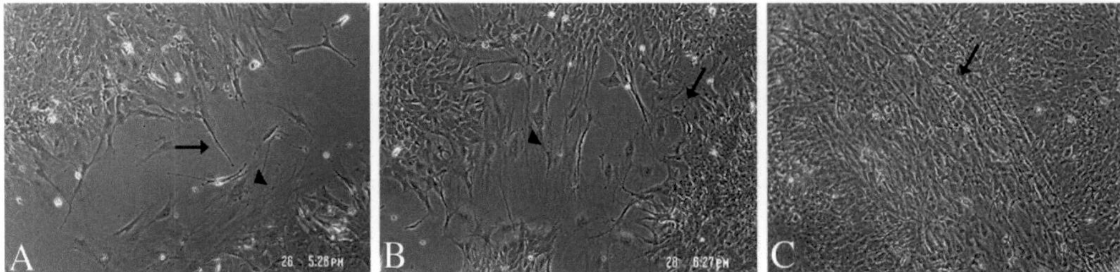

Fig. 11.2. Phase-contrast micrographs of live astrocytes after scratch-wound. Pictures were taken (**A**) 4 days; (**B**) 6 days, and (**C**) 8 days post-scratch-wound, after the cells were scratch-wounded with a sterile pipette tip.

Forty-eight hours after the scratch-wound, 99% of the cells along the scratched area are GFAP-positive; no ED1 or A2B5, a marker for oligodendrocyte progenitors, positive cells are observed.

***7.7. Summary***

The scratch-wound method is a reproducible method and an invaluable tool to study the effects of a mechanical trauma on purified isolated astrocytes in vitro, in particular for assessing astrocyte proliferation and migration in response to a mechanical injury. Although the scratch model has, as all the in vitro techniques, some drawbacks, it is still a useful tool.

## 8. Stretch Injury Model to Study Scar Formation and Reactive Astrogliosis in a Dish

### 8.1. Introduction

The use of effectively purified astrocytes (see above) and the ability to obtain highly differentiated, in vitro "matured" astrocytes are important prerequisites to address cellular and molecular mechanisms of astrocyte reactivity, scar formation, and axon growth inhibition. A new in vitro model of the glial scar has recently been established, which shows astrocyte reactivity and axon growth inhibition (7). We combine two initiating triggers of scar formation, mechanical trauma, and meningeal cell confrontation in this culture model. Astrocytes are grown on deformable membranes and subjected to mechanical trauma, – stretch and pressure – leading to defined, widespread reactivity and making astrocytes a neurite growth inhibitory substrate (7). The cultures used in this model display a rather mature astrocyte phenotype: postmitotic, highly differentiated, process-bearing astrocytes predominate and only few (3–4%) A2B5-positive glial progenitors or immature astrocytes are found (7, 26). These cultures express low levels of another progenitor marker, nestin, but high levels of adult brain extracellular matrix proteins, including tenascins, neurocan, and phosphacan (7). In vitro "maturing" of astrocytes is achieved by culturing them in low density on silastic membranes and, importantly, in the presence of adult-derived equine serum followed by a stepwise serum reduction to finally obtain serum-free cultures. Serum-starved astrocyte or astrocyte-fibroblast co-cultures are subjected to mechanical stretch by an abrupt pressure pulse deforming the culture well. This combination of pressure and biaxial stretch has been correlated with traumatic injury in vivo, where an acute pressure gradient is also translated into tissue strains and deformation (27–29). Mechanical stretching of astrocytes causes transient membrane permeability with membrane integrity being restored within 48 h (30, 31). Mechanical stretch triggers a transient rise in intracellular calcium, release of ATP, and inositol triphosphate signaling in astrocytes cultured on deformable membranes (32–34). Activation of P2 purinergic receptors and extracellular-signal-related kinase (ERK) as well as increased GFAP signals and secretion of S100β have been shown in this stretch-injury model (7, 32–35).

**8.2. Reagents and Equipment**

*General Items:* See **Section 2**

*Additional specific items*

1. Leibowitz' L15, Invitrogen cat # 41300

2. Trypan blue 0.4%, Invitrogen cat # 15250-061

3. Gentamicin, 10 mg/mL distilled, Invitrogen cat # 15710

4. Donor equine serum, Hyclone, cat # SH30074.03

5. Poly-L-lysine in 0.125 borate buffer, Sigma, cat # P2636

6. Rat tail collagen, type 1, BD Bioscience, cat # 354236

7. Bovine plasma fibronectin, Sigma, cat # F1141

8. Hibernate A, Brainbits, cat # HA

9. B27 supplements, Invitrogen, cat # 17504, 50X

10. Glutamax 100×, 200 mM, Invitrogen, cat # 35050

11. DNAse I, Roche, cat # 10104159001

12. Collagenase type I, Worthington, cat # LS004194

13. Papain, Worthington, cat # LS003126

14. Trypsin, 3× crystallized, Worthington, cat # LS003707

15. Percoll (density: 1.130 ± 0.005 g/mL), Sigma, cat # P1644

16. Iodoxal step gradient, Optiprep, Axis-Sheild, cat # 114542

17. 35-mm Petri dish with coverglass bottom, MatTek Corporation, cat # P35G-0-14-C

18. Borosilicate Pasteur pipette reusable glass with bent tip, Bellco, cat # 1273-40004

19. Microscope coverglass 25-mm circle, Fisherbrand, cat # 12-545-102

20. Pre-cleaned frosted microslides 3 × 1", VWR, cat #48312-013

21. Scotch mounting tape, Heavy Duty (3 M, 1 × 50")

22. Scalpel, Becton-Dickinson, cat # 2013-01

23. Cell Injury controller II, CIC II, Custom Design and Fabrication, Virginia Commonwealth University

24. Nitrogen tank, Air Liquide, cat # 255-SCF

25. Silastic membrane culture plates, Bioflex 6-well plates, Flexcell Intl., Untreated # BF-3001U, collagen I-treated, # BF-3001C

26. Microscope with high rising objectives, Telaval 31, Zeiss

27. Beckman centrifuge model J-66, Rotor JS-4.2

28. RC5C Plus centrifuge, Rotor SA600, Sorvall

### 8.3. Maturing Purified Astrocytes in Culture

1. *Coating*: Silastic membranes are coated on the day before passaging with fibronectin or collagen I. For fibronectin, uncoated plates are first incubated for 1 h in 1.5 mL Poly-L-lysine (20 mg/mL). Remove and rinse 3× in $H_2O$. Coat overnight with fibronectin (0.6 µg/mL dPBS) using 1.5 mL/well at 4°C. Coat collagen I-precoated silastic membranes with rat tail collagen I (0.1 mg/mL 0.02 N acetic acid) using 1.5 mL/well overnight at 4°C. Remove solutions completely before seeding cells.

2. *Astrocyte passaging*: After repeated shaking 2–3-week-old confluent astrocyte lawns are freed from microglia and oligodendrocytes and contain $6–8 \times 10^6$ cells/T75 flask. After removing the growth medium incubate cells for 5 min or until cell–cell contacts are loosened in 5–6 mL of versene solution. Appearance of phase-dark cell borders indicates loosening of cell–cell contacts. Remove versene and replace with pre-warmed 2.5 mL of 0.25% trypsin/EDTA to trypsinize cells for 3–5 min at 33°C shaking. When cells are detaching, add 2.5 mL of FBS-containing medium and gently triturate using a burned tip borosilicate pipette. Transfer the single cell suspension into a 50 mL tube. Rinse flask twice with medium to collect remaining cells. Spin cells for 5 min at $200g$ in an IEC clinical centrifuge. Gently remove solution and add 2.5 mL fresh medium to resuspend cells, then add additional 7.5 mL. Determine total yield and concentration per millilitre using a 20 µL aliquot in a hemocytometer. Add medium to achieve a final dilution of 50,000 cells/mL for low-density seeding. About 100,000 cells/2

mL DMEM/F12 + 10% FBS are seeded onto each coated elastic culture well (962 mm$^2$).

3. *Astrocyte differentiation*: Replace medium every other day for 4 days to 1 week until a nearly confluent lawn is established. Always drop medium onto the side wall of the well to minimize astrocyte activation. Switch to differentiation medium: DMEM/F12 + 10% horse serum (donor equine serum). Over the course of the following 2–3 weeks, stepwise reduce horse serum from 10 to 7.5 to 5 to 2.5%. Before stretching (1–5 days), switch to 0.5% horse serum or omit serum entirely (if conditioned medium, CM, is collected). One day before stretching, place exactly 2 mL medium/well assuring reproducible stretch pressure.

**8.4. Stretching**

The cell injury controller CIC II is an electronically controlled pneumatic device, which releases a defined nitrogen gas pressure ("input pressure") that abruptly deforms the elastic culture wells. Cultures are carefully placed on an elevated stand (comes with VCU equipment) and one culture well at a time is sealed with a plug connecting the culture space via tubing to the pressure controller and a nitrogen tank. One or two 50 ms nitrogen pulses between 26-30 psi input pressure are applied to each well, resulting in 3.4–3.8 psi peak injury pressure to the cells stretching the membrane from 100% to 136% (from 35 to 48 mm), with a maximal 10.5 mm deflection of the center. This stimulus intensity corresponds to the "moderate stretch" determined by Ellis and colleagues (27). Cultures are placed back into the incubator.

*For time-lapse imaging*: medium is removed, elastic membranes are cut out and placed upside down on a 35 mm culture dish with coverglass bottom. Same medium is replenished and membrane is fixed with two aluminum blocs and the dish is transferred to a gas- and temperature-controlled time-lapse imaging microscope stage. Unstretched cells grown on elastic membranes serve as "quiescent" controls.

**8.5. Astrocyte-Meningeal Fibroblast Co-cultures**

1. *Isolation of primary meningeal fibroblasts* is done according to Niclou and colleagues with modifications (36). Meninges are peeled from newborn to 5-day-old rat cortices and transferred in cold L15. Meninges are digested for 40 min at 33°C, gently shaking in 0.16% trypsin, 0.25% collagenase type I in Ca$^2$ and Mg$^2$-free HBSS in presence of 40 µg/mL DNAse I. An equal volume of DMEM/F12 + 10% FBS is added and the suspension is dissociated using a burned tip borosilicate pipette. Microglia and blood cells are removed either by leaving the red rim behind when resuspending the pellet or by shaking confluent meningeal cultures or by per-

coll gradient fractionation. For this, the suspension is mixed to obtain a final of 45% percoll in HBSS and spun for 30 min at 4°C, 2300$g$ (Sorvall, Rotor SA600). The fraction below the white rim is used leaving the red pellet of blood cells and undissociated pieces behind. Dilute the cells with twice the volume of medium and spin down for 5 min at 296$g$. Resuspend meningeal cells in DMEM/F12 + 10% FBS and grow in tissue culture-treated flasks until they are confluent within a few days.

2. *Astrocyte-fibroblast co-cultures:* Remove medium from confluent meningeal culture flask, add 5 mL versene, leave it briefly (lawn lifts off easily), and digest for 5 min in 0.25% trypsin/EDTA to achieve a single cell suspension. Add equal volume of medium, triturate, and spin. Prepare a 50,000 cell/mL dilution. Add 100,000 meningeal cells in 2 mL DMEM/F12 + 10% FBS to each confluent astrocyte culture well (962 mm$^2$). Twenty-four hours later switch to DMEM/F12 + 0.5% horse serum and incubate additional 24 h. Forty-eight hours astrocyte–fibroblast co-cultures are stretched as described above.

### 8.6. Regeneration-Relevant Neurite Outgrowth Assay in a Scar-Like Environment

1. *Isolation of postnatal neurons:* One-week-old cortical neurons are used because their neurite regeneration is more relevant to studies of axonal regeneration than that of embryonic neurons (7). Postnatal day 6 rat cortices are dissected, freed from meninges, and neurons are isolated using a protocol from Benson and colleagues with minor modification (37–39). Tissues are trimmed to enrich for somatosensory-motor cortex areas and kept in hibernation solution (Hibernate A, B27 supplements, 0.5 mM glutamate, 50 μg/mL gentamicin). Transfer tissue to a clean lid and slice it using a straight blade scalpel, turn dish and slice across a second time. Transfer cubes into 35-mm culture dish. Digest the tissue 30–40 min in 52 U/mL (2 mg/mL) papain in hibernation solution at 33°C with gentle shaking. Transfer the coarse mixture into a 15-mL tube, spin down 5 min, remove digestion solution, add 2 mL hibernate and slowly triturate 10 times using a burned tip borosilicate pipette. Let undissociated pieces settle for 2 min, then transfer the supernatant into a new tube. Add 2 mL hibernation solution to remaining pieces, triturate 10×, wait 2 min, take supernatant and add to previous. Repeat dissociation a third time. Discard remaining undissociated pieces. The homogenized cell suspension (6 mL) is now carefully placed on top of an iodoxal step gradient (15, 20, 25, 35% prepared using a 1:1 dilution of 60% iodoxal in hibernation medium. Centrifuge for 15 min at 800$g$ at room temper-

ature without brakes (Beckmann, rotor JS-4.2). Remove the top 6 mL and transfer the clear fraction below the dense ring into a new tube omitting the pellet. Add 5 mL of medium to dilute iodoxal and spin down cells at $196g$ for 5 min. Remove supernatant and resuspend cells in 2 mL hibernation medium. The single cell suspension contains circa 50% viable neurons. Store neuron stock solution overnight at 4°C.

2. *Neuron-astrocyte co-cultures.* Determine viable cell concentration and yield of 1 day hibernated neurons, by diluting an aliquot 1:1 with trypan blue, and count 20 μL in a hemocytometer. Determine total cell number and volume of neuron stock solution needed. 100,000 cells (approx. 50,000 neurons) per 962 mm$^2$ astrocyte culture are seeded as follows: Collect 1 mL medium from each culture well and place in tubes keeping stretched and unstretched CM in separate tubes. Add proper volume of neuron suspension to the tubes and mix gently. Replenish each well with 1 mL of the appropriate CM now with 100,000 cells enriched in neurons in each well. Incubate the neuron–astrocyte co-cultures for 24 h.

### 8.7. Critical Steps and Troubleshooting

For the successful outcome it is essential to monitor astrocytes frequently. For monitoring of the cells growing on the recessed, elastic membranes, a microscope with high rising objectives is required. The new generation injury controller, CICII, records the stretch pressure applied to the well as "injury pressure"; this is the part of the pulse not translated into deflection. This reading assures reproducibility of the given mechanical trauma. Cell clustering and astrocyte contraction in response to stretching is substrate-dependent: fibronectin is more adhesive for astrocytes than collagen (7). For CM analysis it is possible to completely omit serum for up to 5 days without loosing cell viability. The neurite regeneration improving potential of various compounds can be effectively tested in this 24 h neuron–astrocyte co-culture assay (I. Wanner, unpublished observation). The compounds are added in double-strength concentration to the tubes containing CM of stretched astrocytes before adding the neurons. Neuron and compound-supplemented CM is added back to the astrocyte cultures, and co-cultures are incubated for 24 h. This procedure assures that factors secreted by stretch-reactive astrocytes remain present during the neurite outgrowth assay.

### 8.8. Analysis Tools and Typical Results

Immediately after the stretch astrocytes still appear as confluent lawn, yet phase-dark cell-cell borders indicate loosening of contacts (**Fig. 11.3** left, time-lapse imaging). Drastic morphological changes occur in stretched astrocytes within 3–16 h after stretching (**Fig. 11.3**, right picture; and **Fig. 11.4**). Flattened, spread astrocytes actively contract and become spindle- or star-shaped with rounded cell bodies and fasciculated, cytoskeleton (**Fig. 11.3**, time-lapse series; I. Wanner, unpublished observation). Shortly after stretching, a small percentage of astrocytes will die. We routinely measure cell death at 48 h after

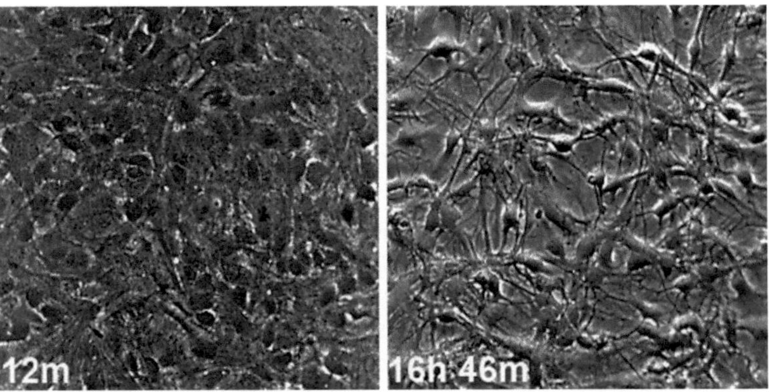


Fig. 11.3. Time-lapse imaging was done using an Olympus IX81 inverted spinning disk confocal microscope with a weather station temperature- and gas-controlled stage. Astrocytes are still flat immediately after the stretch and become reactive over the course of nearly 16 h 45 min after stretching. *Left*: 12 min after stretching the cells form a continuous lawn. *Right*: 16 h 46 min after stretching the lawn is interrupted by little gaps due to astrocyte contraction; reactive astrocytes appear as more phase-contrasted, spindle-shaped, or stellate cells.

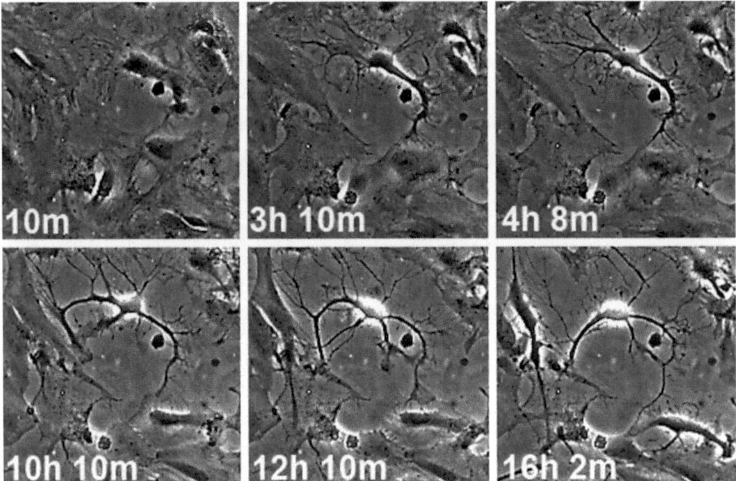


Fig. 11.4. This 16.5 h time-lapse sequence shows the transformation of a flat, quiescent group of astrocytes into reactive, stellate-looking astrocytes upon stretching. *Top*, from *left* to *right*: Cytoplasmic lamellae of the flat astrocytes (at 10 min post-stretching) retract (shown at 3 h 10 min and 4 h 8 min post-stretching). *Bottom*, from *left* to *right*: thin processes appear and the astrocyte cell body becomes rounded thereafter. New processes form with many dynamically protruding and retracting filopodia, once the contracted shape is acquired (between 10 and 16 h).

stretching using a propidium iodide permeability assay (7, 27). Under conditions described here, cell death is typically below 5%. Immunohistochemical analyses of astrocyte reactivity are done 24 h after stretching using double- and triple-labeling protocols (7). Twenty-five mm circular coverglasses are placed over the culture centers (491 mm$^2$), cut out when dried and mounted onto glass slides. Imaging requires little elevation stripes (Scotch mounting tape) to enable a workable objective distance. RNA and protein samples including CM from stretch-reactive astrocytes are harvested using common biochemical isolation protocols (7). Routine quantitative analyses of astrocyte-fibroblast co-cultures showed that 4 day co-cultures prepared the way described above consisted of 40–50% meningeal fibroblasts and 50–60% astrocytes. Mixed lawns show many "reactive sites" and glia limitans-like borders between the two cell types (I. Wanner, unpublished observation and (7). To quantify neurite regeneration, the number of sprouting neurons/fields is determined. Neurite outgrowth is measured using a systematic randomized approach for unbiased sampling. Overall sprouting can be determined using a new semiautomated computational approach based on Igor Pro (Wavemetrix software) (7). In addition, several neurite growth parameters are routinely measured using a software-assisted manual neurite tracing method (Neurolucida, Microbrightfield) (7).

## 9. Summary

One advantage of the pressure-induced stretch injury model is that it produces widespread reproducible reactivity on entire astrocyte lawns. This is particularly useful for larger-scale biochemical analyses of gene expression changes in reactive astrocytes, including real-time PCR analyses and proteomic applications. Thus, this model allows identification of new reactive gliosis markers. Another great advantage of the CIC II–Bioflex plate combination allows versatile imaging including confocal imaging and time-lapse monitoring of astrocyte activation. Lastly, compared to other neurite growth inhibition assays and scar mimicking culture methods, this assay does mimic key cellular and molecular aspects of the scar environment, and neurite growth can be randomly quantified because neurons are placed over widespread scar-like areas. Using stretch as opposed to chemical reagents to cause astrocyte reactivity avoids the problem that such reagents may themselves have an effect on axonal regeneration. In summary, this in vitro injury model offers a unique combination of applications to reproducibly quantify morphological, biochemical, and functional changes of astrocyte reactivity that are shown to occur in scar areas after traumatic injury in vivo.

## References

1. McCarthy KD, de Vellis J. Preparation of separate astroglial and oligodendroglial cell cultures from rat cerebral tissue. J Cell Biol 1980;85(3):890–902.

2. Giulian D, Baker TJ. Characterization of ameboid microglia isolated from developing mammalian brain. J Neurosci 1986;6(8):2163–78.

3. Liva SM, Kahn MA, Dopp JM, de Vellis J. Signal transduction pathways induced by GM-CSF in microglia: significance in the control of proliferation. Glia 1999;26(4):344–52.

4. Holmes E, Hermanson G, Cole R, de Vellis J. Developmental expression of glial-specific mRNAs in primary cultures of rat brain visualized by in situ hybridization. J Neurosci Res 1988;19(4):389–96, 458–65.

5. Saneto RP, de Vellis J. Characterization of cultured rat oligodendrocytes proliferating in a serum-free, chemically defined medium. Proc Natl Acad Sci U S A 1985;82(10):3509–13.

6. Espinosa de los Monteros A, Zhang M, De Vellis J. O2A progenitor cells transplanted into the neonatal rat brain develop into oligodendrocytes but not astrocytes. Proc Natl Acad Sci U S A 1993;90(1):50–4.

7. Wanner IB, Deik M, Torres M, et al. A new in vitro model of the glial scar inhibits axon growth. Glia 2008;56(15):1691–709.

8. Morrison RS, de Vellis J. Growth of purified astrocytes in a chemically defined medium. Proc Natl Acad Sci U S A 1981;78(11):7205–9.

9. Morrison RS, de Vellis J. Differentiation of purified astrocytes in a chemically defined medium. Brain Res 1983;285(3): 337–45.

10. Bressler JP, Cole R, de Vellis J. Neoplastic transformation of newborn rat oligodendrocytes in culture. Cancer Res 1983;43(2):709–15.

11. Bressler JP, de Vellis J. Neoplastic transformation of newborn rat astrocytes in culture. Brain Res 1985;348(1):21–7.

12. Wu VW, Schwartz JP. Cell culture models for reactive gliosis: new perspectives. J Neurosci Res 1998;51(6):675–81.

13. Eng LF, Ghirnikar RS, Lee YL. Glial fibrillary acidic protein: GFAP-thirty-one years (1969–2000). Neurochem Res 2000;25 (9–10):1439–51.

14. Norton WT, Farooq M. Astrocytes cultured from mature brain derive from glial precursor cells. J Neurosci 1989;9(3):769–75.

15. Norton WT, Farooq M, Chiu FC, Bottenstein JE. Pure astrocyte cultures derived from cells isolated from mature brain. Glia 1988;1(6):403–14.

16. Whittemore SR, Sanon HR, Wood PM. Concurrent isolation and characterization of oligodendrocytes, microglia and astrocytes from adult human spinal cord. Int J Dev Neurosci 1993;11(6):755–64.

17. Yu AC, Lee YL, Eng LF. Astrogliosis in culture: I. The model and the effect of antisense oligonucleotides on glial fibrillary acidic protein synthesis. J Neurosci Res 1993;34(3):295–303.

18. Lu EJ, Brown WJ, Cole R, deVellis J. Ultrastructural differentiation and synaptogenesis in aggregating rotation cultures of rat cerebral cells. J Neurosci Res 1980;5(5): 447–63.

19. Ghirnikar RS, Yu AC, Eng LF. Astrogliosis in culture: III. Effect of recombinant retrovirus expressing antisense glial fibrillary acidic protein RNA. J Neurosci Res 1994;38(4): 376–85.

20. Koguchi K, Nakatsuji Y, Nakayama K, Sakoda S. Modulation of astrocyte proliferation by cyclin-dependent kinase inhibitor p27(Kip1). Glia 2002;37(2):93–104.

21. Kornyei Z, Czirok A, Vicsek T, Madarasz E. Proliferative and migratory responses of astrocytes to in vitro injury. J Neurosci Res 2000;61(4):421–9.

22. Hou YJ, Yu AC, Garcia JM, et al. Astrogliosis in culture. IV. Effects of basic fibroblast growth factor. J Neurosci Res 1995;40(3):359–70.

23. He Y, Li HL, Xie WY, Yang CZ, Yu AC, Wang Y. The presence of active Cdk5 associated with p35 in astrocytes and its important role in process elongation of scratched astrocyte. Glia 2007;55(6):573–83.

24. Mandell JW, Gocan NC, Vandenberg SR. Mechanical trauma induces rapid astroglial activation of ERK/MAP kinase: Evidence for a paracrine signal. Glia 2001;34(4): 283–95.

25. Lau LT, Yu AC. Astrocytes produce and release interleukin-1, interleukin-6, tumor necrosis factor alpha and interferon-gamma following traumatic and metabolic injury. J Neurotrauma 2001;18(3):351–9.

26. Schnitzer J, Schachner M. Cell type specificity of a neural cell surface antigen recognized by the monoclonal antibody A2B5. Cell Tissue Res 1982;224(3):625–36.

27. Ellis EF, McKinney JS, Willoughby KA, Liang S, Povlishock JT. A new model for rapid stretch-induced injury of cells in culture: characterization of the model

using astrocytes. J Neurotrauma 1995;12(3): 325–39.

28. Gennarelli TA, Thibault LE. Biological models of head injury. In: Becker DP, Povlishock JT, eds. Central Nervous System Traums Status Report-1985. Bethesda, MD: National Inst. of Health; 1985:pp.391–404.

29. Shreiber DI, Bain AC, Ross DT, et al. Experimental investigation of cerebral contusion: histopathological and immunohistochemical evaluation of dynamic cortical deformation. J Neuropathol Exp Neurol 1999;58(2):153–64.

30. Amruthesh SC, Boerschel MF, McKinney JS, Willoughby KA, Ellis EF. Metabolism of arachidonic acid to epoxyeicosatrienoic acids, hydroxyeicosatetraenoic acids, and prostaglandins in cultured rat hippocampal astrocytes. J Neurochem 1993;61(1):150–9.

31. Lamb RG, Harper CC, McKinney JS, Rzigalinski BA, Ellis EF. Alterations in phosphatidylcholine metabolism of stretch-injured cultured rat astrocytes. J Neurochem 1997;68(5):1904–10.

32. Floyd CL, Rzigalinski BA, Sitterding HA, Willoughby KA, Ellis EF. Antagonism of group I metabotropic glutamate receptors and PLC attenuates increases in inositol trisphosphate and reduces reactive gliosis in strain-injured astrocytes. J Neurotrauma 2004;21(2):205–16.

33. Floyd CL, Rzigalinski BA, Weber JT, Sitterding HA, Willoughby KA, Ellis EF. Traumatic injury of cultured astrocytes alters inositol (1,4,5)-trisphosphate-mediated signaling. Glia 2001;33(1):12–23.

34. Neary JT, Kang Y, Willoughby KA, Ellis EF. Activation of extracellular signal-regulated kinase by stretch-induced injury in astrocytes involves extracellular ATP and P2 purinergic receptors. J Neurosci 2003;23(6):2348–56.

35. Willoughby KA, Kleindienst A, Muller C, Chen T, Muir JK, Ellis EF. S100B protein is released by in vitro trauma and reduces delayed neuronal injury. J Neurochem 2004;91(6):1284–91.

36. Niclou SP, Franssen EH, Ehlert EM, Taniguchi M, Verhaagen J. Meningeal cell-derived semaphorin 3A inhibits neurite outgrowth. Mol Cell Neurosci 2003;24(4):902–12.

37. Benson MD, Romero MI, Lush ME, Lu QR, Henkemeyer M, Parada LF. Ephrin-B3 is a myelin-based inhibitor of neurite outgrowth. Proc Natl Acad Sci U S A 2005;102(30):10694–9.

38. Brewer GJ. Serum-free B27/neurobasal medium supports differentiated growth of neurons from the striatum, substantia nigra, septum, cerebral cortex, cerebellum, and dentate gyrus. J Neurosci Res 1995;42(5):674–83.

39. Brewer GJ. Isolation and culture of adult rat hippocampal neurons. J Neurosci Methods 1997;71(2):143–55.

# Chapter 12

# Oligodendrocyte Progenitor Cell Culture

Akiko Nishiyama, Ryusuke Suzuki, Hao Zuo, and Xiaoqin Zhu

## Abstract

Oligodendrocyte progenitor cells represent the largest proliferating progenitor population in the post-natal central nervous system. They not only give rise to oligodendrocytes but also have the potential to differentiate into astrocytes or revert to multipotential neural stem-like cells under certain conditions. Dissociated cultures of these cells provide a useful tool to study their function and dissect the molecular mechanisms that regulate their behavior. This chapter describes and compares three methods of purifying oligodendrocytes progenitor cells: purification by differential adhesion, immunopanning, and fluorescence-activated cell sorting.

**Key words:** Oligodendrocyte, immunopanning, cell sorting, rat.

## 1. Introduction

Oligodendrocyte progenitor cells (OPCs) are the largest proliferating progenitor pool in the postnatal central nervous system (CNS) (1, 2). Because of their ability to give rise to myelin-forming oligodendrocytes, the role for OPCs in myelin repair in demyelinating diseases such as multiple sclerosis has been extensively studied (3). Cultures of OPCs provide a tool for examining the mechanisms that affect their proliferation, migration, and differentiation. Despite the limitation in reproducing the in vivo environment, dissociated cultures of OPCs have been widely exploited for the ease of observation and manipulation. Dissociated OPCs can be co-cultured with other types of cells to study cellular interactions. For example, myelinating cultures, while technically challenging, can provide a powerful tool for dissecting the mechanisms that affect the myelination process (4, 5). Recently, the fate of OPCs has become an intensely investigated

L.C. Doering (ed.), *Protocols for Neural Cell Culture*, Springer Protocols Handbooks,
DOI 10.1007/978-1-60761-292-6_12, © Humana Press, a part of Springer Science+Business Media, LLC 1992, 1997, 2001, 2010

area of research, fueled by the debate of whether OPCs are multi-potential cells (6–8). In addition to studying the fate potential of cells isolated in culture, their in vivo behavior can be studied after transplanting them into the brain. These studies require improved methods for obtaining a highly purified and defined population of cells.

There are several methods to isolate OPCs from the perinatal rodent central nervous system (CNS). The advantages and disadvantages are listed in **Table 12.1**. The shake-off method was first introduced by McCarthy and de Vellis (9) and uses differential adhesive properties of astrocytes, microglia, and OPCs. The least adhesive OPCs can be physically separated from astrocytes and microglia by vigorously shaking flasks of mixed glial

**Table 12.1**

**Comparison of the three methods for purifying OPCs**

|  | Advantages | Disadvantages |
| --- | --- | --- |
| Shake-off method | • Both astrocytes and OPCs can be isolated | • Relatively low purity, 90–95% |
|  | • Relatively simple and inexpensive | • Cells are aged for at least 7 days while they are maintained as a mixed glial culture |
|  | • High yield |  |
| Immunopanning | • Relatively simple and inexpensive | • Limited to cell surface markers that are resistant to proteolytic enzymes used for cell dissociation |
|  | • Can combine positive and negative selection | • Takes a few days to prepare immunopanning plates |
|  | • High purity | • Relatively low yield |
|  | • Cell can be purified directly from the brain |  |
| FACS | • Highest purity | • Low yield |
|  | • Cells can be isolated directly from the brain | • Requires instrumentation |
|  |  | • Requires a transgenic mouse line with a cell type-specific fluorescent tag or an antibody to a cell surface antigen that is resistant to proteolytic cleavage |

cell culture and collecting detached cells. More recently, this method was optimized to enhance the survival of OPCs (10). The immunopanning method was developed by Raff and colleagues, who isolated OPCs from perinatal rat optic nerves based on the expression of the A2B5 ganglioside on their cell surface (11, 12). Fluorescence-activated cell sorting (FACS) is another way to isolate OPCs for culture. This method applies a laser beam on a stream of single cells to separate OPCs that are expressing a fluorescent tag from non-fluorescent cells. The fluorescent tag may be provided by using a transgenic mouse line that specifically expresses a fluorescent protein in OPCs or by labeling dissociated OPCs with a fluorescent antibody. This method is becoming more widely used as different transgenic mouse lines are generated (8, 7).

In this chapter, these three commonly used methods for purifying OPCs from perinatal CNS are described. The first section (**Section 2**) describes the procedure for dissociating cells from perinatal CNS, which is common to all three purification methods. This is followed by a description of the shake-off method, immunopanning, and FACS.

## 2. Protocol: Dissociation of Glial Cells from Perinatal Rodent Brain

### 2.1. Animals: Postnatal Day 2–4 (P2-4) Mice or Rats

### 2.2. Materials

#### 2.2.1. Dissection Tools

1. Scissors. (1) straight, 13 cm. Scissors. (2) fine, 9–12 cm.

2. Forceps. (1) straight, fine, serrated, 10 cm.

3. Forceps. (2) 45° angled, fine (Dumont-type, no. 5–45) (**Fig. 12.1**).

4. Tools are sterilized by submerging them in 70% ethanol for 10 min and thoroughly drying them.

5. Stereomicroscope

6. Water bath

#### 2.2.2. Reagents

1. $Ca^{2+}$-$Mg^{2+}$-free minimum essential medium (S-MEM) (Invitrogen, cat. no. 11380-037).

2. Dulbecco's modified eagle medium (DMEM) (Invitrogen, cat. no. 11960-044).
   Add penicillin, streptomycin, and L-glutamine into DMEM before use.

3. Penicillin streptomycin solution (100×) (Invitrogen, cat. no. 15140–122).

4. L-Glutamine (200 mM; 100×) (Invitrogen, cat. no. 25030-081).

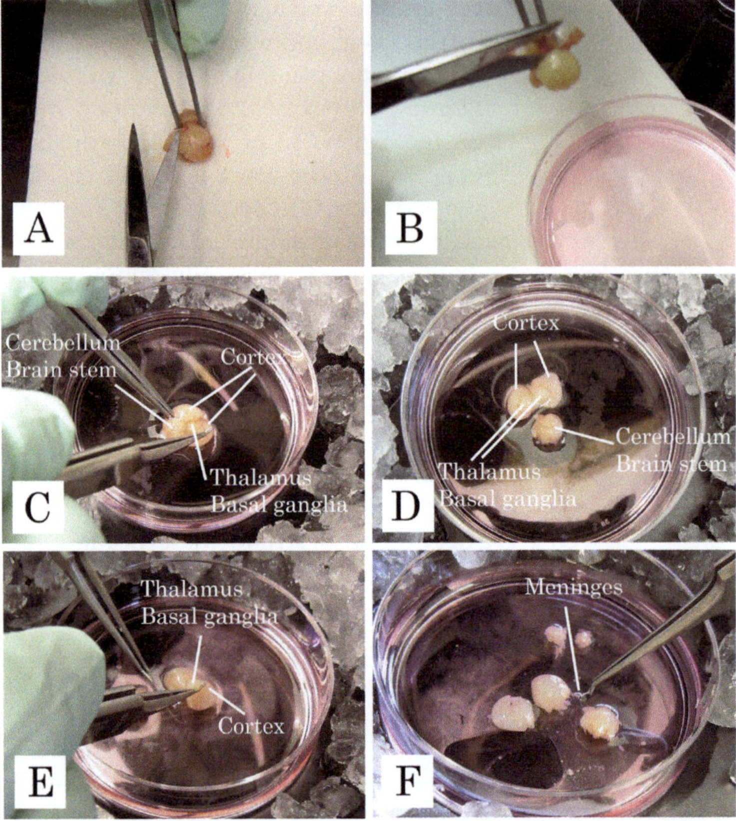



Fig. 12.1. Dissection of cerebral cortical tissue from P3 mouse brain. The skull is removed to expose the brain (**A**). Olfactory bulbs are removed and the brain is dropped into a Petri dish with prechilled SMEM (**B**). The cerebellum and brain stem are held by one pair of tweezers. A sagittal cut is made halfway through the interhemispheric sulcus, and then the cortex is flipped over by the other tweezers (**C**). The cerebellum and brain stem are separated from the brain, and the thalamus and basal ganglia are exposed (**D**). The thalamus and basal ganglia are pinched by tweezers and detached from the cortex (**E**). The meninges are peeled off from the surface of brain parenchyma (**F**).

5. Fetal bovine serum (FBS) (Invitrogen, cat. no. 26140-079).

6. FBS is heated in a water bath at 56°C for 30 min before use to inactivate complement and stored in aliquots at –20°C.

7. Trypsin (Sigma, cat. no. T9935). Store aliquots of 1% solution (in saline, filtered) at –20°C.

8. Deoxyribonuclease I (Sigma, cat. no. D4527). Store aliquots of 5 mg/mL solution (in saline, filtered) at –20°C.

9. Kynurenic acid.

*2.2.3. Culture Supplies*

1. Serological pipettes, sterile (10 mL)

2. Pasteur pipettes

3. Petri dishes, 60 mm, plastic, sterile

4. 100 mm tissue culture dishes (Corning)

5. 75 cm$^2$ tissue culture flasks

### 2.3. Methods

*2.3.1. Dissection of Cortical Tissues from Perinatal Rodent Brains*

Decapitation of the animals and dissection of cerebral cortex are carried out under sterile conditions.

Prepare 60 mm Petri dishes containing 2 mL of ice-cold S-MEM. Use one dish for every two brains (i.e., 5 dishes for 10 brains) and keep them chilled on ice. Prepare another 60 mm Petri dish with 5 mL of S-MEM, which will be used to collect the dissected and cleaned tissue. If the number of brains exceeds 8 for mice or 6 for rats, pool the tissues in two dishes. Sometimes, 0.05 mg/mL kynurenic acid is added to S-MEM to increase cell survival

1. Anesthetize the animals in accordance with the protocols approved by the institution. We anesthetize perinatal mice or rats by hypothermia.

   Decapitate and spray 70% ethanol on the head. With scissors, make an incision in the scalp to expose the skull. Then using sturdy scissors make one transverse cut in the skull above the cerebellum and two sagittal cuts along the sides of the brain. The skull flap can now be lifted from the posterior end toward the anterior end of the brain in one piece (**Fig. 12.1A**). Using tweezers, lift the brain from the base of the skull, severing the olfactory bulbs and cranial nerves with scissors (**Fig. 12.1B**).

2. Once the brain is freed from the cranium, drop it into the dish with 2 mL of S-MEM.

3. Under a stereomicroscope and with two forceps with 45° angled fine tips, remove the brainstem and cerebellum (**Fig. 12.1C**).

4. Flip over each side of the cerebrum from posterior or medial side and remove the thalamus, hypothalamus, and striatum (**Fig. 12.1, D and E**). The remaining tissue will comprise the cerebral cortex and subcortical white matter. We will refer to this as the cortical tissue.

5. Remove the meninges and choroid plexus from the cortical tissue. Ideally, the meninges should come off in a single piece (**Fig. 12.1F**). Pool the collected tissues in the dish with 5 mL S-MEM and keep the dish on ice until all cortical tissues are cleaned and collected. Proceed with cell dissociation described below.

*2.3.2. Dissociation of Cells from the Brains*

1. Take the Petri dishes with the dissected and cleaned cortical tissues into a tissue culture hood. Tilt the dish and transfer the S-MEM into a sterile 15 mL centrifuge tube.

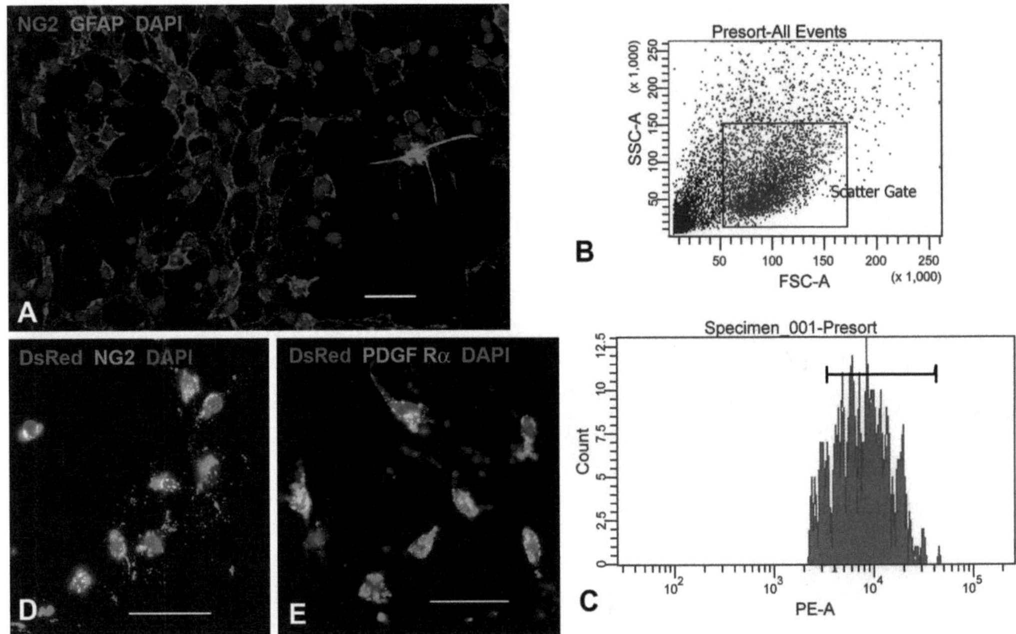

Fig. 12.2. Examples of purified OPCs and FACS gate setting. (A). Mouse OPCs purified by the shake-off method and maintained for 2 days in NBM + B27 + PDGF, immunolabeled with rabbit anti-NG2 (*red*) and mouse anti-GFAP (*green*; glial fibrillary acidic protein) antibodies. The culture contains a small number of GFAP+ astrocytes. (B). Scatter gate setting for FACS to eliminate small dead cells (*left side*, FSC <50) and granular macrophages (*upper half*). (C). Fluorescence gate setting to ensure 100% purity of the sorted cells. (D). Sorted cells immunolabeled with rabbit anti-NG2 antibody (*green*) 1 day after sorting. (E). Sorted cells immunolabeled with rabbit anti-PDGF Rα antibody (*green*; alpha receptor for platelet-derived growth factor) 1 day after sorting. Red color in **D** and **E** represent DsRed fluorescence. *Blue* in **A**, **D**, and **E** represent DAPI. Scale bars = 20 μm.

2. Mince the tissues with small scissors into pieces that are small enough to be taken up by a Pasteur pipette.

3. Suspend the minced tissue with the S-MEM saved from Step 1 and transfer the tissue pieces into the 15 mL tube.

4. Add trypsin to a final concentration of 0.1–0.2%. Higher concentrations of trypsin are used for older brains. Incubate in a 37°C water bath for 15 min. Shake the tube rigorously once during the incubation time to disperse the tissue.

5. After the incubation, add an equal volume (5 mL) of DMEM containing 10% FBS (10F-DMEM) to neutralize trypsin.

6. Add DNaseI to a final concentration of 0.05–0.1 mg/mL.

7. Triturate the cells by pipetting up and down 10 times with a 10 mL serological pipette.

8. Let stand for 2–3 min to allow undissociated tissue pieces to come down to the bottom of the tube.

9. Transfer the supernatant to a fresh 15 mL centrifuge tube and centrifuge at 1,100 rpm (200 $g$) for 3 min to spin down cells.

10. Remove the supernatant and suspend the cells in fresh 10F-DMEM.

11. Use the cells for OPC isolation described in **Sections 3, 4 and 5** below.
    **Note:**
    1. Best results are obtained from P2 to 3 rodent brains. Dissociation becomes more difficult with increasing age of the animal. *See* for example, reference 17 for dissociating OPCs from adult optic nerves.

    2. Approximately $2 \times 10^6$ cells are obtained from cortical tissues of one mouse brain and $1 \times 10^7$ cells from one rat brain.

## 3. Purification of OPC by the Shake-Off Procedure

The shake-off method to purify OPCs was first reported by McCarthy and de Vellis (9). This method is based on the fact that OPCs are more loosely attached to the bottom of flasks than other cell types such as astrocytes. We describe a modified version of this protocol with improved OPC survival (10).

### 3.1. Materials

1. Neurobasal medium (NBM) (Invitrogen, cat. no. 21103–049). Add penicillin, streptomycin, and L-glutamine into NBM before use.

2. Penicillin streptomycin solution (100×) (Invitrogen, cat. no. 15140–122).

3. L-Glutamine (200 mM; 100×) (Invitrogen, cat. no. 25030–081).

4. Poly-L-lysine (Sigma, cat. no. P1524, MW>300,000). Dissolve in sterile water in a 50 mL sterile polypropylene tube at a concentration of 3 mg/mL by heating the entire tube in a boiling water bath for 15 min. Dilute a portion of this solution to a working concentration of 100 μg/mL with sterile water. Store both solutions at 4°C.

5. B27 serum-free supplement (50×) (Invitrogen, cat. no. 17504–044).

6. PDGF-AA (R&D Systems, cat. no. 221-AA). Add HCl and bovine serum albumin before use according to the manufacturer's recommendation.

7. 75 cm² tissue culture flasks (Falcon, cat. no. 353024). Coat the flasks with 100 μg/mL poly-L-lysine overnight

at room temperature or >2 h at 37°C. Rinse them with sterile $dH_2O$ twice and let them dry before use.

8. Tissue culture dishes, 100 mm, plastic, sterile (Corning, cat. no. 430293).

9. Serological pipettes, plastic, sterile, (5, 10, and 25 mL).

10. A humidified $CO_2$ incubator, set at 37°C with 5% $CO_2$.

11. A 37°C incubator with a shaker.

### 3.2. Preparation of Flasks of Dissociated Mixed Glial Cell Culture

1. Plate dissociated mixed glial cells obtained from above (*see* **Section 2**) into 75 cm$^2$ tissue culture flasks with 8–10 mL 10F-DMEM. Cortical tissues from 3 to 4 mouse brains or 2 to 3 rat brains are cultured in one 75 cm$^2$ flask.

2. Culture the cells in a humidified incubator at 37°C with 5% $CO_2$.

3. Feed the cells with fresh 10F-DMEM (8–10 mL/flask) once every 3–4 days.

### 3.3. Harvesting and Plating OPCs

1. OPCs can be harvested after 7–10 days of incubation. Coat culture surfaces such as tissue culture dishes, multiwell plates, or glass coverslips with 100 μg/mL poly-L-lysine at 4°C overnight or >2 h at 37°C.

2. Add 8–10 mL of fresh 10F-DMEM to the flasks of mixed glial cultures and equilibrate in the $CO_2$ incubator for at least 2 h before starting the shaking.

3. Close the caps tightly and tape the flasks to the incubated shaker. Shake at 250 rpm at 37°C for 15–18 h.

4. On the following day, increase the speed to 350 rpm and shake the flasks for an additional 30 min.

5. Collect the supernatant containing OPCs and preplate them on uncoated tissue culture dishes and incubate at 37°C for 30 min to allow microglia to attach. Swirl the plate once after 15 min. We have obtained the best results using Corning tissue culture dishes (cat. no. 430293) for this step.

6. Add fresh 10F-DMEM to the original flasks and return them to the incubator. A new batch of OPCs can be harvested in 5–7 days. OPCs can be harvested up to three times from each flask.

7. Harvest the non-adherent cells from the Corning preplating dishes. Most macrophages/microglia should have been adhered to the bottom of the dish.

8. Centrifuge the harvested cells at 1,100 rpm (200 *g*) for 3 min. Remove the supernatant and add fresh 10F-DMEM.

9. Plate the harvested OPCs into tissue culture dishes (35 or 60 mm), multiwell plates, or on glass coverslips in multiwell

plates, depending on the nature of the experiments to be performed. The culture surface should be coated with 100 µg/mL poly-L-lysine as described above. It is recommended that a portion of cells be plated on 12 mm coverslips in a 24-well plate to check the purity of OPC by immunostaining.

**3.4. Culturing OPCs**

1. Approximately 1 h after plating, OPCs should be adhered to the bottom. Change the medium to serum-free medium (NBM containing B27 supplements, and 10–15 ng/mL PDGF-AA). This will cause OPCs to proliferate for a few days, but they will eventually differentiate. The best OPC population can be obtained 2 days after plating, when the majority of the cells are NG2 + OPCs (**Fig. 12.2A**), with a small number of contaminating GFAP + cells.

2. To obtain differentiated oligodendrocytes, continue to culture the expanded OPCs in the absence of PDGF AA and add 30 ng/mL thyroid hormone (T3).

   **Note:**

   1. An average of $2 \times 10^5$ mouse OPCs or $1 \times 10^6$ rat OPCs can be obtained from each flask immediately after shaking.

   2. Low cell density ($<1 \times 10^4$ cells/mm$^2$) can reduce the viability of OPCs.

---

## 4. Purification of OPCs by Immunopanning

In the late 1970s and early 1980s, various cell surface antigens were identified, which were specifically expressed in various cell types (13, 14). Many of these cell surface antigens are glycolipid antigens and are resistant to proteolytic cleavage. Thus, these antigens could be exploited for purifying the desired cell population, as they are not lost upon enzymatic treatment during cell dissociation. Positive and negative selection procedures can be combined to yield a highly purified cell population. The following method describes the procedure for isolating A2B5 + OPCs. This method is used most effectively for purifying rat OPC from the white matter such as optic nerves.

**4.1. Materials**

The following are commonly used antibodies for immunopanning oligodendrocyte lineage cells:

*4.1.1. Antibodies*

1. A2B5 (IgM; clone 105): detects a glycolipid antigen on the surface of OPCs and some neuronal cells (Eisenbarth et al. 13)

2. O4 (IgM; detects sulfatides on the surface of oligodendrocyte lineage cells at different developmental stages (14, 15). The timing of appearance of the O4 antigen differs in different species as shown below.
   - Chick: OPCs (similar to A2B5 + cells in the rodent)
   - Mouse: immature oligodendrocytes
   - Rat: late progenitor cells and premyelinating oligodendrocytes

3. O1 (IgM; 14): detects galactocerebroside on newly differentiated oligodendrocytes

4. Ran-2 (rat neural antigen-2; IgG2; (16): detects an antigen on flat astrocytes and is commonly used to negatively select astrocytes when purifying oligodendrocyte lineage cells

*4.1.2. Other Materials*

1. Petri dishes (100 mm diameter)
2. Glass coverslips and tissue culture dishes coated with 100 μg/mL poly-L-lysine.
3. A2B5 hybridoma cells (ATCC, CRL-1520, glycolipid antigen clone 105, IgM)
4. Antibody-coating buffer, 50 mM TrisHCl, pH 9.5, autoclaved
5. Goat IgG fraction to mouse IgM immunoglobulins (Cappel, cat. no. 55460)
6. S-MEM
7. Culture medium for hybridoma cells: DMEM and FBS
8. D-MEM for OPCs

**4.2. Harvesting Hybridoma Supernatants**

Grow A2B5 hybridoma cells in 10F-DMEM in 100 mm tissue culture dishes. Even though these cells are known to grow in suspension culture, they weakly adhere to tissue culture dishes. Therefore, the medium can be changed by simply aspirating the culture medium from the dishes with cells.

Change medium every 3 days.

When the cells reach confluence, passage the cells as follows. Aspirate the medium gently. Add 2–5 mL of fresh medium. Dislodge the cells from the surface of the dish with a Pasteur pipette. Split the suspended cells into 3–4 new 100 mm tissue culture dishes.

When the cells reach 70–80% confluence, discard old medium and add 8 mL of fresh culture medium. Culture for 3 days.

Transfer the medium into a sterile 50 mL conical tube. Centrifuge at 1,100 rpm (200 $g$) for 5 min to remove cellular debris. Transfer the supernatant to a fresh 50 mL tube. This is the A2B5 hybridoma supernatant and can be stored at –20°C.

**4.3. Preparation of Immunopanning Dishes**

Coat 100 mm Petri dishes (do not use tissue culture dishes) with 100 μg/12 mL of goat anti-mouse IgM antibodies diluted in antibody-coating buffer overnight at 4°C. Seal with parafilm.

On the following day, rinse the dishes 3 times with 6 mL of sterile PBS. Add 9 mL of A2B5 hybridoma supernatant in 10F-DMEM at 4°C overnight or at 37°C for more than 2 h. We seem to obtain better results when the incubation is done overnight.

Immediately before adding the cells, rinse the dishes 3 times with sterile PBS. Leave last rinse on the dish until ready to add the cells.

**4.4. Dissection and Dissociation of Brains from P2-4 Rats or Mice**

Prepare dissociated mixed glial cells as described above (*see* 2. Preparation of dissociated mixed glial cells for the following procedures). Suspend the cells in 6 mL/$2\times10^6$ cells of 10F-DMEM.

**4.5. Immunopanning**

1. Add the suspended cells to the washed immunopanning dishes. Incubate at 37°C for 30 min at 37°C. Swirl the dishes once after 15 min.

2. Aspirate the medium and non-adherent cells. Rinse vigorously (by rocking the dishes) with PBS two times to remove loosely adhering cells.

3. Add 2–3 mL of 10F-DMEM and pipette off the adhered cells using a Pasteur pipette. Instead of mechanical removal, gentle trypsinization (2–5 min in 0.1% trypsin) can be used at this step.

4. Collect the cells by centrifugation at 1,100 rpm (200 $g$) for 3 min.

5. Plate the cells on poly-L-lysine-coated surface.
   **Note:**
   1. This method can be extended to other antigens, such as NG2, expressed on the surface of OPCs. However, since NG2 is cleaved by trypsin, which is commonly used to dissociate cells, cells must be allowed to re-express NG2 at the surface. Usually, it takes 2–4 h for a sufficient amount of NG2 antigen to re-appear on the surface. Using papain instead of trypsin may reduce the extent of cleavage of NG2.

   2. In our hands, the purity of immunopanned cells from the spinal cord is usually higher than that from the brain. We get a large number of astrocytes mixed with OPC in A2B5-immunopanned cells from mouse brains.

## 5. Fluorescence-Activated Cell Sorting

We describe an example of FACS purification of OPCs from NG2DsRedBAC transgenic mice (Jackson Laboratory, Stock# 008241 STOCK Tg(Cspg4-DsRed.T1)1Akik/J), which express the red fluorescent protein DsRed under the regulation of the *ng2* gene. This technique can be applied to sorting cells from other transgenic lines or from mixed cells after immunofluorescently labeling them with an antibody to an OPC antigen. We describe the method we use for Becton-Dickinson FACSAria, but it can be adopted for other cell sorters.

### 5.1. Animals

Brain tissue from P2 to P3 mice (wild-type and NG2DsRedBAC transgenic mice) is used for the preparation of cells.

### 5.2. Materials

1. Glass coverslips coated with 100 μg/ml poly-L-lysine (PLL) as described above
2. Dissociation medium: S-MEM containing 0.05 mg/mL kynurenic acid (Fluka or Sigma).
3. Culture medium: 10F-DMEM and NBM containing 15 ng/mL PDGF-AA or other media, depending on the experiment.
4. 40-μm cell strainer (BD Falcon, cat. no. 352340).

### 5.3. Methods

1. Dissociate cells from cortical tissues of wild-type and transgenic mice as described above. Approximately $1–2 \times 10^6$ cells are obtained from one mouse brain.
2. Resuspend cells in DMEM-10F at a concentration of $1–3 \times 10^6$ cells/mL, and filter the cell suspension through a 40-μm cell strainer. Collect the filtered cell suspension in the sorting tubes appropriate for the cell sorter. Prepare several collection tubes with 200 μL of PBS containing 30% FBS to collect the sorted cells.

### 5.4. FACSAria Start-Up and Sorting

1. Run approximately 10,000 cells from wild-type cell preparation. Obtain a cell distribution plot for wild-type or control cells.
2. Run approximately 10,000 cells from tg cell preparation. Obtain a similar cell distribution plot from transgenic cells.
3. Compare two plots and set the sort gate.
4. First set the scatter gate as shown in **Fig. 12.2B**. Small cells on the forward scatter axis (FSC-A) are excluded, as they likely represent dead cells. Cells with high granularity (high numbers on the side scatter axis (SSC-A) are excluded, as they likely represent macrophages.

5. Set the red fluorescence (PE-A axis) gate as shown in **Fig. 12.2C**. It is important to empirically optimize this gate to obtain a 100% pure population after sorting.

6. An additional gate can be set to exclude cells that exhibit FITC fluorescence, as they are likely to be macrophages with autofluorescence.

7. Place the sample tube and collection tube in position and start sorting. The flow rate is determined by the sorting rate and the concentration of the input cell suspension. We use a sorting rate of 3,000 events/second to achieve adequate cell separation with good cell survival. The rate can be modified to fulfill different requirements. The expected yield is 80,000–120,000 cells per mouse brain.

8. After sorting, it is recommended that approximately 10,000 sorted cells be analyzed by the cells sorter to verify the sorted cells.

9. In addition, we routinely check the quality and purity of the sorted cells by immunofluorescence microscopy of the sorted cells plated on coverslips (*see* below).

**5.5. Recovery of the Sorted Cells**

1. To collect the sorted cells, transfer the suspension to a 15 mL tube and centrifuge at 1,100 rpm (200 *g*) for 3 min.

2. Add 1 mL of 10F-DMEM and resuspend the cells.

3. Plate the cells on top of PLL-coated coverslips and incubate cells in $CO_2$ incubator for 1–2 h.

4. After the cells have attached to the coverslips, change the medium to NBM containing PDGF-AA.

5. Culture the cells in the appropriate medium.
   **Note:** We routinely stain sorted cells 4 h after plating to check the purity of the sorted cells. Only batches of sorted cells that are 100% DsRed+ cells are used in experiments. Of these cells, 98.5% are NG2+, and 1.5% are O1+ (Zhu et al. 8). Note in **Fig. 12.2C** that in order to obtain 100% pure cells, a portion of cells with a low level of DsRed must be discarded. **Figure 12.2D and E** shows the results of NG2 and PDGF Rα staining of the sorted cells 1 day after sorting.

## References

1. Dawson MR, Polito A, Levine JM, and Reynolds R. NG2-expressing glial progenitor cells: an abundant and widespread population of cycling cells in the adult rat CNS. Mol Cell Neurosci 2003; 24: 476–488.

2. Nishiyama A. Polydendrocytes. NG2 cells with many roles in development and repair of the CNS. Neuroscientist 2007; 13:62–76.

3. Franklin RJ. Why does remyelination fail in multiple sclerosis? Nat Rev Neurosci 2002; 3(9):705–714.

4. Bunge MB, Bunge RP, Peterson ER, and Murray MR. A light and electron microscope study of long-term organized cultures of rat dorsal root ganglia. J Cell Biol 1967; 32(2):439–466.

5. Demerens C, Stankoff B, Logak M, Anglade P, Allinquant B, Couraud F, Zalc B, and Lubetzki C. Induction of myelination in the central nervous system by electrical activity. Proc Natl Acad Sci U S A 1996; 93: 9887–9892.

6. Kondo T and Raff M. Oligodendrocyte precursor cells reprogrammed to become multipotential CNS stem cells. Science 2000; 289: 1754–1757.

7. Belachew S, Chittajallu R, Aguirre AA, Yuan X, and Kirby M, Anderson S and Gallo V. Postnatal NG2 proteoglycan-expressing progenitor cells are intrinsically multipotent and generate functional neurons. J Cell Biol 2003; 161:169–186.

8. Zhu X, Bergles DE, and Nishiyama A. NG2 cells generate both oligodendrocytes and gray matter astrocytes. Development 2008; 135:145–157.

9. McCarthy KD and de Vellis J. Preparation of separate astroglial and oligodendroglial cell cultures from rat cerebral tissue. J Cell Biol 1980; 85:890–902.

10. Yang Z, Watanabe M and Nishiyama A. Optimization of oligodendrocyte progenitor cell culture method for enhanced survival. J Neurosci Methods 2005; 149:50–56.

11. Raff MC, Miller RH and Noble M. A glial progenitor cell that develops in vitro into an astrocyte or an oligodendrocyte depending on culture medium. Nature 1983; 303: 390–396.

12. Barres BA, Hart IK, Coles HSR, Burne JF, Voyvodic JT, Richardson WD, and Raff MC. Cell death and control of cell survival in the oligodendrocyte lineage. Cell 1992; 70: 31–46.

13. Eisenbarth GS, Walsh FS, and Nirenberg M. Monoclonal antibody to a plasma membrane antigen of neurons. Proc Natl Acad Sci U S A. 1979; 6(10): 4913–4917.

14. Sommer I, and Schachner M. Monoclonal antibodies (O1 to O4) to oligodendrocyte cell surfaces: an immunocytological study in the central nervous system. Dev Biol 1981; 83(2):311–327.

15. Bansal R and Pfeiffer SE. Novel stage in the oligodendrocyte lineage defined by reactivity of progenitors with R-mAb prior to O1 anti-galactocerebroside J Neurosci Res. 1992; 32:309–316.

16. Bartlett PF, Noble MD, Pruss RM, Raff MC, Rattray S, and Williams CA. Rat neural antigen-2 (RAN-2): a cell surface antigen on astrocytes, ependymal cells, Müller cells and lepto-meninges defined by a monoclonal antibody. Brain Res 1981; 204(2):339–351.

17. Shi J, Marinovich A, and Barres BA. purification and characterization of adult oligodendrocyte progenitor cells from the rat optic nerve. J Neurosci 1998; 18(12):4627–4636.

# Chapter 13

# Isolation of Microglia Subpopulations

## Makoto Sawada and Hiromi Suzuki

## Abstract

We describe a technique to isolate two major phenotypically different populations of microglia (type I and type II) in mixed brain culture. Type I microglia have surface markers for mature monocytic cells and produce microglia-specific cytokines. Type II microglia have cell-surface characteristics exclusively observed in immature bone marrow cells but not in mature monocytic cells, and do not express the hallmarks of mature microglia such as C3b receptors, CD14, or cytokine mRNAs.

**Key words:** Mouse, microglia subtype, antibody, flow cytometry, RT-PCR.

## 1. Introduction

Microglia, macrophage-like cells in the central nervous system (CNS), are multifunctional cells; they play important roles in the development, differentiation, and maintenance of neural cells via their phagocytic activity and production of enzymes, cytokines, and trophic factors (1). Microglial cells show a rather uniform distribution of cell numbers throughout the brain, with only minor populations in some brain regions. Their in situ morphologies, however, may vary markedly from elongated forms observed in apposition with neuronal fibers to spherical cell bodies with sometimes extremely elaborated branching. Furthermore, human fetal microglia display heterogeneity in phenotype identified by CD68 (2). This heterogeneity gave rise to the hypothesis that these cells are differentially conditioned by their microenvironment and, therefore, also display specific patterns of differential gene expression. For example, mRNAs of TNF-alpha, CD4, and Fc gamma receptor II are differentially expressed in microglia isolated from different areas of the brain (3). It has been reported that production of IL-12 (4) and histamine (5) is different in the subtypes of

L.C. Doering (ed.), *Protocols for Neural Cell Culture*, Springer Protocols Handbooks,
DOI 10.1007/978-1-60761-292-6_13, © Humana Press, a part of Springer Science+Business Media, LLC 1992, 1997, 2001, 2010

microglia. Furthermore, the subtypes of microglia show distinct acidification profiles by treatment with pepstatin A (6) and selective clearance of oligomeric beta-amyloid peptide (7).

In the present protocol, we indicate the method to prepare two distinct subpopulations of microglia from mouse mixed brain cultures.

## 2. Reagents and Equipment

### 2.1. Reagents

1. PKH26 (Zynaxis, Malvern, PA)
2. Diluent B (a phagocytic cell-labeling solution, Zynaxis, Malvern, PA)
3. FITC-labeled IgG (Cappel, West Chester, PA)
4. Anti-MacI antibody (1:100 dilution, BMA Biomedical, Switzerland)
5. Anti-ER-MP12 antibody (1:50 dilution, BMA Biomedical, Switzerland)
6. FITC-labeled anti-rat IgG (1:100 dilution, Cappel, West Chester, PA)
7. Phosphate-buffered saline (pH 7.2) containing 0.02% EDTA
8. Microglial cell culture medium: Eagle's MEM supplemented with 10% fetal calf serum, 5 mg/mL bovine insulin, and 0.2 % glucose

### 2.2. Animals

1. Neonatal C57BL/6 mice (Jackson Laboratory, Bar Habor, MA)

### 2.3. Equipment

1. Culture flask: F75 culture flask (Falcon 3024, Becton-Dickinson Japan, Tokyo, Japan)
2. Purification plate: noncoated plastic dishes (Falcon 1001, Becton-Dickinson Japan, Tokyo, Japan)
3. EPICS Elite fluorescence-activated cell sorter (Courter, Hialeah, FL)

## 3. Protocol

### 3.1. Preparation of Mixed Brain Cell Culture

Primary mixed glial cultures were prepared as a previous report (8).

1. A brain is isolated from a pup carefully.
2. The meninges are removed carefully in chilled Hanks' balanced salt solution.

3. Each brain is dissociated by nylon mesh (75 micron mesh) in chilled Hanks' balanced salt solution.

4. Dissociated cells are washed with Hanks' balanced salt solution.

5. The cell suspension is triturated with a fire-stretched glass pipette and plated in a F75 culture flask in 10 mL of microglial cell culture medium.

6. Medium was changed every 3 days.

### 3.2. Isolation of Two Types of Microglia

#### 3.2.1. Type I Microglia

1. One fraction of microglia is liberated by mechanical agitation at 150 rpm for 120 min in an orbital shaker at 37°C on the 14th day.

2. Liberated cell suspension is plated on a noncoated plastic dish (Falcon 1001 plate) for 30 min.

3. Remove nonattached cells with culture medium and wash gently with pre-warmed medium twice.

4. Attached cells (type I microglia) are liberated by a rubber policeman in 1 mL of microglial cell culture medium.

5. An aliquot (7 $\mu$L) of cell suspension is counted; an average yield of the cells was $1.7 \times 10^6$ cells/brain in a typical experiment.

6. The purity of type I microglia is determined by immunostaining with FITC-labeled IgG (used in 1:100 dilution); about 3000 cells in 100 $\mu$L medium is placed and attached on a glass cover slip briefly, then stained with diluted FITC-IgG for 10 min. In a typical experiment, the purity of type I microglia is more than 99%.

#### 3.2.2. Type II Microglia

Another fraction of microglia is liberated by trypsinization following mechanical agitation:

1. A mixed brain cell culture in a F75 flask is washed with 10 mL of phosphate-buffered saline (pH 7.2) twice.

2. Digest by 2 mL of trypsin–EDTA solution for 15 min at 37°C.

3. Reaction is terminated by adding 10 mL of microglial cell culture medium and flash by pipetting.

4. The cell suspension is triturated with a fire-stretched glass pipette and plated on a noncoated plastic dish (Falcon 1001 plate) for 30 min.

5. Remove nonattached cells with culture medium and wash gently with pre-warmed medium five times.

6. Attached cells (type II microglia) are liberated by a rubber policeman in 1 mL of microglial cell culture medium.

7. An aliquot (7 μL) of cell suspension is counted; an average yield of the cells was $1–2 \times 10^6$ cells/brain in a typical experiment.

8. The purity of type II microglia is determined by immunostaining with FITC-labeled IgG (used in 1:100 dilution); about 3000 cells in 100 μL medium is placed and attached on a glass coverslip briefly, then stained with diluted FITC-IgG for 10 min. In a typical experiment, the purity of type II microglia was about 90–95%.

**3.3. Fluorescent Dye Staining**

Astrocytes, microglia, and mixed glial cultures were stained with the fluorescent dye PKH26 (Zynaxis, Malvern, PA) as follows:

1. Cells in 10 cm id. plastic culture dishes were incubated with PKH26 staining solution containing 10 mM of PKH26, 50% Diluent B (a phagocytic cell-labeling solution, Zynaxis, Malvern, PA), and 50% culture medium for 15 min.

2. Washed with 10 mL serum-containing medium for three times.

3. PKH26-stained cells were harvested using a rubber policeman in 2 mL of ice-cold phosphate-buffered saline (pH 7.2) containing 0.02% EDTA.

4. Washed with 5 mL ice-cold phosphate-buffered saline (pH 7.2) by centrifugation three times.

5. Analyzed with an EPICS Elite fluorescence activated cell sorter (Courter, Hialeah, FL).

**3.4. Immunostaining**

Type I and type II microglia were labeled with anti-MacI antibody (1:100 dilution, BMA Biomedical, Switzerland) or anti-ER-MP12 antibody (1:50 dilution, BMA Biomedical, Switzerland) at 0°C for 30 min, stained with FITC-labeled anti-rat IgG (1:100 dilution, Cappel, West Chester, PA), then analyzed with an EPICS Elite fluorescence-activated cell sorter (FACS) (Courter, Hialeah, FL).

*3.4.1. RNA Preparation and RT-PCR*

PKH26-stained microglia were sorted into two fractions: one with a high forward scattering (FS) intensity (type I microglia) and the other with a low FS intensity (type II microglia). Total RNA was extracted from approximately $1 \times 10^5$ microglia of both fractions by a modified acid phenol–guanidine–chloroform method (9). PCR was performed as described previously (10) with the following primers: IL-1 β sense, 5′-ATGGCAACTGTTCCTGAACTCAACT, antisense, 5′-CAGGACAGGTATAGATTCTTTCCTTT; class I MHC sense, 5′-ACATGGAGCTTGTGGAGACC, antisense, 5′-AGTCGGA-GAGACATTTCAGAGC; CD14 sense, 5′-CCTTAGTCACAA-TTCACTGCGG, antisense, 5′-ATCAGGGGTCAAGTTTGC;

and β-actin sense, 5′-.GTGGGCCGCTCTAGGCACCAA, antisense 5′-CTCTTTGATGTCACGCACGATTTC. Aliquots of PCR product were subjected to electrophoresis in 2% agarose gels in TBE buffer. The gels were then stained with ethidium bromide and photographed.

## 4. Typical Protocol Results

### 4.1. Mixed Glial Culture

Mixed glial culture from the neonatal mouse brain is a model system for glial cell differentiation; the three major types of glial cells, astrocytes, oligodendrocytes, and microglia, sequentially differentiate and increase in number as they do in vivo (11). At 10–14 days in vitro (DIV) in normal mouse mixed glial cultures, many phase-bright round cells appeared on top of the astrocyte monolayer; and these cells show a burst-forming proliferation (**Fig. 13.1A**), which paralleled microglial growth in vivo (11). There are two types of microglia in the 14 DIV mixed glial culture; one is positive for Mac I antibody staining and shows round shape, and another is positive for ER-MP12 antibody staining and shows ramified shape.

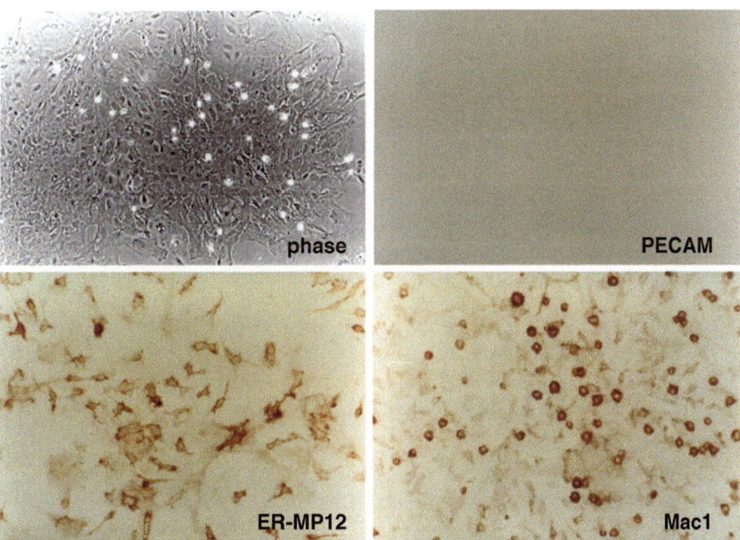


Fig. 13.1. Mixed glial cultures at 14 days in vitro (DIV) from normal brains had many phase-bright cells on top of the astrocyte layer (**A**). Mixed glial cultures were stained with PE-CAM (**B**), ER-MP12 (**C**), and Mac I (**D**).

### 4.2. Type I Microglia

Since the phase-bright round cells could be easily liberated from the astrocyte monolayer by mechanical agitation (8), we isolated these cells from mixed glial culture, stained them with rhodamine-labeled IgG to examine Fc receptor binding, which is a hallmark of microglia, and counted rhodamine-positive cells. More than 99% of the liberated cells from normal mouse mixed glial culture were Fc receptor-positive microglia.

### 4.3. Type II Microglia

Fc receptor-positive microglia were also observed in an astrocyte monolayer, and these cells could be discriminated and purified from astrocytes and other remaining cells by trypsinization followed by allowing the cells to attach to noncoated plastic dishes. We found similar numbers of Fc receptor-positive microglia in the collected adherent cells.

### 4.4. FACS

We analyzed characteristics of microglia in whole cells of the mixed glial cultures from mouse brains with a FACS with a fluorescent dye specific for phagocytic cells, PKH26 (12, 13). PKH26 stained microglia efficiently; purified microglia were stained with intensity at least two orders higher than purified astrocytes, which is the major cell population in mixed glial cultures (indicated by an arrow in **Fig. 13.2**). PKH26-stained microglia from normal mouse brains were divided into two populations by the intensity of FS (**Fig. 13.2**). The type I microglia liberated from the astrocyte layer by mechanical agitation were enriched in high-FS intensity cells, while those which were not liberated by agitation but separated from astrocytes by trypsinization and non-coated plastic dish treatment (type II) were enriched in low-FS intensity cells. When mixed glial cultures from normal mice were treated with human recombinant M-CSF, the high-FS population increased in number while the low-FS population decreased. On the other

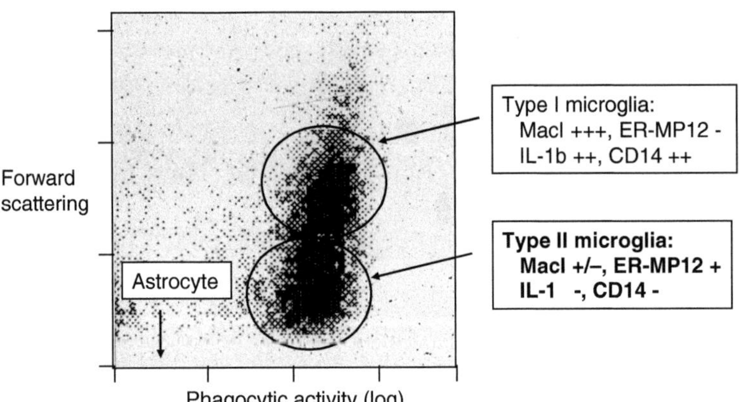


Fig. 13.2. Fluorescence flow cytometric analysis of PKH26-labeled microglia. Two-dimensional histograms of PKH26-stained mixed glial cultures from normal brain indicating the existence of two populations of microglia; high-FS and low-FS populations. Histogram of microglia purified by mechanical agitation (type I) indicating the similar FS intensity of high-FS population of microglia in mixed glial cultures from normal brain. That of microglia purified by trypsinization (type II) indicating the similar FS intensity of low-FS population of microglia in mixed glial cultures from normal brain.

hand, these two subpopulations of microglia had similar phagocytic capacities because both populations had the same fluorescent intensity following PKH26 incorporation (**Fig. 13.2**).

To characterize the differences between the two subpopulations of microglia further, we compared surface antigen expression on both types of cells. Type I microglia were stained with an anti-Mac1 antibody, which recognizes complement C3b receptor on the surface of cells of the monocyte lineage (14). Type II microglia were stained faintly with this antibody. On the other hand, staining with ER-MP12 and ER-MP20 antibodies gave the opposite results.

Distinct phenotypes of the two microglial subpopulations were also observed at the level of mRNA expression. We fractionated the high- and low-FS microglial subpopulations by sorting with a FACS and analyzed these cells by RT-PCR (10). The sorted type I microglia showed comparable levels of interleukin-1 β, class I MHC, and CD14 mRNA expression. The sorted type II microglia, however, did not express interleukin-1 β or CD14 mRNAs, and expressed small amounts of class I MHC mRNA (**Fig. 13.3**). Contamination with other types of glial cells was negligible because both types of microglia showed phagocytic activity indicated by PKH26 fluorescent dye staining (data not shown).

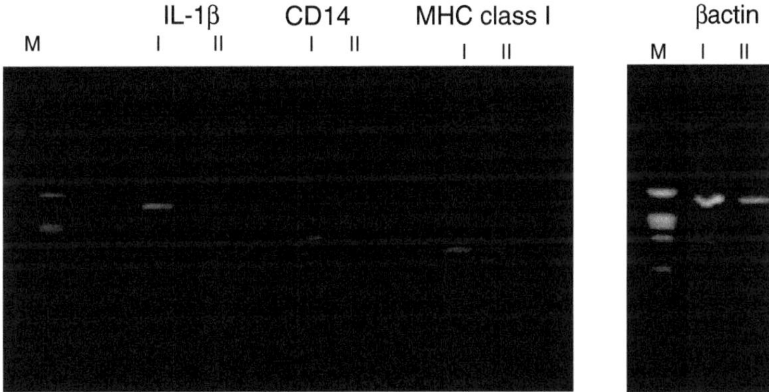


Fig. 13.3. Distinct mRNA expression in microglia subpopulations. mRNA expression of Interleukin 1 β (IL-1 β, lane 2 and 3), CD14 (lane 4 and 5), class I MHC (lane 6 and 7), and β -actin (lane 9 and 10) were examined. Lane 2, 4, 6, and 9 indicate RT-PCR products from type I microglia. Lane 3, 5, 7, and 10 indicate those from type II microglia. lane 1 and 8, Hae III digested pUC119 as a molecular weight marker.

## 5. Conclusions

Type II microglia did not express CD14 mRNA, which is a receptor for LPS–LPS-binding protein complex and one of the major signal transduction molecules involved in LPS stimulation (15), suggesting they might be less responsive to LPS stimulation than type I microglia. Furthermore, type II microglia express less class I MHC mRNA than type I. From these observations, the two distinct types of microglia seemed to be functionally different. Further studies should be performed to elucidate the in vivo functions of both types of microglia. We have found that type II, but not type I, microglia were observed in the M-CSF-deficent *op/op* mice. A previous study showed that post-traumatic microglial proliferation was not observed in *op/op* mouse brain (16), suggesting type I microglia may have a role in degeneration/repair of brain injury.

The type II microglia have a novel surface characteristic in that they are immunoreactive for ER-MP 12 and 20, antibodies to which react with immature bone marrow cells but not mature monocytes, macrophages, or other tissue-resident macrophages (17, 18). We previously reported that ER-MP12 and 20 positive microglia exist in the brain of young-adult mice (19). These findings do not support the current hypothesis (20) that microglia arise from blood monocytes, which are thought to be a source of brain macrophages observed in blood–brain barrier rupture.

## Acknowledgments

This study was supported in part by Grants-in-Aid for Scientific Research from the Japanese Ministry of Education, Science and Culture, from the Japanese Ministry of Health and Welfare.

## References

1. Sawada, M., Suzumura, A. and Marunouchi, T. 1995. Cytokine network in the central nervous system and its roles in growth and differentiation of glial and neuronal cells. Int. J. Dev. Neurosci. 13:253–264.

2. Rezaie, P., Patel, K. and Male, D.K. 1999 Microglia in the human fetal spinal cord; patterns of distribution, morphology and phenotype. Dev. Brain. Res. 115:71–78.

3. Ren, L., Lubrich, B., Biber, K. and Gebicke-Haerter, P.J. 1999 Mol. Brain. Res. 65:198–205.

4. Suzumura, A., Sawada, M. and Takayanagi, T. 1998. Production of interleukin-12 and expression of its receptors by murine microglia. Brain Res. 787:139–142.

5. Katoh, Y., Niimi, M., Yamamoto, Y., Kawamura, T., Morimoto-Ishizuka, T., Sawada, M., Takemori, H. and Yamatodani, A. 2001. Neurosci. Lett. 305:181–184.

6. Okada, M., Irie, S., Sawada, M., Urae, R., Urae, A., Iwata, N., Ozaki, N., Akazawa, K. and Nakanishi, H. 2003. Pepstatin A induces extracellular acidification distinct from aspartic protease inhibition in microglial cell lines. Glia. 43:167–174.

7. Shimizu, F., Kawahara, K., Kajizono, M., Sawada, M. and Nakayama, H. 2008. Interleukin-4-induced selective clearance of oligomeric beta-amyloid peptide1-42 by rat primary type-2 microglia. J. Immunol. 181:6503–6513.

8. Sawada, M., Suzumura, A., Yamamoto, H. and Marunouchi, T. 1990. Activation and proliferation of the isolated microglia by colony stimulating factor-1 and possible

involvement of protein kinase C. Brain Res. 509:119–124.

9. Sawada, M., Suzumura, A. and Marunouchi, T. 1992. Down regulation of CD4 expression in cultured microglia by immunosuppressants and lipopolysaccharide. Biochem. Biophys. Res. Commun. 189: 869–876.

10. Sawada, M., Itoh, Y., Suzumura, A. and Marunouchi, T. 1993a. Expression of cytokine receptors in cultured neuronal and glial cells. Neurosci. Lett. 160:131–134.

11. Suzumura, A., Mezitis, S. G. E., Gonatas, N. K. and Silberberg, D. H. 1987. MHC antigen expression on bulk isolated macrophage-microglia from newborn mouse brain: induction of Ia antigen expression by g-interferon. J. Neuroimmunol. 15:263–278.

12. Horan, P. K. and Slezak, S. E. 1989. Stable cell membrane labelling. Nature. 340:167–168.

13. Melnicoff, M. J, Morahan, P. S., Jensen, B. D., Breslin, E. W. and Horan, P. K. 1988. In vivo labeling of resident peritoneal macrophages. Leukoc Biol. 43:387–397.

14. Griffin, F. J., Bianco, C. and Silverstein, S. C. 1975. Characterization of the macrophage receptor for complement and demonstration of its functional independence from the receptor for the Fc portion of immunoglobulin G. J. Exp. Med. 141:1269–1277.

15. Lee, J. D., Kato, K., Tobias, P., Kirkland, S. T. N. and Ulevitch, R. J. 1992. Transfection of CD14 into 70Z/3 cells dramatically enhances the sensitivity to complexes of lipopolysaccharide (LPS) and LPS binding protein. J. Exp. Med. 175: 1697–1705.

16. Raivich, G., Moreno, F. M., Moller, J. C. and Kreutzberg, G. W. 1994. Inhibition of post-traumatic microglial proliferation in a genetic model of macrophage colony-stimulating factor deficiency in the mouse. Eur. J. Neurosci. 6:1615–1618.

17. Leenen, P. J., Melis, M., Slieker, W. A. and Van, E. W. 1990a. Murine macrophage precursor characterization. II. Monoclonal antibodies against macrophage precursor antigens. Eur. J. Immunol. 20:27–34.

18. Leenen, P. J., Slieker, W. A., Melis, M. and Van, E. W. 1990b. Murine macrophage precursor characterization. I. Production, phenotype and differentiation of macrophage precursor hybrids. Eur. J. Immunol. 20:15–25.

19. Tanaka, M., Marunouchi, T. and Sawada, M. 1997 Expression of Ly-6C on microglia in the developing and adult mouse brain. Neurosci. Lett. 239:17–20.

20. Ling, E. A. and Wong, W. C. 1993. The origin and nature of ramified and amoeboid microglia: a historical review and current concepts. Glia. 7:9–18.

# Chapter 14

# Microglia from Progenitor Cells in Mouse Neopallium

**Sergey Fedoroff and Arleen Richardson**

## Abstract

This chapter outlines a procedure for initiation of mouse neopallium (cerebral cortex and its underlying white matter) cell cultures and their development into nearly pure microglia cultures. The cytokine CSF-1 is required for the development, survival, and differentiation of microglia. In cultures, astroglia secrete CSF-1 into the medium (Hao et al. 1990, J. Neurosci. Res. 27, 314–323). Therefore, a simple protocol is described, originally developed by 12 (Hao et al. 1991, Int J. Dev. Neurosci. 9, 1–14), which utilizes CSF-1-secreting astroglia cultures for production of better than 99% pure cultures of microglia.

**Key words:** Mouse, neopallium, CSF-1, conditioned media.

## 1. Introduction

Brains of newborn rats or mice are generally used as the source of tissue for glial cultures. Only about 1% of cells survive the cell disaggregation process and culture environment, and neurons that survive die within the first few days of culturing. Such cultures contain progenitor cells and glia cells in different stages of differentiation (1, 2). Terminally differentiated glia cells, which can no longer divide, are overgrown by the proliferating, immature cells. How the cultures will develop and which cell types (astroglia, oligodendroglia, ependymal cells, or microglia) will predominate depend on the culture medium and the physical conditions under which the cells are grown. The medium can be modified either by changing its chemically defined components or by adding or deleting serum. It is also possible to add growth factors and/or cytokines either to the culture medium in pure recombinant form or as medium conditioned by cells that produce and secrete growth factors/cytokines into the medium (the latter is considerably cheaper).

L.C. Doering (ed.), *Protocols for Neural Cell Culture*, Springer Protocols Handbooks,
DOI 10.1007/978-1-60761-292-6_14, © Humana Press, a part of Springer Science+Business Media, LLC 1992, 1997, 2001, 2010

The addition of cytokines to cultures can have a dramatic effect on the morphology and function of cells. It should be noted that cytokines and growth factors may affect more than one cell type and may initiate variable effects in different cell types. The growth factors/cytokines may interact with other factors in the medium synergistically, additively, or in an inhibitory way. It is important to remember that the half-life of cytokines is short. However, the cytokines/growth factors can be added to the culture in microcapsules, which release the factors at a constant rate over a long period of time, thus assuring a constant concentration of a given factor to culture medium over long periods of time (3–5).

## 2. Protocol

### 2.1. Neopallium Cultures: Preparation of Newborn Mouse for Dissection of Neopallium

#### 2.1.1. Materials

1. Mouse, newborn.
2. Forceps (2) curved, 12 cm.
3. Forceps (1) straight, blunt, 14 cm.
4. Forceps (2) Irex no. 5.
5. Forceps (1) fine, curved, 10 cm.
6. Scissors (1) large, 15 cm.
7. Scissors (1) straight, 10 cm.
8. Scissors (1) curved, 10 cm.
9. Needles, 25-gauge, sterile.
10. High-sucrose phosphate buffer.
11. Wax dissecting dish (1) sterile.
12. Petri dishes (3) glass, sterile, 100 mm.
13. Ethanol, 70% in Coplin jar.

#### 2.1.2. Preparation of Materials

1. Place several discrete drops of high-sucrose phosphate buffer around the periphery of the Petri dish. (One cerebral hemisphere will be placed in each drop.) In the center of the Petri dish, place a number of drops of high-sucrose phosphate buffer in which the neopallia will be collected.
2. Sterilize all instruments. When the instruments have been sterilized, place them into two sterile glass Petri dishes and keep the ends covered with a lid.

#### 2.1.3. Preparation of Newborn Mouse

1. Deeply anesthetize a newborn mouse with Halothane$^{TM}$ or Metofane$^{TM}$.
2. Using large straight forceps, pick up the mouse and briefly submerge it in the container of 70% alcohol.

3. Using the large scissors, decapitate the mouse so that the head falls onto the wax dissecting dish.

4. Using a sterile needle, pin the head through the nose, dorsal side up, to the wax dish (**Fig. 14.1A**).

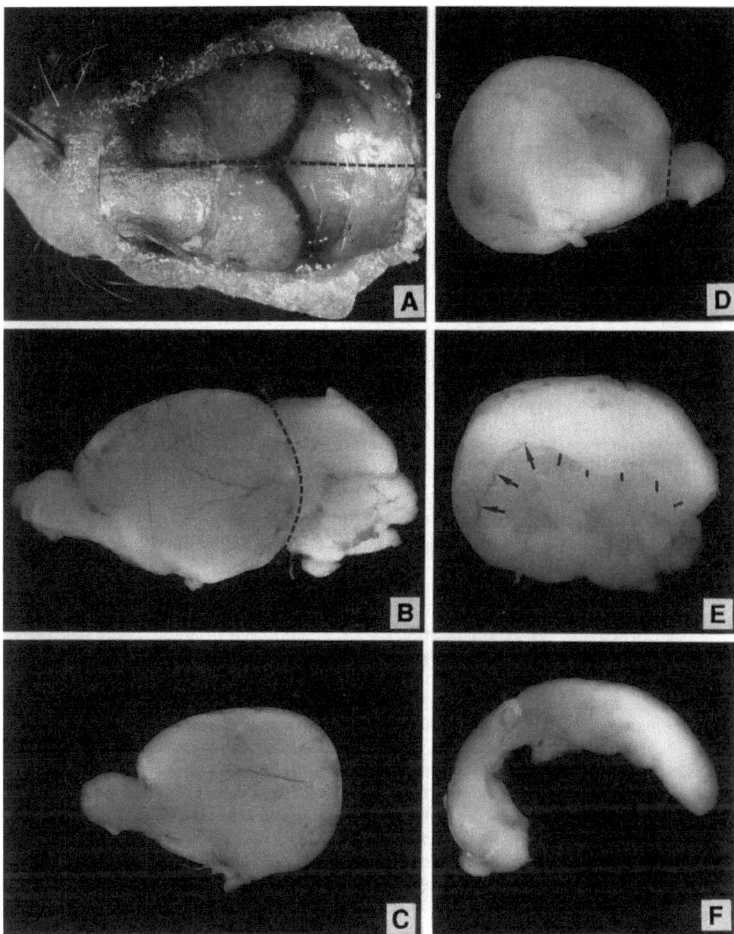

Fig. 14.1. Dissection of newborn mouse neopallium. (**A**) Head of mouse pinned through snout, using sterile needle, skin removed. *Dotted line* indicates midline cut through skull and brain. (**B**) Half of brain removed from skull, showing cerebral hemisphere, cerebellum, and olfactory bulb. *Dotted line* indicates cut for removal of cerebellum. (**C**) One cerebral hemisphere after removal of cerebellum (dorsal view). (**D**) The same cerebral hemisphere seen in (**C**), with *dotted line* indicating cut for removal of olfactory bulb. (**E**) Bow-shaped hemisphere after the removal of basal ganglia. *Arrows* depict the site of dissection to remove hippocampus. (**F**) Isolated neopallium.

**2.2. Dissection of the Cerebral Hemispheres**

1. Using sterile curved forceps and curved scissors, remove the skin over the dorsal surface of the skull. This is easily accomplished by beginning the cut ventral to the ears, continuing toward the snout, then across the head just above the eyes, and finally back toward the neck, and, at the same time, by gently pulling the skin from the skull with forceps (**Fig. 14.1A**).

2. Using straight sterile scissors, divide the skull and cerebral hemispheres from posterior to anterior, i.e., by bringing the bottom blade of the scissors up through the brain. Do not cut down initially through the skull (**Fig. 14.1A**) because the pressure will mutilate the soft brain tissue.

3. Remove the cerebral hemispheres:
   a. Using a pair of sterile curved forceps, grasp the snout of the mouse and squeeze gently. This serves to hold the head in place as well as to open the skull.

   b. Next, using the fine curved forceps, lift the brain out of the skull and place each hemisphere into a drop of medium in the sterile Petri dish.

   c. Using the Irex forceps, remove the delicate, almost transparent meninges. Begin at the olfactory bulbs and peel meninges off, by pulling gently posteriorly. Try to avoid tearing the meninges.

   d. Pinch off the cerebellum (**Fig. 14.1B**) and olfactory bulb from the cerebral hemisphere (**Fig. 14.1C, D**), using the Irex forceps.

### 2.3. Dissection of the Neopallium

1. Place the cerebral hemisphere with the ventral surface facing up (**Fig. 14.1D**). Scoop out the basal ganglia, using the fine forceps. With the removal of the basal ganglia, a bowl-shaped hemisphere is left (**Fig. 14.1E**).

2. Remove the hippocampus using the Irex forceps (**Fig. 14.1E**). The remaining part of the hemisphere is neopallium (**Fig. 14.1F**).

3. Transfer each neopallium into the drops in the center of the Petri dish.

4. Cut each neopallium into two or three fragments.

### 2.4. Preparation of Neopallial Cultures

#### 2.4.1. Materials

1. Puck's balanced salt solution (Puck's BSS).

2. Medium (30 mL), modified Eagle's minimum essential medium (mMEM) containing 5% horse serum (HS) (Invitrogen, Canada, Burlington, ON).

3. Tissue culture flask (1) 75 cm$^2$.

4. Beaker (1) covered with 75 μm Nitex$^{TM}$ mesh.

5. Centrifuge tube (1) 50 mL, sterile.

6. Pasteur pipette (1) sterile, with bent tip.

7. Rubber bulb (1) to fit Pasteur pipette.

#### 2.4.2. Disaggregation of Neopallia

1. Place fragments of neopallia on top of 75 μm Nitex mesh, which is stretched over a beaker.

2. Place a few drops of growth medium over the neopallia to keep them moist.

3. Gently roll the fragments of neopallia over the mesh using the back side of a curved Pasteur pipette and, at the same time, add medium that has been taken up in the curved Pasteur pipette. Avoid adding air bubbles. Add 5 mL medium/brain to the mesh, and continue rolling the fragments until the neopallia are completely disaggregated and the cells have passed through the mesh. Rinse the mesh with the medium.

4. Remove the mesh and transfer the cells in the beaker to a centrifuge tube. Use a pipette to resuspend the cells.

5. Determine the cell number:
   (a) Prepare a sample from the cell suspension for counting in a hemocytometer. Use 0.05 or 0.1 mL 0.3% Nigrosin (Sigma Aldrich, Oakville, ON) as a viability indicator (viable cells exclude the dye) and 0.2 or 0.4 mL cell suspension, respectively.

   (b) Fill the hemocytometer and count the number of viable cells (unstained) in at least four large corner squares of one chamber.

   (c) Calculate the number of viable cells per milliliter.

*2.4.3. Preparation of Neopallial Cultures*

1. Before plating, make sure that the cells are well suspended. Plate $3–5 \times 10^6$ viable cells in a 75 cm$^2$ tissue culture flask in a final volume of 12 mL mMEM with 5% HS.

2. Incubate culture flasks in a 37°C highly humidified atmosphere containing 5% $CO_2$ in air.

3. Feed cultures every 2–3 days for 10 days.

**2.5. Comments**

1. Instead of flasks, Petri dishes of various sizes can be used. A convenient way to prepare neopallial cultures is to place coverslips into a nonculture Petri dish and to plant cells as described above. The coverslips can be removed from the Petri dish at any stage of culturing and used for sequential analyses. Cells grown on coverslips are convenient for immunocytochemistry, histochemistry, morphometry, or electron microscopy.

2. When large numbers of neopalial cells are required for chemical analysis or for conditioning medium, large-sized flasks, roller bottles, or large flasks filled with beads can be used to increase the surface area.

## 3. Microglia Cultures

**3.1. Materials**

1. Culture medium, mMEM with 5% HS (horse serum)

2. LM cell-conditioned medium.

3. Hanks' balanced salt solution (HBSS).

### 3.2. Preparation of Highly Enriched Microglia Cultures

1. Prepare neopallial cell cultures as described in **Section 2.4**, and incubate 10–12 days.

2. Examine the cultures under the microscope to ensure that they have reached confluency, but are not too dense.

3. Wash and re-feed the cultures, and incubate for another 10–12 days without medium change.

4. Remove the medium and wash cultures with HBSS. Repeat the washing until all dead and floating cells have been removed.

5. Re-feed the cultures with mMEM with 5% HS, and add 20% LM cell-conditioned medium, which contains CSF-1, or add 1000 IU/mL recombinant CSF-1.

6. Incubate cultures at 37°C in a highly humidified atmosphere containing 5% $CO_2$ in air.

### 3.3. Comments

The first indication of the formation of microglia is the appearance of phase-dark cells with poorly defined, thick, cytoplasmic processes abutting the underlying neopallial cells. A few days later, the phase-dark cells show cytoplasmic vacuoles and become attached to the substratum. Subsequently, they assume a morphology characteristic of amoeboid cells (vacuoles, irregular shape, and pseudopodia), and eventually form process-bearing cells that resemble parenchymal, ramified microglia. In the presence of CSF-1, microglia proliferate (6).

### 3.4. Subculturing of Microglia

The cultures of microglia (**Fig. 14.2**) can be used directly or subcultured. Microglia attach firmly to the substratum. To remove them from the substratum, microglia can be treated with trypsin or cold versene.

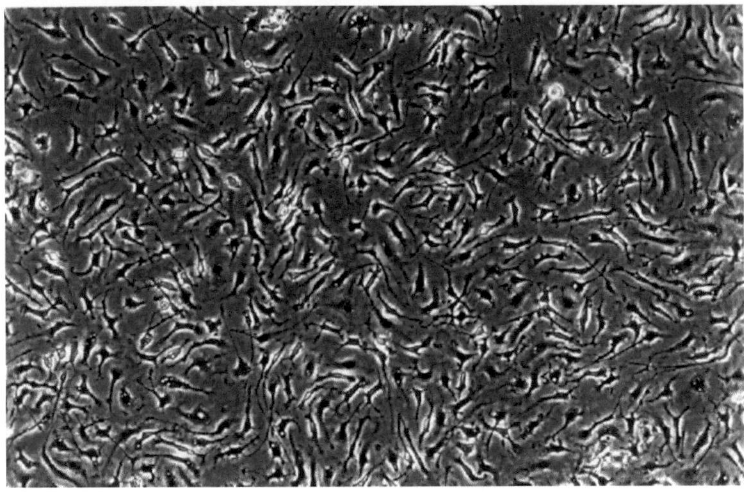

Fig. 14.2. Microglia from mouse neonatal neopallial cell cultures.

1. Trypsin: Treat cultures with 0.25% trypsin for 10 min at room temperature. The trypsin can then be inhibited by the addition of medium containing serum or soy bean trypsin inhibitor. Transfer the resulting disaggregated cell suspension into another culture vessel.

2. Cold versene: Treating cultures with cold versene is a gentler and more effective way to remove microglia from the substratum, without altering their cell surface antigenic characteristics.
   a. Cool a culture of microglia at 4°C for 10 min.
      **Note**: Microglia round up when subjected to low temperature, thus decreasing the area of the cell surface attached to the plastic substratum.
   b. Gently triturate culture 30 times with a Pasteur pipette.
   c. Transfer the cell suspension into a test tube and centrifuge at 200$g$ for 5 min.
   d. Remove the versene solution and resuspend microglia in fresh medium.
   e. Transfer the resulting disaggregated cell suspension into another culture vessel.

## 4. Comments

1. Gebicke-Haerter et al. (7) reported that bacterial lipopolysaccharide (LPS), present in culture medium, inhibits the formation of brain macrophages. It should be noted that most commercially available fetal bovine serum contains some LPS (8, 9) and thus may inhibit the formation of microglia. It is therefore advisable to use horse serum, which normally has no detectable LPS. Moreover, it is advisable to use the embryos of C3H/HeJ mice, which are resistant to LPS (10).

2. This method for isolation of microglia from mouse neopallial cell cultures can also be used to obtain microglia from rats or humans.

3. To avoid pH change, use media in which the buffer calibrates with the atmospheric air $CO_2$, e.g., high-sucrose phosphate buffer, HBSS, or Hibernate A medium.

## 5. Reagents

1. Medium: MEM is modified (mMEM) to contain a fourfold concentration of vitamins, a double concentration of amino acids except glutamine, and 7.5 mM glucose (Invitrogen

Canada, custom media, Burlington, ON). Before using the modified MEM for culturing, 5.2 mL 1 $M$ sodium bicarbonate and 2.5 mL 200 mM glutamine are added to 200 mL mMEM, and the pH is adjusted to 7.2 by bubbling 5% $CO_2$ into the medium. The prepared modified MEM is supplemented with 5% HS.

2. Puck's BSS:
   a. Composition:
      NaCl 8.00 g/L
      KCl 0.40 g/L
      $Na_2HPO_4 \cdot 2H_2O$ 0.06 g/L
      $KH_2PO_4$ 0.06 g/L
      Glucose 1.00 g/L
      1% Phenol red 2.00 mL

   b. Procedure for making 1 L Puck's BSS:
      i. Dissolve all components, in order (except phenol red), in a beaker containing 400 mL triple-distilled water.
      ii. Add phenol red to a 1 L volumetric flask. Add other components to the flask. Rinse beaker used for solutions with triple-distilled water, and add to the volumetric flask. Add enough water to make the volume up to 1 L.
      iii. Stir on magnetic stirrer for 30 min (minimum time).
      iv. Filter through 0.22 μm filter.
      v. Label and store stock solution at 4°C.

3. High-sucrose phosphate buffer: This maintenance medium is for freshly dissected parts of the brain. It contains physiological ion concentrations (Na+ and K+), some buffering action, and an energy source. Sucrose is the best compound for adjustment of osmolality, because it is neither taken up by the cells nor extracellularly degraded.

   Component g/L

   NaCl 8.00

   KCl 0.40

   $Na_2HPO_4$ 0.024

   $KH_2PO_4$ 0.03

   Glucose 0.90

   Sucrose 20.0

   a. Dissolve chemicals in 1 L high-quality distilled water.

   b. Sterilize by filtration through a 0.22 μm filter, aliquot, and store at 4°C.

### 5.1. Preparation of LM Cell-Conditioned Medium

For the preparation of conditioned medium containing CSF-1, LM cells (American Type Culture Collection, cat. no. CCL-1.2) are used. The advantage of using LM cells is that they can grow well in suspension and in serum-free medium. By using suspension flasks, a large amount of conditioned medium can easily be prepared. LM cells can also be grown in large stationary flasks or roller bottles. The cells are grown in Medium 199 with 0.5% Bacto-peptone, without the addition of serum.

Suspension cultures are usually inoculated with $6 \times 10^5$ cells/mL, and grown for a week or 10 days without feeding, or until they reach about $2 \times 10^6$ cells/mL. By then, the culture medium appears orange in color and cloudy because of the high cell concentration. The cells are allowed to settle and the supernatant is removed and centrifuged at $200g$ for 15 min. The supernatant medium is then filtered through a stack of 1.2, 0.8, and 0.45-$\mu$m pore-size filters, and sterilized by filtration through a 0.2-$\mu$m pore-size filter. The sterile conditioned medium is aliquoted, labeled, and stored at $-80°C$. The conditioned medium can also be concentrated (Amicon stirred ultrafiltration cells, UF discs YM-10; Millipore, Billerica.Ma). It is important that an aliquot of the conditioned medium be used for determination of the CSF-1 content in the medium, using the radioreceptor assay (11, *as modified by* 6). The biological activity of CSF-1 can be determined by using the sodium XTT colorimetric population growth assay of 5/10.14 and DA1.K cells (12).

### 5.2. Nitex Mesh-Covered Beakers

*See* Chapter 11 for suppliers of Nitex mesh and the preparation of Nitex mesh-covered beakers.

### 5.3. Wax Dissecting Dishes

#### 5.3.1. Preparation

1. Break nine sticks of Kerr Brand blue inlay casting wax, regular type II (Sybron, Emeryville, CA) into pieces. Place wax pieces in 100 mm Petri dishes.
2. Place the dishes containing the wax in a pan and sterilize by dry heat in an oven for 2 h at 250°F.
3. The dishes may be allowed to cool overnight in the oven. If removed while hot, be careful not to tilt the dishes so that wax reaches the cover of the Petri dish.
4. When cool, wrap wax dishes in aluminum foil and store until needed.

#### 5.3.2. Cleaning Used Wax Dissecting Dishes

1. Clean the wax dissecting dishes by rinsing with water. Dry with a soft, lint-free tissue.
2. Wash the wax dishes with 70% ethanol and dry with a soft, lint-free tissue. Close the lids and sterilize as above for new wax dishes.

## References


1. Alliot, F., Lecain, E., Grima, B., and Pessac, B. (1991), Microglial progenitors with a high proliferative potential in the embryonic and adult mouse brain. *Proc. Natl. Acad. Sci., USA,* **88,** 1541–1545.
2. Fedoroff, S., Zhai, R., and Novak, J. P. (1997), Microglia and astroglia have a common progenitor cell. *J. Neurosci. Res.* **50,** 477–486.
3. Maysinger, D., Jalsenjak, I., and Cuello, C. (1992), Microencapsulated nerve growth factor: effects on the forebrain neurons following devascularizing cortical lesions. *Neuroscience Lett.* **140,** 71–74.
4. Maysinger, D., Berezovskaya, O., and Fedoroff, S. (1996a), The hematopoietic cytokine colony stimulating factor 1 is also growth factor in the CNS: (II) Microencapsulated CSF-1 and LM-10 cells as delivery systems. *Exp. Neurol.* **141,** 47–56.
5. Maysinger, D., Krieglstein, K., Filipovic-Grcic, J., Sendtner, M., Unsicher, K., and Richardson, P. (1996b), Microencapsulated ciliary neurotrophic factor physical properties and biological activities. *Exp. Neurol.* **138,** 177–188.
6. Hao, C., Guilbert, L. J., and Fedoroff, S. (1990), Production of colony-stimulating factor-1 (CSF-1) by mouse astroglia in vitro. *J. Neurosci. Res.* **27,** 314–323.
7. Hao, C., Richardson, A. and Fedoroff, S. (1991), Macrophage-like cells originate from neuroepithelium in culture: characterization and properties of the macrophage-like cells. *Int. J. Dev.Neurosci.* **9,** 1–14.
8. Gebicke-Haerter, P. J., Baker, J., Schobert, A., and Northoff, H. (1989), Lipopolysaccharide-free conditions in primary astrocyte cultures allow growth and isolation of microglial cells. *J. Neurosci. Res.* **9,** 183–194.
9. Northoff, H., Gluck, D. Wolpl, A., Kubanek, B., and Galanos, C. (1986a), Lipopolysaccharide induced elaboration of interleukin-1 (IL-1) by human monocytes: Use for the detection of LPS in serum and influence of serum-LPS interactions. *Rev. Infect. Dis.* 9 (Suppl. 5), 599–602.
10. Northoff, H., Kabelits, D., and Galanos, C. (1986b), Interleukin 1 production for detection of bacterial polysaccharide in fetal calf sera and other solutions. *Immunol. Today* **7,** 126–127.
11. Sultzer, B. M., Castagna, R., Bandekar, J., and Wong, P. (1993), Lipopolysaccharide nonresponder cells: TheC3H/HeJ defect. *Immunobiol.* **187,** 257–271.
12. Das, S. K., Stanley, E. R., Guilbert, L. J., and Farman, L. W. (1980), Determination of a colony stimulating factor subclass by a specific receptor on a macrophage cell line. *J. Cell Physiol.* **104,** 359–366.
13. Branch, R. D. and Guilbert, L. J. (1986), Practical in vitro assay systems for the measurement of hematopoietic growth factors. *J. Tissue Culture Methods* **10,** 101–108.


## Further Reading


Abd-El-Basset, E. and Fedoroff, S. (1995), Effect of bacterial wall lipopolysaccharide (LPS) on morphology, motility and cytoskeletal organization of microglia in cultures. *J. Neurosci. Res.* **41,** 222–237.

Abd-El-Basset, E. and Fedoroff, S. (1994), Dynamics of actin filaments in microglia during Fc receptor-mediated phagocytosis. *Acta Neuropathol.* **88,** 527–537.

Fedoroff, S. (1995), Development of microglia, in *Neuroglia,* Kettenmann, H. and Ransom, B. R., eds., Oxford University Press, New York, pp. 162–181.

Fedoroff, S., Hao, C., Ahmed, J. and Guilbert, L. J. (1993), Paracrine and autocrine signaling in regulation of microglia survival, in: *Biology and Pathology of Astrocyte-Neuron Interactions,* Fedoroff, S., Juurlink, B. H. J., and Doucette, R., eds. Plenum, New York, pp. 247–261.

Floden, A. M. and Combs, C.K. (2007), Microglia repetitively isolated from in vitro mixed Glial cultures retain their initial phenotype. *J Neurosci Methods* **164,** 218–224.

Gingras, M., Gagnon, V., Minotti, S., Durham, H.D. and Berthod, F. (2007), Optimized protocols for isolation of primary motor neurons, astrocytes and microglia from embryonic mouse spinal cord.

Giulian, D. and Bauer, T. J. (1986), Characterization of ameboid microglia isolated from developing mammalian brain. *J. Neurosci.* **6,** 2163–2178.

Hayes, G. M., Woodroofe, M. N. and Cuzner, M. L. (1988), Characterization of microglia isolated from adult human and rat brain. *J. Neuroimmunol.* **19,** 177–189.

Neuhaus, J. and Fedoroff, S. (1994), Development of microglia in mouse neopallial cell cultures. *Glia* **11,** 11–17.

Novak, J. P. and Fedoroff, S. (1999), Model of the dynamics of a branching system of the glial cell lineages in vitro. *J. Biol. Systems* **7,** 429–447.

Rieske, E., Graeber, M. B., Tetzlaff, W., Czlonkowska, A. Streit, W. J. and Kreutzberg, G. W. (1989), Microglia and microglia-derived brain macrophages in culture: generation from axotomized rat facial nuclei, identification and characterization in vitro. *Brain Res.* **429,** 1–14.

Richardson, A., Hao, C. and Fedoroff, S. (1993), Microglia progenitor cells: a subpopulation in cultures of mouse neopallial astroglia. *Glia* **7,** 25–33.

Yokoyama, A., Yang, L. Itoh, S., Mori, K. and Tanaka, J. (2004), Microglia, a potential source of neurons, astrocytes, and oligodendrocytes. *Glia* **45,** 96–104.

# Chapter 15

# Primary Schwann Cell Cultures

## Haesun A. Kim and Patrice Maurel

## Abstract

The most widely used method (the Brockes' method) for preparing primary Schwann cell culture uses neonatal rat sciatic nerves as the primary source of Schwann cells. The procedure is relatively simple and yields a highly purified population of Schwann cells in a short period of time. The method has also been used to prepare Schwann cells from mice, however, with limitation. For example, the Brockes' method is not applicable when the genotypes of mouse neonates are unknown or if the mouse mutants do not develop to term. In the first part of the chapter, we described a method ideal for preparing Schwann cells in a transgenic/knockout mouse study. The method uses embryonic dorsal root ganglia as the primary source of Schwann cells and allows preparing separate, highly purified Schwann cell cultures from individual mouse embryos in less than 2 weeks. In the second part of the chapter, we describe a procedure for preparing Schwann cells from neonatal rat sciatic nerves using the Brockes' method, including protocols for Schwann cell purification, expansion, and storage.

**Key words:** Rat Schwann cells, mouse Schwann cells, dorsal root ganglia, sciatic nerves.

## 1. Introduction

The study of Schwann cell development and myelination has been facilitated by the availability to isolate and establish pure population of primary Schwann cells. In addition, transgenic mouse models and naturally occurring mouse mutants serve as increasing important tools for the study of Schwann cell biology. The most widely used method of preparing Schwann cell culture was originally developed by Brockes et al. (1), which uses sciatic nerves of new-born rats as the primary source of Schwann cells. The procedure is relatively simple and yields a highly purified population of Schwann cells in a short period of time (described in **Section 3**). A similar method can be used to prepare Schwann

L.C. Doering (ed.), *Protocols for Neural Cell Culture*, Springer Protocols Handbooks,
DOI 10.1007/978-1-60761-292-6_15, © Humana Press, a part of Springer Science+Business Media, LLC 1992, 1997, 2001, 2010

cells from mouse sciatic nerves (2–6); however, the use of multiple neonatal nerves limits the application of the method only to wild-type mice or mouse mutants that develop to term.

In this chapter, we describe a method for establishing a purified population of Schwann cells from a single mouse embryo (E12.5–13.5). The method was originally described in studies defining abnormalities in Schwann cells derived from mice deficient in Neurofibromatosis 1 (*Nf1*) gene locus (7, 8). The *Nf1*-null mutant mice die in utero between E11 and E15 prior to the formation of mature peripheral nerves (9, 10), making it impossible to use the Brockes' method. The Kim's method allows to establish separate Schwann cell cultures from individual mouse embryos and generates highly purified (>99.5%) mouse Schwann cells in large quantity in less than 2 weeks.

## 2. Preparation of Schwann Cells from a Single Mouse Embryo

In this method, embryonic dorsal root ganglia are used as the main source for Schwann cells, a concept originally developed by Wood and Bunge (11). The procedure includes three steps: (1) dissection and dissociation of the embryonic DRG; (2) expansion of Schwann cell precursors, followed by mechanical separation of the Schwann cell–neuronal network from the underlying fibroblasts; (3) purification of Schwann cells from the associated neurons and subsequent expansion of the purified Schwann cells. A schematic illustration of the procedure is shown in **Fig. 15.1**.

### 2.1. Day 1: Drg Dissection and Plating of the Dissociated Drg

#### 2.1.1. Media and Solutions

1. 70% Ethanol.
2. Leibovitz's L-15 medium (Gibco, 11415-064).
3. Trypsin, 0.25% in Hanks' balanced salt solution without $Ca^{++}$ and $Mg^{++}$ (Invitrogen, 15050-057).
4. *L-15/10%FBS:* L-15 supplemented with 10% fetal bovine serum (FBS) (Mediatech, 35-010-CV).
5. *DRG plating medium:* Dulbecco's modified Eagle's medium (DMEM) with high glucose (Mediatech, 10-017-CV) (DMEM), 10% FBS, 2 mM L-glutamine, nerve growth factor (NGF 2.5S) (Harlan Bioproduct, BT-5025), 50 ng/mL. Pre-warmed to 37°C before use.

#### 2.1.2. Materials

1. Timed pregnant mouse: E12.5–E13.5.
2. Scissors (1), 16.5 cm, standard, straight sharp/blunt tips (Fine Science Tools [FST], 14001-16).
3. Forceps (1), 12 cm, narrow pattern, straight (FST, 11002-12).

Step 1: DRG dissection and dissociation

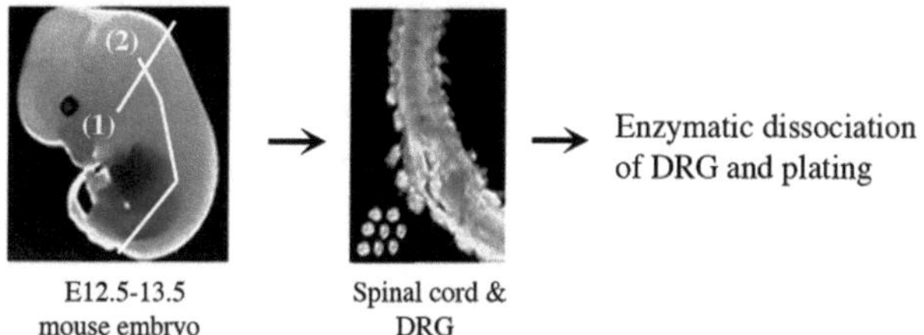



Step 2: Separation of Schwann cell-neuron network

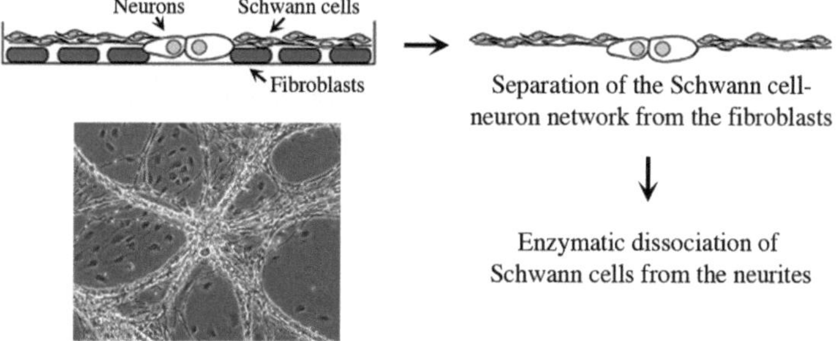



Step 3: Purification and expansion of Schwann cells

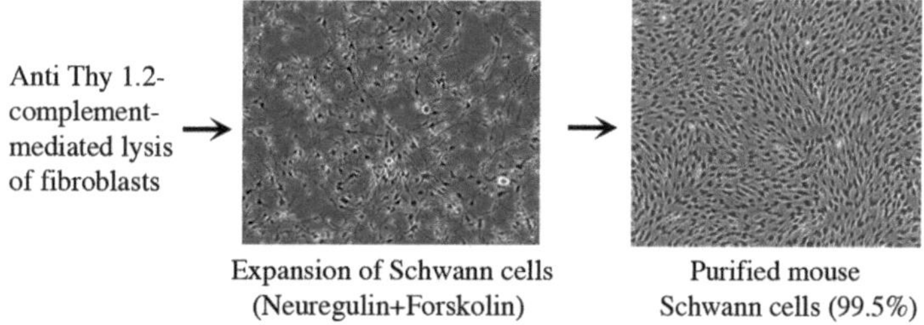



Fig. 15.1. A schematic diagram of a procedure for isolating Schwann cells from a mouse embryo.

4. Fine Iris scissors (1), 8.5 cm, straight (FST, 14090-09).

5. Graefe forceps (1), 10 cm, straight (FST, 11050-10).

6. Spring scissors (1), 8.5 cm (FST, 15009-08).

7. Forceps (1), Dumont curved tip, #7 (FST, 11271-30).

8. Forceps (2), Dumont fine tip, #5 (FST, 11254-20).

9. Dissecting microscope.

10. 100 mm Petri dishes (2).

11. 35 mm culture dishes (three per each embryo harvested).

12. Eppendorf tubes (one per each embryo harvested).

13. 15 mL conical tubes (one per each embryo harvested).

14. Sterile glass Pasteur pipettes, cotton-plugged.

15. Air incubator, 37°C.

16. Tissue culture incubator, 37°C, 90% humidity, 10% $CO_2$.

*2.1.3. Procedure*

2.1.3.1. Preparation
of Embryos

1. Sterilize forceps and scissors by immersing them into 70% ethanol for 20 min. Air dry. Alternatively, insert the tips of the instruments into a hot bead sterilizer for 10–20 s.

2. Euthanize the pregnant mouse by $CO_2$ exposure. Place the mouse on its back and clean the abdomen area with 70% ethanol.

3. Using a set of scissors (16.5 cm) and forceps (12 cm), cut across the lower abdomen through the skin and the muscle layer. While being careful not to touch the internal organs, continue cutting up the left and right side to expose the internal cavity.

4. Using a set of fine Iris scissors and Graefe forceps, remove the uterine horn and transfer in a sterile 100 mm Petri dish.

5. With spring scissors and curved forceps, cut across a uterine sac and push the embryo out. Separate the embryo from the placenta and amniotic membrane and transfer it into a 100 mm Petri dish containing L-15. Continue until all the embryos are removed.

6. Label two sets of 35 mm culture dishes according to the number of embryos harvested (for example, label the dishes E1, E2, E3, E4, E5 for five embryos removed). Fill each dish with 3 mL of L-15. Also prepare 1.5 mL Eppendorf tubes labeled accordingly.

7. Place each embryo in a separate 35 mm dish containing L-15.
**Note**: From this point on embryos are processed individually.

8. Place a dish containing an embryo under a dissecting microscope. Decapitate the embryo with spring scissors (**Fig. 15.1** Step 1, line 1) and transfer the head to a pre-labeled 1.5 mL Eppendorf tube prepared in Step 6. Repeat the step for all the embryos. Place the Eppendorf tubes at 4°C until ready to process the heads for preparing genomic DNA for genotyping.

9. Place the embryo on its side. Insert the tips of spring scissors vertically into the rostral end (where head used to be), just anterior to the vertebral column and begin making 3–4 sequential cuts from rostral to caudal (from head to tail) direction (**Fig. 15.1**, Step 1, line 2). Remove the ventral half portion of the embryo including all the limbs.

10. Lay the remaining dorsal tissue on its back so that the ventral side is facing up. Using two sets of fine forceps, remove any remaining organs (lungs, kidneys, aorta) from the ventral surface. The vertebral body should be visible in the middle of the dorsal plate.

11. Repeat Steps 9–10 for all the embryos.
    **Note**: If L-15 medium in the dish becomes too cloudy with blood, transfer the dorsal plate into a new 35 mm dish with fresh L-15.

**2.1.3.2. Removal of Spinal Cord**

1. While gently holding the dorsal plate at the tail region with a set of forceps, use another set to make ventral cuts along each side of the vertebral body from head-to-tail direction. Remove the vertebral body to expose the underlying spinal cord. Gently lift up the rostral end of the spinal cord and pull it up, freeing it from the vertebral canal. Dorsal root ganglia attached along the sides of the cord should be visible (**Fig. 15.1**, Step 1, right panel).

2. Transfer the cord to the second pre-labeled 35 mm dish prepared in Step 6.

3. Repeat Steps 1–2 for all the embryos

**2.1.3.3. Removal of DRG**

1. While gently holding the cord between forceps, use the other set to pinch off the ganglia from the dorsal root starting from one end of the spinal cord. Be careful not to remove any meninges from the spinal cord. Remove all the ganglia (approximately 37–42 ganglia/spinal cord/embryo) and discard the cord. Repeat the step for all the spinal cords.

**2.1.3.4. DRG Dissociation**

1. Gently swirl each dish to collect all ganglia toward the center of the dish. Under a dissecting microscope, remove the L-15 medium from the dish using a cotton-plugged glass Pasteur pipette while being careful not to remove any ganglia.

2. Add 1.5 mL of 0.25% trypsin to each dish. Place the dishes into a 37°C air incubator for 15–20 min. Swirl dishes every 10 min. Meanwhile, label 15 mL conical tubes according to the number of embryos and add 10 mL of L-15/10% FBS into each tube.

3. Transfer ganglia collected from individual embryo into a 15 mL conical tube containing 10 mL L-15/10% FBS.

4. Centrifuge the ganglia at 50 $g$ for 5 min.

5. Remove the supernatant while being careful not to disturb the pellet. Add 10 mL of L-15/10% FBS. Centrifuge for 5 min as above.

6. Remove the supernatant and add 1 mL of DRG plating medium to each tube. Prepare narrow-bored glass Pasteur pipettes by flaming the tips. Triturate the pellet 10–15 times through the tip of a pipette or until an even cell suspension is obtained. Be careful not to generate too many air bubbles while triturating. Repeat the step for all the tubes and make sure to use separate pipette for each pellet.

7. Add additional 1 mL of DRG plating medium to the cell suspension.

8. Plate dissociated cells (2 mL) derived from a single embryo in an uncoated 35 mm culture dish. Place dishes in a tissue culture incubator.

   **Note**: Dissociated DRG cultures are composed of neurons, fibroblasts, and the Schwann cell precursors. It is important to use uncoated plastic culture dishes since use of any substrate (collagen, Matri-gel)-coated dishes makes the fibroblast become bundled up with neurons and the associated Schwann cells, making it difficult to separate the Schwann cell–neuron network from the underlying fibroblast layer.

### 2.2. Day 2–8: Expansion of Schwann Cell Precursors and Separation of Schwann Cell–Neuron Network from the Fibroblasts

#### 2.2.1. Media and Solutions

1. *N2/NGF:* 1:1 ratio of DMEM and F-12 (Invitrogen, 11765-054) and N2 supplement (GIBCO, 17502-048). Add NGF (50 ng/mL). Pre-warmed to 37°C before use.

#### 2.2.2. Materials

1. Dissecting microscope.
2. $27_G 1/2$ needles, sterile.
3. 15 mL conical tubes.
4. Sterile glass Pasteur pipettes, cotton-plugged.

#### 2.2.3. Procedure

##### 2.2.3.1. Expansion of Schwann Cell Precursors

1. Sixteen to eighteen hours after plating of the dissociated DRG, replace the medium with 2 mL of N2/NGF. Under the serum-free condition, fibroblast growth is suppressed while axon-associated Schwann cell precursors continue to proliferate. There is no need to change the medium.

2. Meanwhile, prepare genomic DNA from each embryo head collected in Step 8 of the previous **Section 2.1.3.1 Step 8** and genotype each embryo by PCR. Mark each dish with the appropriate genotype of the embryo from which the cells were derived.

   **Note**: Expected morphologies of the dissociated DRG culture after plating.

   *24 hours* Most cells should have attached to the plate.

   *48 hours* Many neurons extend neurites. Schwann cell precursors beginning to make contact with neurites should be visible. However, the precursors do not show the characteristic spindle-shape morphology of Schwann cell at this point. They look rather flat with fibroblasts-like morphology.

   *4 days* Extensive neuronal network develops. Schwann cells greatly increase in number as they proliferate in contact with neurons. Fibroblasts appear as large flat cells with no obvious association with neurons (**Fig. 15.2A**).

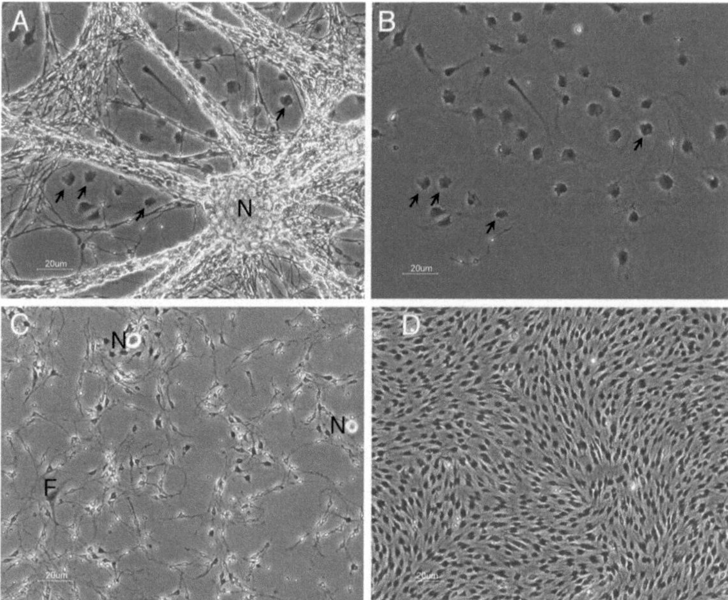

Fig. 15.2. Mouse Schwann cell purification. (**A**) Phase-contrast micrographs of mouse dissociated DRG culture on day 7. Neuronal cell bodies (N) and extended axons are seen. Schwann cell bodies are seen tightly associated with the axons. Large flat fibroblasts (*arrows*) are also present with no obvious association with the neurons. (**B**) Most of the fibroblasts are left behind after removal of the Schwann cell–neuron network. (**C**) Mouse Schwann cells plated after being dissociated from the neurons. Neuronal cell bodies (N) and contaminating fibroblasts are also seen in the culture. (**D**) After being maintained for 2 days in a serum-free condition in the presence of neuregulin and Forskolin, the culture is populated with Schwann cells with the characteristic bipolar, spindle-shaped cell bodies.

If everything goes well as it should, fibroblasts should remain quiescent without obvious increase in number under the serum-free condition.

*6–7 days* Virtually, the entire neurite network should be occupied by Schwann cells. Schwann cells begin to acquire spindle-shape morphology as they mature in culture. DRG axons extend out to the periphery of the dish and begin to bundle up along the curvature of the dish.

2.2.3.2. Separation of Schwann Cell–Neuron Network from Fibroblasts (6–7 Days Following Plating of the Dissociated DRG)

1. Under a dissecting microscope, neuron-associated Schwann cells are visible as beaded structures along neurites. Axon bundles along the periphery of the dish should be visible under a dissecting scope. Starting from a corner of the plate where axonal bundles have formed, lift up the Schwann cell–neuron network using a $27_G1/2$ needle, and slowly peel it off toward the center of the plate. Usually, the neuronal network that is being peeled off can be distinguished from the fibroblast layer left behind under the microscope (**Fig. 15.2B**). If DRGs are initially dissociated

well, almost complete separation of the Schwann cell-neuron network should be achieved from the underlying fibroblast layer.

**Note**: If dense populations of fibroblasts are observed in the direction from which neurons are being peeled off, cut off the neuronal network at that point and start lifting the remaining network on the plate from a new point. Sometimes, fibroblasts become condensed heavily underneath neuronal cell bodies, making the separation difficult at the region. In this case, the fibroblast-dense area can be excised from the plate using a surgical razor blade before proceeding with the separation procedure. It is crucial to obtain Schwann cell–neuron networks that are free of contaminating fibroblasts at this point, because removal of fibroblasts from the later Schwann cell cultures is more difficult.

2. After successfully detaching Schwann cell–neuron network from the plate, transfer the culture medium containing the floating neuronal network into a 15 mL conical tube. Collect all the Schwann cell–neuron networks from embryos of the same genotype into the same tube.

## 2.3. Day 8–14: Mouse Schwann Cell Purification and Expansion

### 2.3.1. Media and Solutions

1. *Trypsin–Collagenase Solution:* 0.25% trypsin (GIBCO) solution supplemented with 0.1% collagenase type 1 (Worthington Biochemical).

2. *Monoclonal anti-Thy 1.2 solution:* Dilute anti-Thy 1.2 antibody (Serotec, MCA1474T) 1:1000 in DMEM and store 1 mL aliquots at –80°C.

3. *Rabbit complement solution:* Reconstitute lyophilized rabbit complement (GIBCO, 640-9200AN) in 5 mL of ddH$_2$O. Add 35 mL of DMEM and filter sterilize. Store 1 mL aliquots at –80°C.

4. *Schwann cell medium:* DMEM, 10% FBS, 2 mM L-glutamine. Pre-warm to 37°C before use.

5. *Schwann cell growth medium:* Schwann cell medium supplemented with recombinant human neuregulin-1-β1 EGF domain (R&D, 396-HB-050) (10 ng/mL) and Forskolin (Sigma, F-6886) (2.5 μM). Pre-warmed to 37°C before use.

6. *N2-Schwann cell growth medium:* N2 (*see* **Section 2.2.1.**) supplemented with recombinant human neuregulin-β-1 EGF domain (10 ng/mL) and Forskolin (2.5 μM). Pre-warmed to 37°C before use.

7. *Poly-L-lysine solution:* Poly-L-lysine (Sigma, P-7890), 0.05 mg/mL prepared in 0.15 M sodium borate pH 8.0, filter sterilized.

8. Phosphate-buffered saline (PBS).

*2.3.2. Materials*

1. Poly-L-lysine-coated 100 mm culture dishes.
2. Glass Pasteur pipettes, cotton-plugged.

*2.3.3. Procedure*

1. After collecting Schwann cell–neuron networks from all the genotypes, pellet the cells by centrifugation at 200 $g$ for 5 min.

2. Remove the supernatant and resuspend the pellet in trypsin–collagenase solution in a volume of approximately 0.5 mL per embryo. For example, if cells from six embryos of the same genotype are pooled, 3 mL of the enzyme solution is used to dissociate Schwann cells. However, if cells from a single embryo are processed, a minimum amount of 1 mL is used.

3. Incubate cells at 37°C for 30 min. Swirl tubes every 10 min.

4. Stop the enzyme reaction by adding Schwann cell medium. Pellet the cells as above.

5. Remove the supernatant. Resuspend the pellet in 1 mL of anti-Thy 1.2 antibody solution. Incubate for 30 min at 37°C.

6. Pellet the cells by centrifugation and remove the supernatant. Resuspend the cells in 1 mL of rabbit complement solution. Incubate for 30 min at 37°C.

7. Count the cell number. An average of $0.3 \times 10^6$ cells are obtained from a single embryo. Thus, if cells were pooled from three embryos, there should be about $1 \times 10^6$ cells. Plate cells from the same genotype onto a poly-L-lysine-coated 100 mm culture dish (see below) in Schwann cell medium. If less than $0.5 \times 10^6$ cells are obtained, plate the cells onto a 60 mm culture dish.

   *Preparing poly-L-lysine-coated plates:*
   a. Add PBS to a tissue culture dish, just enough to wet the bottom of the dish. Remove the PBS

   b. Add poly-L-lysine solution enough to cover the surface of the dish (3 mL for 100 mm dish). Leave at room temperature for 30 min.

   c. Remove the poly-L-lysine. Rinse the dish with PBS three times. After the final wash, leave PBS in the dish until ready to use.

8. Next day, bipolar, spindle-shaped Schwann cells should be visible with few neuronal cell bodies (**Fig. 15.2C**). Change the medium to N2-Schwann cell growth medium. Neuregulin and Forskolin contained in the growth medium start to stimulate Schwann cell proliferation while growth

of any contaminating fibroblasts is suppressed under the serum-free condition. Neurons should die off within 2 days in the absence of NGF in the medium.

9. There is no need to change the medium until the cultures become confluent (6–7 days after the plating), at which point the cultures should contain 99–99.5% mouse Schwann cells (**Fig. 15.2D**). Replate the cells onto new poly-L-lysine-coated plates at a density of $1 \times 10^6$ cells/100 mm plate. From this point on, the cells are grown in Schwann cell growth medium.

   **Note**: As the passage number increases, mouse Schwann cells tend to become less responsive to growth factors and begin to senesce. The demise of cells occurs faster in serum-free media than in serum containing media. After switching to Schwann cell growth medium, the mouse Schwann cells can be grown up to passage 6–7, or frozen down and stored in liquid nitrogen for future use (*see* **Section 3.5.2.** for freezing cells).

## 3. Preparation of Schwann Cells from Neonatal Rat Sciatic Nerves

This method, developed from the original protocol described by Brockes et al. (1), with modifications from Porter et al. (12), is commonly used for the preparation of pure populations of Schwann cells. Sciatic nerves from postnatal day 1–3 (P1-3) Sprague-Dawley rat pups are used as the source of Schwann cells. The procedure includes three basic steps: (1) dissecting and dissociating the sciatic nerves; (2) purification of Schwann cell with antimetabolic agents, and by complement-mediated killing of contaminating fibroblasts; (3) expansion of the purified Schwann cells and preparing frozen aliquots.

### 3.1. Culture Media and Solutions

1. *Schwann cell medium: see* **Section 2.3.1.**

2. *Schwann cell growth medium: see* **Section 2.3.1.**

3. *Trypsin–Collagenase in L-15:* Add 1 mL of 2.5% trypsin (Gibco 15090-046) and 1 mL of 1% collagenase, type 1 into 8 mL pre-warmed L-15.

4. *Poly-L-lysine solution*: see **Section 2.3.1.**

5. *Cytosine-β-arabino furanoside hydrochloride (Ara-C)* (Sigma C-6645): Prepare a 1 mM stock solution in culture-grade sterile distilled water; filter sterilize; store 1 mL aliquots at –80°C.

6. *Rabbit complement*: Lyophilized powder (Calbiochem 234400), reconstitute with 5 mL of culture-grade sterile

distilled water. Prepare one-time use 400 μL aliquots; keep at –80°C.

7. *Cell freezing medium:* 10% DMSO, 90% FBS; filter sterilize; keep and use at 4°C.

8. 70% ethanol.

9. Leibovitz's L-15 medium (Gibco 11415-064).

10. L-15/10% FBS.

11. Hanks' balanced saline solution (HBSS), calcium- and magnesium-free (Gibco 14170-112).

12. Penicillin/streptomycin (100× stock; Gibco 15140-0122).

13. 0.25% Trypsin–EDTA solution (Gibco 25200-056).

14. Anti-Thy-1.1 antibody (Serotec MCA04G).

### 3.2. Materials

1. Sprague-Dawley rat pups, postnatal day 1–3, 2 litters (about 20–25 pups).

2. Dissecting microscope.

3. Scissors (1), 16.5 cm, standard, straight sharp/blunt tips (FST, 14001-16).

4. Scissors (1), 10.5 cm, shallow curved (FST, 14370-23).

5. Forceps (1), 12 cm, narrow pattern, straight (FST, 11002-12).

6. Spring scissors (1), 8.5 cm (FST, 15009-08).

7. Forceps (2), Dumont fine tip, #5 (FST, 11254-20).

8. Needles $27_G1/2$ and $21_G11/2$ (Becton Dickinson 305109 and 305167, respectively).

9. Tissue culture dishes, 60 and 100 mm.

10. Conical tubes, 15 and 50 mL.

11. Serological pipettes, 2 mL.

12. Cryotubes, 2 mL (Corning 430658).

13. Tissue culture incubator, 37°C, 90% humidity, 10% $CO_2$.

### 3.3. Day 1: Harvesting Sciatic Nerves and the Dissociation

#### 3.3.1. Preparation

1. Sterilize dissection tools by immersing them into 70% ethanol for 20 min. Air dry.

2. Prepare 3 × 60 mm tissue culture plates coated with poly-L-lysine as described in **Section 2.3.3.Step** 7 Add 2 mL of Schwann cell medium supplemented with penicillin/streptomycin (1×) per plate. Keep the plates in a tissue culture incubator at 37°C, 90% humidity and 10% $CO_2$, until ready to use.

3. Prepare 15 mL conical tubes with 10 mL of L-15 medium; keep chilled on ice.

*3.3.2. Dissection (One Pup at a Time)*

1. Euthanize a rat pup by decapitation using the large set of scissors (16.5 cm). Place the animal on the dissecting board, dorsal side up, and pin it into place with $21_G 11/2$ and $27_G 1/2$ needles, splaying the hind legs into an inverted V shape (**Fig. 15.3A**). Spray the body with 70% ethanol.

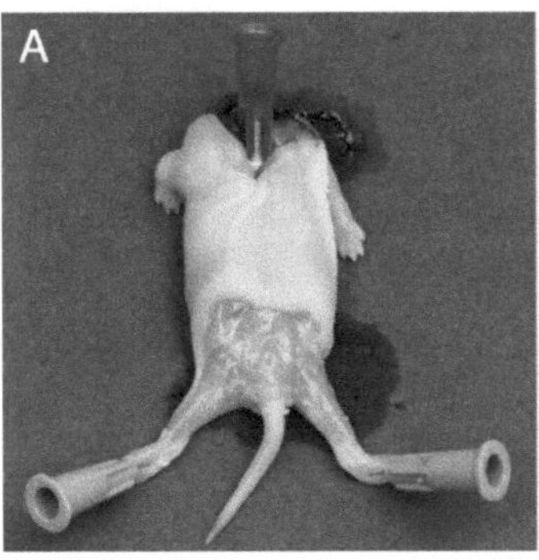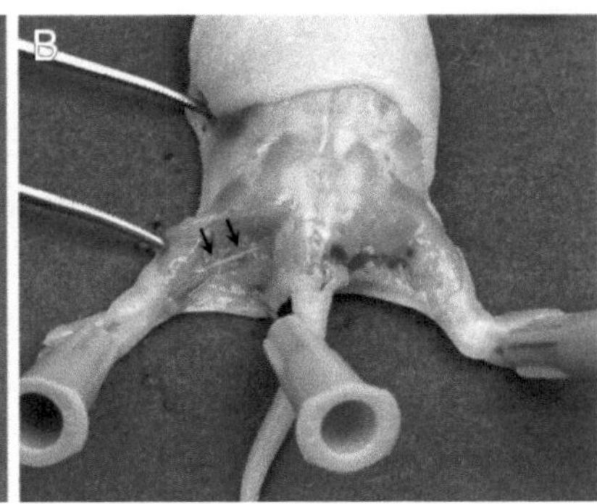

Fig. 15.3. Dissection of the sciatic nerve from rat neonates. (**A**) Euthanized pup is pinned into place with one $21_G 11/2$ and two $27_G 1/2$ needles, splaying the hind legs into an inverted V shape. The skin has been removed from the lower back and hind limbs. (**B**) The hamstrings muscles at the back of the legs have been teased away, revealing the sciatic nerve (*arrows*).

2. Using the set of small curved scissors and forceps, cut the skin away from the hind limbs and the lower back (**Fig. 15.3A**).

3. With the #5 Dumont forceps, gently tease apart the hamstrings muscles from the knee to the hipbone (**Fig. 15.3B**).

4. Locate the sciatic nerve, which runs posterior and parallel to the femoral bone as shown in Figure 3B. Gently grasp the sciatic nerve with one of the Dumont forceps and cut the nerve at each end with the spring scissors. The dissected nerves will be about 1 cm (0.5 in.) in length.

5. Transfer the sciatic nerves immediately into ice-cold L-15 medium in a 15-mL conical tube.

6. Repeat Steps 1–5 for all the remaining pups.

*3.3.3. Dissociation*

1. Once all the sciatic nerves have been harvested, centrifuge the tube for 5 min at 50 $g$ at room temperature.

2. Gently decant the supernatant and resuspend the sciatic nerves in 10 mL of pre-warmed trypsin–collagenase solution. Incubate the tube in a 37°C water bath for 30 min, inverting it every 5–10 min.

3. Centrifuge for 5 min as above. Discard the supernatant and gently resuspend the pelleted sciatic nerves in 10 mL of Schwann cell medium. Repeat centrifugation.

4. Gently aspirate the supernatant and add 3 mL of Schwann cell medium, supplemented with penicillin/streptomycin (1×) to the sciatic nerves; using a 2 mL serological pipette, triturate the sciatic nerves until they are completely dissociated; it should take about 10–20 pipettings.

5. Plate 1 mL of cell suspension per 60 mm poly-L-lysine-coated tissue culture dish (3 dishes total); place the dishes in a tissue culture incubator. The culture should consist of Schwann cells and fibroblasts (**Fig. 15.4A**)

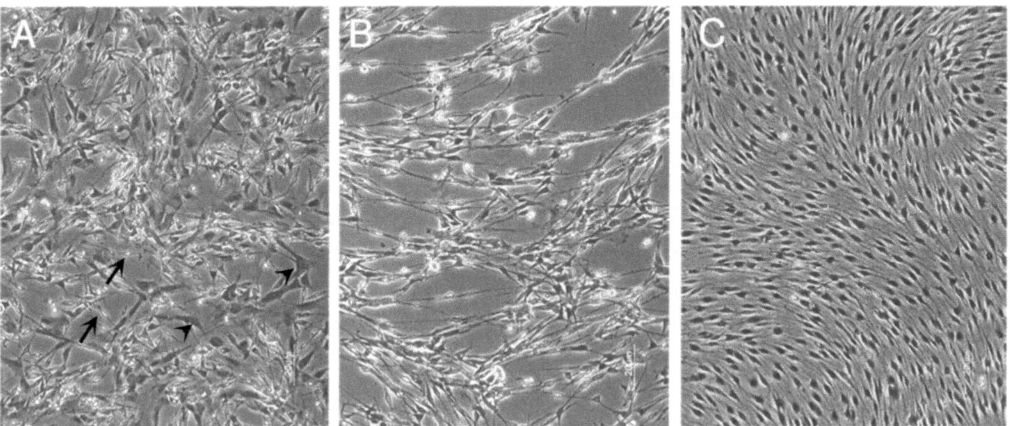

Fig. 15.4. Rat Schwann cell purification. (**A**) Cells 24 h after dissociation of the sciatic nerves. Schwann cells have a bipolar, spindle-shaped morphology (*arrows*), distinguishable from the fibroblasts' flattened and spread-out morphology (*arrowheads*). (**B**) After antimetabolic treatment most of the fibroblasts have been eliminated. (**C**) After expansion Schwann cells form a monolayer of swirling, confluent *spindle-shaped* cells.

### 3.4. Day 2–10: Rat Schwann Cell Purification

#### 3.4.1. Removal of Fibroblasts by Antimetabolic Treatment

In the absence of growth factors, such as neuregulin-1-β1, Schwann cells do not proliferate under serum condition (13), however, fibroblasts do. Cytosine-ß-arabino furanoside hydrochloride (Ara-C) is an antimetabolic agent that, after conversion into cytosine arabinoside triphosphate, impairs DNA synthesis and kill rapidly dividing cells, such as fibroblasts. This enables selective removal of fibroblasts while leaving the Schwann cells intact.

1. Wash cells twice with 3 mL of HBSS.

2. After the last wash, add 3 mL of Schwann cell medium supplemented with Ara-C at the final concentration of 10 μM (1:100 dilution from stock).

3. Place the cells in the tissue culture incubator.
   **Note**: After 2–3 days, there should be many dead cells, mainly fibroblasts, floating in the medium. The spindle-shaped Schwann cells should remain attached to the plates (**Fig. 15.4B**).

4. After 3 days of Ara-C treatment, wash the cells twice with 3 mL of HBSS and replace the medium with 3 mL of fresh Schwann cell medium. Place the cells in the incubator for another 2 days.

*3.4.2. Removal of the Remaining Contaminating Fibroblasts by Complement-Mediated Killing*

Thy 1.1 is a murine alloantigen that is displayed at the surface of rat fibroblasts, but not of rat Schwann cells. Using an anti-Thy 1.1 antibody in conjunction with serum complement, it is therefore possible to selectively kill the remaining contaminating fibroblasts by complement-mediated cytolysis.

1. Wash cells once with 3 mL of HBSS.

2. Wash cells once with Schwann cell medium.

3. Add 3 mL of Schwann cell medium supplemented with 60 µl of the anti-Thy-1.1 antibody. Incubate at 37°C for 15 min.

4. Add 400 µl of rabbit complement; mix by gently swirling the dishes. Incubate for 30–45 min at 37°C; monitor cells under the microscope every 10 min to assess the death of the fibroblasts as well as the health of the Schwann cells.

5. Remove medium and wash cells twice with 3 mL of HBSS.

6. Add Schwann cell growth medium and place the dishes in the tissue culture incubator. Under the growth condition, Schwann cells will begin to divide.

7. After 2 days, repeat Steps 1 through 6.
   **Note**: Observe the cultures for fibroblasts contamination and Schwann cell viability. Schwann cells should have a bipolar, spindle-shaped morphology, distinguishable from the fibroblasts flattened and spread-out morphology. At this stage, Schwann cell purity should be greater than 99%. If it is not, the complement-mediated killing should be repeated before expanding the cells.

8. Culture cells until they form a 100% confluent monolayer (**Fig. 15.4C**).

*3.5. Day 11–25: Schwann Cell Expansion and Storage*

*3.5.1. Expansion*

1. When Schwann cells form a confluent monolayer, wash the cells twice with 3 mL of HBSS.

2. Add 2 mL of 0.25% trypsin–EDTA solution; incubate at 37°C for 2–5 min. Observe the cells under light microscopy, while gently shaking the culture dish.

3. When the cells start to come off the dish, add 2 mL of L-15/10% FBS per dish. Collect the cells from all 3 × 60 mm plates in a 15 mL tube.

4. Centrifuge for 5 min at 200 $g$, at room temperature.

5. Discard supernatant and gently resuspend the cells in 9 mL of Schwann cell growth medium.

6. Plate cells onto 9 × 100 mm tissue culture dishes (poly-L-lysine-coated), 1 mL per dish. This 1:3 split corresponds to passage # 1.

7. Incubate cells until they form a confluent monolayer (about 5–7 days).
**Note**: When the Schwann cells form a confluent monolayer, the cell density should be approximately 4–5 × $10^6$ cells per 100 mm dish. At this point, the cells can be trypsinized and replated at a desired density for experiments. Alternatively, the cells can be expanded further and stored in liquid nitrogen for future use.

8. Trypsinize the cells with 5 mL trypsin–EDTA per dish. Add 5 mL of L-15/10% FBS and transfer the cells in a 15 mL tube, one tube per plate (nine tubes). Centrifuge.

9. Discard supernatant and gently resuspend the cells in 5 mL (per tube) of Schwann cell growth medium. Collect cells from all nine tubes into a 50 mL conical tube.

10. Plate cells onto 45 × 100 mm tissue culture dishes (poly-L-lysine-coated), 1 mL per dish. This 1:5 split corresponds to passage # 2. Place the cells in the incubator until 80–90% confluent (5–7 days).

*3.5.2. Freezing for Storage*

1. When cells are 80–90% confluent, wash cells twice with 5 mL of HBSS.

2. Add 5 mL of 0.25% trypsin–EDTA solution; incubate at 37°C for 2–5 min.

3. Trypsinize the cells and add 5 mL of L-15/10% FBS. Collect cells in 50 mL tubes (25 mL per tube). Do not process all the plates; work 10 plates at a time.

4. Centrifuge for 10 min, at 200 $g$, at room temperature.

5. Discard supernatant and gently resuspend the cells in 10 mL of L-15/10% FBS; count cells using a hematocytometer.

6. Centrifuge and discard supernatant. Gently resuspend the cells in ice-cold freezing medium with a final density of 2 × $10^6$ Schwann cells/1 mL.

7. Make 1 mL aliquots in cryotubes, and freeze overnight at –80°C.

8. Next day, transfer frozen aliquots to liquid nitrogen for long-term storage.

*3.5.3. Thawing Schwann Cells*

1. Quickly thaw a frozen aliquot in a 37°C water bath.

2. Transfer the cell suspension to a 15 mL tube containing 9 mL of pre-warmed (37°C) L-15/10% FBS. Centrifuge.

3. Discard supernatant and gently resuspend the cells in 8 mL of Schwann cell growth medium.

4. Plate cells onto one 100 mm poly-L-lysine-coated tissue culture dish. Expand the cells as above.

## References


1. Brockes JP, Fields KL, Raff MC. Studies on cultured rat Schwann cells. I. Establishment of purified populations from cultures of peripheral nerve. Brain Res 1979;165(1):105–18.
2. Manent J, Oguievetskaia K, Bayer J, Ratner N, Giovannini M. Magnetic cell sorting for enriching Schwann cells from adult mouse peripheral nerves. J Neurosci Methods 2003;123(2):167–73.
3. Seilheimer B, Schachner M. Regulation of neural cell adhesion molecule expression on cultured mouse Schwann cells by nerve growth factor. Embo J 1987;6(6):1611–6.
4. Shine HD, Sidman RL. Immunoreactive myelin basic proteins are not detected when shiverer mutant Schwann cells and fibroblasts are co-cultured with normal neurons. J Cell Biol 1984;98(4):1291–5.
5. Stevens B, Tanner S, Fields RD. Control of myelination by specific patterns of neural impulses. J Neurosci 1998;18(22):9303–11.
6. Zhang BT, Hikawa N, Horie H, Takenaka T. Mitogen induced proliferation of isolated adult mouse Schwann cells. J Neurosci Res 1995;41(5):648–54.
7. Kim HA, Rosenbaum T, Marchionni MA, Ratner N, DeClue JE. Schwann cells from neurofibromin deficient mice exhibit activation of p21ras, inhibition of cell proliferation and morphological changes. Oncogene 1995;11(2):325–35.
8. Kim HA, Ling B, Ratner N. Nf1-deficient mouse Schwann cells are angiogenic and invasive and can be induced to hyperproliferate: reversion of some phenotypes by an inhibitor of farnesyl protein transferase. Molecular & Cellular Biology 1997;17(2):862–72.
9. Jacks T, Shih TS, Schmitt EM, Bronson RT, Bernards A, Weinberg RA. Tumour predisposition in mice heterozygous for a targeted mutation in Nf1. Nat Genet 1994;7(3):353–61.
10. Brannan CI, Perkins AS, Vogel KS, et al. Targeted disruption of the neurofibromatosis type-1 gene leads to developmental abnormalities in heart and various neural crest-derived tissues. Genes Dev 1994;8(9):1019–29.
11. Wood PM, Bunge RP. Evidence that sensory axons are mitogenic for Schwann cells. Nature 1975;256:662–4.
12. Porter S, Clark MB, Glaser L, Bunge RP. Schwann cells stimulated to proliferate in the absence of neurons retain full functional capability. J Neurosci 1986;6(10):3070–8.
13. Brockes JP, Fields KL, Raff MC. A surface antigenic marker for rat Schwann cells. Nature 1977;266(5600):364–6.

# Chapter 16

# Primary Dissociated Astrocyte and Neuron Co-culture

Shelley Jacobs and Laurie C. Doering

## Abstract

We describe a method to isolate and co-culture dissociated hippocampal neurons and cortical astrocytes from young mice (E17 and newborn, respectively). This protocol is useful to investigate the effects of astrocytes on the developmental biology of neurons. By independently isolating the astrocytes and neurons (with different genetic backgrounds) from different mice, an in vitro environment can be created where one cell type is deficient in a gene or protein of interest. For example, this method is well suited to examine the effect of a genetic mutation in astrocytes on the development of neuronal processes and synapses.

**Key words:** Mouse, astrocyte, neuron, co-culture, development.

## 1. Introduction

To facilitate the study of interactions between cells of the central nervous system (CNS), in vitro preparations are often the method of choice by many neuroscientists. When neurons are isolated from their in vivo niche, it becomes possible to manipulate specific aspects of their development, maturation, and signaling. In vitro cellular preparations can also be used to probe the effects of a genetic mutation (knockout mouse strain) on a specific CNS cell type, without the complications and restrictions imposed by the in vivo environment.

Hippocampal neurons can be isolated from embryonic or adult tissue. *See* also the chapter by Nault and De Koninck. However, the isolation of cells from the embryonic CNS is easier from a technical viewpoint. Neurons and their precursors can

L.C. Doering (ed.), *Protocols for Neural Cell Culture*, Springer Protocols Handbooks,
DOI 10.1007/978-1-60761-292-6_16, © Humana Press, a part of Springer Science+Business Media, LLC 1992, 1997, 2001, 2010

be dissociated from late-stage embryos (i.e., E17 mouse), with higher viable yields due in part to the immature neural networks. Primary hippocampal neurons grown in vitro develop extensive dendritic arbors studded with characteristic dendritic spines. When seeded in sufficient density, these neurons form networks mediated by synaptic connections (1).

The cellular homogeneity exploited in a number of in vitro paradigms, where neurons are the primary cell type of interest, has prompted a number of laboratories to develop protocols for their isolated growth. However, it appears that unlike their peripheral counterparts, CNS neurons require astroglial support for normal development. This support can come in a variety of forms. Some researchers grow neurons in chemically defined media that have been preconditioned by astrocytes and, therefore, contain the soluble astrocyte-derived factors required for neuron growth *(2, 3)*. Others have developed a protocol where neurons can be grown in a "sandwich" culture, where the neurons are grown on a coverslip, inverted on a monolayer of astrocytes (4). Alternatively, and perhaps the least complicated, is the use of mixed cultures of neurons and astrocytes from the same tissue source, in which the astrocytes quickly form a monolayer beneath the neurons (3, 5).

Given that CNS neurons require glial interactions for growth, it is important to have methods to investigate the molecular interactions between neurons and glial cells. Here we describe a protocol that is relatively simple from a technical point of view, yet provides a means for investigating the effects of non-neuronal cells on neuronal development. Specifically, this protocol examines neuronal maturation and synaptic development when combined with astrocytes carrying a specific genetic mutation. Here, the neurons are isolated from the hippocampus of a late embryonic mouse and co-cultured directly in apposition to cortical astrocytes from a newborn mouse pup. The co-culture is maintained on glass coverslips to facilitate immunocytochemical studies. This protocol therefore provides a key means to investigate the effects of astrocytes on neuronal growth and synapse formation in genetic mouse models of disease.

Our protocol is divided into four steps: cortical astrocyte isolation, preparation of coverslips for co-culture, plating cortical astrocytes for co-culture, and isolating and plating hippocampal neurons for co-culture (**Fig. 16.1.**) It is important to note that the steps of this protocol are dependent on the coordinated timing of astrocytes in culture with a timed-pregnant mouse at the appropriate gestational age (E17). Here, E17 is determined with the day of the sperm plug counted as E1. For the purposes of outlining the timing of the protocol, the day of neuron isolation will be considered the pivotal step, and the timing of all other steps will be counted from that day.

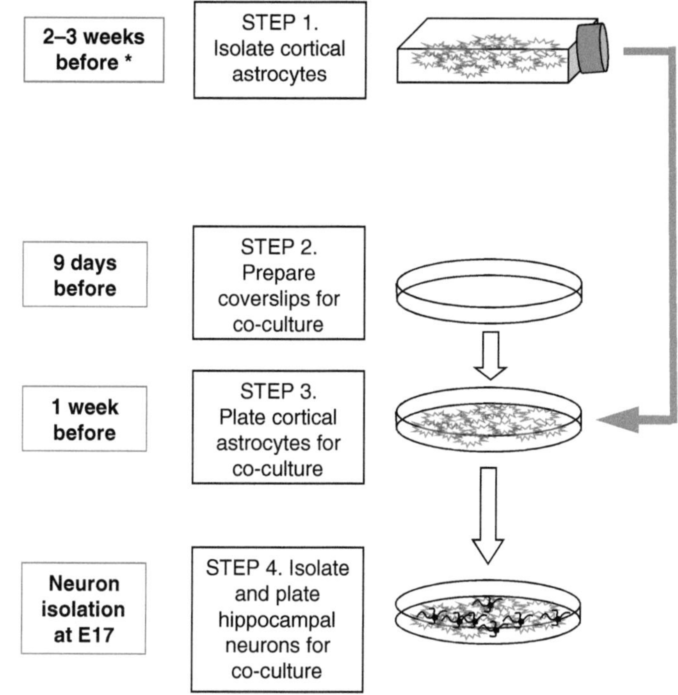


Fig. 16.1. Schematic diagram illustrating the sequential steps for preparing neurons and astrocytes in a co-culture system.

### 1.1. Approximate Time to Perform Each Step

| | |
|---|---|
| STEP 1. Isolating cortical astrocytes for co-culture | 1.5–2 h |
| STEP 2. Preparing coverslips for co-culture | DAY 1: 30 min–1 h<br>DAY 2: 1 h |
| STEP 3. Plating cortical astrocytes for co-culture | 1 h |
| STEP 4. Isolating and plating hippocampal neurons for co-culture | 2–3 h |

### 1.2. General Equipment

1. Biological hood
2. Water bath (37°C)
3. Dissecting microscope
4. Centrifuge
5. Cell culture incubator, 37°C, 5% $CO_2$

### 1.3. Solutions

1. Calcium- and magnesium-free Hanks buffered saline solution (CMF-HBSS) (100 mL)

| 10 mL | 10× HBSS (Invitrogen, Cat. No. 14185-052) |
| 1 mL | 1 M HEPES (Invitrogen, Cat. No. 15630-080) |
| 89 mL | dH$_2$O |

Filter through 0.22 μm filtration system (Stericup, Cat. No. SCGPU01RE/ SCGPU05RE)

2. Glial media (GM) (100 mL)

| 10 mL | Horse Serum (Invitrogen, Cat. No. 16050-122) |
| 2 mL | 30% (w/v) D-glucose (filtered through 0.22 μm filter) (Sigma, Cat. No. G8769) |
| 88 mL | Minimal essential media with Earl salts and L-glutamine (MEM) (Invitrogen, Cat. No. 11095-080) |

*Can be stored at 4°C for up to a week.

3. Phosphate-buffered saline (PBS) (500 mL)

| 50 mL | 0.1 M Phosphate buffer |
| 450 mL | 0.9% (w/v) NaCl in dH$_2$O (Sigma, S6191) |

Filter through 0.22 μm filtration system (Stericup, Cat. No. SCGPU01RE/ SCGPU05RE).

4. Neural maintenance media (NMM) (100 mL)

| 1 mL | N2 Supplement (100×)(Invitrogen, Cat. No. 17502-048) |
| 1 mL | 100 mM sodium pyruvate solution (Invitrogen, Cat. No. 11360-070) |
| 2 mL | 30% (w/v) D-glucose (filtered through 0.22 μm filter, Stericup, Cat. No. SCGPU01RE) (Sigma, Cat. No. G8769) |
| 96 mL | MEM (Invitrogen, Cat. No. 11095-080) |

*Make fresh (can be stored for 24 h at 4°C).

## 2. Isolating Cortical Astrocytes for Co-culture

### 2.1. Reagents and Equipment

1. Four newborn mouse pups (postnatal day 0 or 1) (Four pups will prepare enough astrocytes for 1 T75 flask.)

2. Dissecting tools sterilized with 95% EtOH: fine dissecting forceps (two pair), curved forceps, curved scissors, small scissors

3. 60 mm × 15 mm tissue culture dishes (Falcon, Cat. No. 353002)

4. 15 mL polystyrene Falcon tubes (Falcon Cat. No. 352095)

5. 50 mL Falcon tubes (Falcon, Cat. No. 352070)

6. 70 μm cell strainer (Falcon, Cat. No. 352350)

7. T75 culture flask (Sarstedt, Cat. No. 41111066)

8. 5 mL serological pipettes

9. 12 mL serological pipettes

10. 1.5 mL Eppendorf tubes (for tail samples for genotyping if desired) (Eppendorf, Cat. No. 022364120)

11. No. 20 scalpel blades

12. CMF-HBSS (*see* **Section 1.3**)

13. DNase I (Roche Applied Science, Cat. No. 10104159)

14. One aliquot (1.5 mL) 2.5% (10×) trypsin (Invitrogen, Cat. No. 15090-046)

15. 10% Glial media (GM) (*see* **Section 1.3**)

**2.2. Protocol**

*Preparation*

1. Turn on water bath (37°C) and biological hood.

2. Place one aliquot 2.5% trypsin in water bath.

3. Sterilize dissecting tools and area.

4. Prepare DNase I.
    (i) Weigh 15 mg of DNase into 15 mL Falcon tube.
    (ii) In hood, add 1.5 mL sterile CMF-HBSS.

5. In hood, add 4 mL CMF-HBSS to each 60 mm culture dishes (four required).

6. In hood, add 8 mL CMF-HBSS to a 50 mL Falcon tube.
   *Cell isolation:*

7. Isolate brain from pup and place into one dish with CMF-HBSS.

8. Repeat with three other pups
   (Optional – clip tail for genotyping, place in 1.5 mL Eppendorf tubes and store at –20°C)

9. Under dissecting microscope, remove hemispheres from hindbrain and midbrain, and place hemispheres into second 60 mm dish.

10. Remove meninges and place into third 60 mm dish.
    **Caution:** Removal of meninges must be complete. Meninges contain fibroblasts which can overgrow the astrocyte culture.

11. Remove hippocampi and place isolated cortices into final 60 mm dish.

12. Mince cortices with No. 20 scalpel blades.

13. In biological hood, transfer tissue and CMF-HBSS to 50 mL Falcon with 8 mL CMF-HBSS.

14. Rinse culture dish with 1.5 mL DNase solution and add to 50 mL Falcon tube.

15. Rinse culture dish with 1.5 mL 2.5% trypsin and add to 50 mL Falcon tube.

16. Swirl tube and place in 37°C water bath for 5 min.

17. Triturate with 12 mL serological pipette 10 times, expelling cell suspension against side of tube, and return to 37°C water bath for 5 min.

18. Triturate with 5 mL serological pipette 10–15 times, and return to water bath for 5 min.

19. Remove from water bath and add 15 mL of 10% GM and triturate well.

20. Pass cell suspension through cell strainer into second 50 mL falcon tube.

21. Centrifuge for 5 min ($150$–$200g$).

22. Resuspend in 20 mL of 10% GM and plate in T75 flask.
    **Tip:** The cortices of four mouse pups should yield on average $8 \times 10^6$ cells, which is an appropriate quantity to seed in one T75 flask.

23. Place in incubator 37°C, 5% $CO_2$.

24. Replace with fresh GM after 24 h.
    **Tip:** Do not be alarmed if there appears to be few cells attached to flask. The astrocytes will proliferate quickly. Seeding the astrocytes at too high a density may result in an increased number of microglia in the culture. (In high numbers these are toxic to neuron growth.)

25. Replace half the GM every 2–3 days.
    **Tip:** While maintaining the culture, never change the media completely; the conditioning of the media promotes astrocyte growth.
    **Tip:** When changing media, slap the flask 5–10 times on the palm of your hand before changing the media. This will help remove any loosely bound cells and reduce the numbers of microglia in the culture.
    **Tip:** Astrocytes usually reach an appropriate level of confluence to be used for co-culture after 10–14 days in culture.

## 2.3. Preparing Coverslips for Co-culture

### 2.3.1. Reagents and Equipment

**DAY 1**

1. 24-well tissue culture plates (Falcon, Cat. No. 353047)

2. 13 mm round glass coverslips (Bellco Glass, Cat. No. 1943-00012)

3. 95% EtOH in a small beaker

4. Forceps

5. EtOH burner

6. Serological pipettes

7. Sterile Pasteur pipettes

8. Sterile filtered $dH_2O$ (filtered with 0.2 $\mu$m filter)

9. Sterile filtered poly-L-lysine (PLL) (Sigma, Cat. No. P1399 1 mg/mL) (filtered before use with 0.2 $\mu$m filter)

*2.3.2. Protocol*

1. In hood, using forceps, dip coverglass into 95% EtOH and pass through flame of burner.

2. Allow to cool for a few seconds and then place into one well of 24-well plate.

3. Repeat as necessary for the number of wells desired.

4. Rinse each coverslip with 0.5 mL $dH_2O$.

5. Apply 0.5 mL PLL per well, making sure the coverslip is fully submerged.
   **Tip:** You can use the pipette tip or drop more PLL on the coverslip to ensure full coverage.

6. Cover each 24-well plate with a sheet of paper towel and place into fridge (4°C) overnight.

## DAY 2

*2.3.3. Reagents and Equipment*

1. Sterile filtered $dH_2O$ (filtered with 0.2 $\mu$m filter)

2. Sterile phosphate-buffered saline (PBS) (*see* **Section 1.3**)

3. Natural mouse laminin (Invitrogen, Cat. No. 23017-015), aliquoted and stored at –80°C.

*2.3.4. Protocol*

1. Remove PLL.
   **Tip:** PLL can be re-used. Store at 4°C and filter before each use.

2. Wash two times with $dH_2O$.
   **Caution:** Do not allow coverslips to dry out before unbound PLL is completely rinsed off. This could cause decreased cell survival.

3. Allow plates to air-dry *completely* in hood (approximately 30 min).

4. Dilute laminin in sterile PBS and add 0.5 mL per well. (Optional – Wrap edges with parafilm.)

5. Incubate for at least 4 h, or more commonly overnight, at room temperature (RT).

**Tip:** PLL/laminin coating is stable for up to 2 days at RT.
**Note:** Before plating cells, the laminin needs to be removed, the wells washed two times with MEM and filled with 10% GM. The plates can then be stored in the cell culture incubator until use. This is done the day of astrocyte plating. *(STEP 3. Plating Cortical Astrocytes for Co-Culture)*

---

## 3. Plating Cortical Astrocytes for Co-culture

### 3.1. Reagents and Equipment

1. Previously PLL/laminin prepared coverslips in 24-well plates.
   *(STEP 2. Preparing Coverslips for Co-Culture)*

2. 1 T75 flask of astrocytes (isolated approximately 2 weeks prior).
   *(STEP 1. Isolating Cortical Astrocytes for Co-Culture)*

3. Minimal essential media (MEM) (Invitrogen, Cat. No. 11095-080)

4. 10% glial medium (GM)

5. Neural maintenance medium (NMM) (*see* **Section 1.3**)

6. One aliquot (10 mL) 0.05% trypsin–EDTA (Invitrogen, Cat. No. 25300-062)

7. 15 mL polystyrene Falcon tube (Falcon, Cat. No. 352095)

8. 0.5 mL Eppendorf tube (Eppendorf, Cat. No. 0030124.502)

9. 200 μL yellow tips

10. Hemacytometer (for counting cells)

11. Trypan blue (to determine viability if desired)

### 3.2. Protocol

*Preparation*

1. Wash prepared coverslips two times with 0.5 ml MEM.
   **Caution:** Be careful not to let the coverslips dry between washes. This is toxic to cells.

2. Replace MEM with 0.5 mL per well 10% GM, and place in incubator 37°C, 5% $CO_2$ to equilibrate.

3. Remove media from T75 flask.

4. Wash with 5 mL 0.05% trypsin–EDTA.

5. Add second half of trypsin aliquot, and place into incubator (37°C, 5% $CO_2$) for 5 min (approx). (Check flask to make sure the majority of cells are rounded up and floating.)

6. Add 5 mL 10% GM to flask, and rinse five times.

7. Move cell suspension into 15 mL Falcon tube and centrifuge 150–200$g$ for 5 min.

8. Remove supernatant and resuspend cells in 2 mL 10% GM.

9. Count cells to determine cell density in suspension, and calculate the number of microliters needed to result in 9000 cells/well.

   For example, cell concentration in suspension is 1245 cells/$\mu$L. Want 9000 cells/well … 9000/1245 is 7 $\mu$L. Therefore, add 7 $\mu$L of cell suspension to each well of the 24-well plate previously equilibrated in the incubator.

10. Replace half of GM after 3 days.

    **Tip:** Do not replace all of the media. The astrocytes require the conditioned media for optimal growth.

11. At 6 days in culture replace 10% GM with neural maintenance medium (NMM) in preparation for *STEP 4. Isolation and Plating Hippocampal Neurons for Co-Culture*.

---

## 4. Isolating and Plating Hippocampal Neurons for Co-culture

### 4.1. Reagents and Equipment

1. Timed Pregnant Dam Mouse: four mouse pups at E17 (Day of plug is E1).
   (Four pups will prepare enough neurons for one experiment.)

2. Previously plated astrocytes (With NMM-conditioned media, i.e., NMM added 24 h prior to neuron isolation *(STEP 3. Plating Astrocytes for Co-Culture)*)

3. Dissecting tools sterilized in 95% EtOH: fine dissecting forceps (two pairs), curved forceps, curved scissors, small scissors, three pairs of forceps, and three pairs of scissors (one set for each tissue level of the dissection)

4. 4100 mm tissue culture dishes (Falcon, Cat. No.353003)

5. 460 mm × 15 mm tissue culture dishes (Falcon, Cat. No. 353002)

6. 115 mL polystyrene Falcon tube (Falcon, Cat. No. 352095)

7. 5 mL serological pipettes

8. Sterile Pasteur pipettes

9. Sterile fire-polished Pasteur pipettes (polished to remove sharp edges)

10. Sterile fire-polished Pasteur pipette with borehole diameter reduced to half original size – no less

11. 41.5 mL Eppendorf tubes (for tail samples for genotyping if desired) (Eppendorf, Cat. No. 022364120)

12. 0.5mL Eppendorf tubes (for counting cells) (Eppendorf, Cat. No. 0030124.502)

13. Hemacytometer (for counting cells)

14. Sterile filtered saline (0.9% w/v NaCl in $dH_2O$) (Sigma, Cat No. S6191) (Filtered through 0.22 μm filter (Stericup Cat. No. SCGPU05RE)

15. CMF-HBSS

16. One aliquot (0.5 mL) 2.5% trypsin (Invitrogen, Cat. No. 15090-046)

17. NMM (*see* **Section 1.3**)

**4.2. Protocol**

*Preparation*

1. Turn on water bath (37°C) and hood.

2. Place one aliquot 2.5% trypsin in water bath.

3. Sterilize dissection tools and dissecting area.

4. In hood, add 4 mL CMF-HBSS to each of 460 mm culture dishes.

5. Add 15 mL sterile saline to each of 4100 mm culture dishes.

*Pup isolation*

6. Euthanize pregnant dam.

7. Remove uterus and place into one 100 mm culture dish.

8. Rinse blood from uterus and move into second 100 mm culture dish.

9. Remove four pups randomly from uterus and place into third 100 mm dish.

*Hippocampal isolation*

10. Using curved scissors, decapitate pup and place in fourth 100 mm dish.

11. Remove brain and place into one dish with CMF-HBSS.
    **Tip:** When dissecting out the brain, it can help to stabilize the skull with forceps placed through the eye sockets.

12. Repeat with three other pups.
    (Optional – Clip tail for genotyping, place in 1.5 mL Eppendorf tubes and store at –20°C.)

13. Under dissecting microscope, remove hemispheres from hindbrain and midbrain and place hemispheres into second 60 mm dish.

14. Remove meninges and place cleaned cortices into third 60 mm dish.
    **Caution:** The removal of meninges needs to be complete.

15. Isolate the hippocampi and place into final 60 mm dish.
    **Tip:** When isolating the hippocampus, handle it as little as

possible to avoid damaging the tissue. To minimize damage to the hippocampus, it may help to manipulate the brain by grasping the *cortex* with a pair of forceps.

*Cell isolation*

16. In biological hood, transfer tissue and CMF-HBSS to 15 mL Falcon tube and add 0.5 mL CMF-HBSS (to bring volume to 4.5 mL).

17. Add 0.5 mL 2.5% trypsin to 15 mL Falcon tube.

18. Swirl tube and place in 37°C water bath for 15 min.

19. Remove supernatant (as much as possible), and add 5 mL CMF-HBSS. Swirl tube and let stand for 5 min.

20. Repeat two times.
    **Caution:** Care must be taken to ensure hippocampal tissue is not removed with the supernatant.

21. Remove supernatant and add 2 mL NMM.

22. Triturate 10–15 times with a fire-polished glass pipette.

23. Switch to a reduced borehole fire-polished pipette and triturate another 10–15 times.
    **Caution:** Avoid generating air bubbles. These will reduce the neuron yield. Four mouse pups should yield $\sim 4 \times 10^6$ cells, with a viability of $> 85\%$. Expel suspension against the wall of the tube to minimize the generation of air bubbles.
    **Tip:** When the cells are fully isolated, the suspension will appear cloudy and will contain no visible tissue.

24. Determine cell suspension density and the number of microliter required to yield 20,000 cells per well.

25. Add the appropriate measure of cell suspension to each well of previously plated and conditioned astrocytes.

26. Place in incubator at 37°C, 5% $CO_2$.

27. Replace one third of media with fresh NMM every 7 days in culture, as necessary.
    **Tip:** Do not replace all of the media. The neurons require conditioned media from the astrocytes for optimal growth.
    **Note:** Many other neuron–astrocyte co-culture protocols include the addition of a glial inhibitor, such as cytosine arabinoside (AraC), to the medium a few days after the co-culture is established. We prefer not to use inhibitors in consideration that glial inhibitors can alter cell physiology (6–8). Instead, we rely on the knowledge that our starting population, E17 mouse hippocampus, consists of a relatively low population of glial cells, and the NMM used in this protocol does not promote glial proliferation. Under these conditions we obtain a relatively low level of glial contamination in our neuron population.

## 5. Critical Steps

1. *Removal of meninges.* This is important to ensure minimal contamination of culture with fibroblasts which do not support neuron growth.

2. *Plating density of the astrocytes.* Astrocytes should be seeded at no more than $1 \times 10^7$ cells/T75 flask ($\sim$7.5–8 $\times$ $10^6$ cells/T75 flask is optimal). Increasing the density can result in increased growth of microglia that, in high numbers, are toxic to neuron growth.

3. *Cooling coverslips.* Ensure coverslips are cooled sufficiently. Coverslips that are still warm may stick to the plastic of the 24-well plate.

4. *Rinsing coverslips with $dH_2O$.* The EtOH needs to be rinsed completely. Residual EtOH is toxic to cells.

5. *Applying PLL to coverslip.* Check all coverslips – Do not assume coverslips are completely immersed in PLL. Sometimes air bubble can accumulate and lift the coverslip so that some, or all, are exposed beyond the PLL. If the coverslip is not submerged, the astrocytes may not adhere to and cover the entire surface of the coverslip.

6. *Washing after PLL.* Do not allow coverslips to dry before all unbound PLL is removed. Residual unbound PLL is toxic to cells.

7. *Air drying of coverslips before laminin.* Ensure coverslips are completely dry prior to the addition of laminin. Residual moisture can prevent the laminin from coating the coverslips effectively and can result in poor adherence of the cells.

8. *Washing after laminin.* Do not allow coverslips to dry before all unbound laminin is removed. Residual unbound laminin is toxic to cells.

9. *Washing with 0.05% trypsin–EDTA.* Rinsing with trypsin by incubation at RT for approximately a minute, and discarding this portion, helps remove any loosely bound cells (i.e., contaminating microglia).

10. *Conditioning NMM.* The conditioning of the NMM is essential for optimal neuron growth. The GM on the plated astrocytes should be replaced with NMM at least 24 h before the addition of the neurons to the culture. If neuron survival is not adequate, increase the conditioning time to 48 h.

11. *Washing of hippocampi after trypsin.* Care needs to be taken when performing this step to ensure that hippocampal tissue is not removed with the supernatant. During the ini-

tial washes the tissue may not fully settle into the bottom of the tube. Remove only as much supernatant as possible with each wash. The removal of the supernatant becomes easier with each successive wash. If the supernatant cannot be adequately removed after three washes, wash a fourth time. At this point the tissue should settle more completely into the bottom of the tube. Do not centrifuge to expedite this process; centrifugation will decrease neuron yield.

12. *Trituration of hippocampal tissue.* Avoid the generation of air bubbles as this will reduce neuron yield. The expulsion of the suspension against the side of the tube will reduce the generation of air bubbles. The cells will be adequately released from the tissue when the suspension appears cloudy with no visible pieces of tissue.

## 6. Typical Protocol Results

Typical cell yield

| Source | Number of cells |
| --- | --- |
| Mouse cortex (newborn: P0-P1) (four pups) | $\sim 8 \times 10^6$ cells |
| Mouse cortical astrocytes (one flask) | $\sim 1–1.5 \times 10^6$ cells |
| Mouse hippocampus (E17) (four pups) | $\sim 4 \times 10^6$ cells |

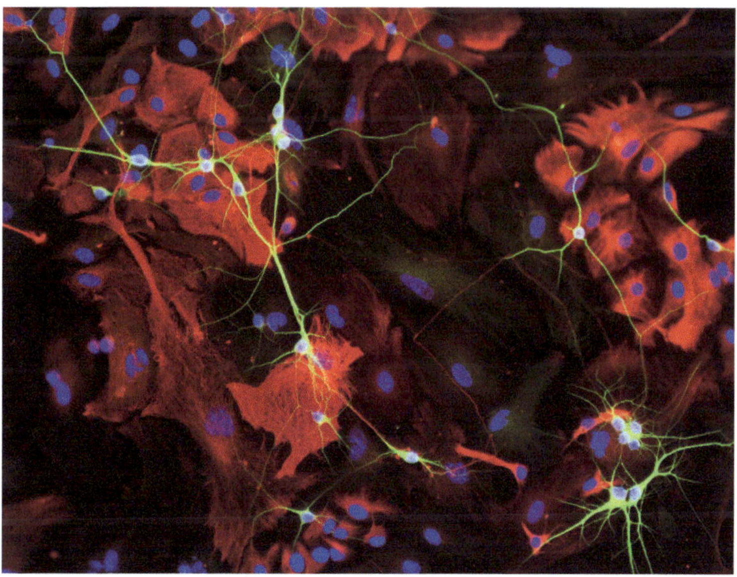

Fig. 16.2. Mouse hippocampal neuron–cortical astrocyte co-culture. Typical appearance of astrocytes identified with an antibody to GFAP (*red*) and neurons marked with an antibody to MAP2 (*green*). All cell nuclei counterstained with DAPI (*blue*). Co-culture photographed after 7 days in vitro.

## Table 16.1
## Troubleshooting guide

| Problem | Potential cause | Corrective measures |
| --- | --- | --- |
| Yield normal, but astrocytes die shortly after plating | Failure of astrocytes to adhere to substrate | Ensure coverslips are washed well during coating process |
| Sticky mass develops while isolating cells from tissue | Triturating too vigorously<br>Pipette tip too small<br>Pipette tip too sharp | Ensure pipette is appropriately fire-polished and work to triturate more steadily. (Trituration takes practice to determine the mechanics for optimal cell yield.) |
| Neuron yield low | Triturating too vigorously | Triturate at a steady rate, and expel against the wall of the tube |
| | Borehole of pipette too small | Ensure that the borehole of the reduced borehole pipette is no less than half the original diameter |
| | Pipette tip too sharp | Ensure that glass pipettes are fully fire-polished.<br>The tip of a 1 ml plastic pipette is too sharp for optimal release of neurons. |
| Slow initial development of neurons | Insufficient conditioning of NMM | Add NMM to glial cultures at 48 h prior to plating neurons |
| | Media exposed too long to air while plating | Ensure plates are out of incubator for minimum time possible. Add cells to plates only 12 wells at a time, and return to incubator between each set. |
| Suboptimal neuron survival or maturation | Glial feeder layer has high numbers of microglia (toxic to neurons), or fibroblasts (do not support neuron growth) | Confirm there is a high (>95%) of GFAP-positive cells. Discard glial cultures with a high percentage of microglia (small, round, phase bright, loosely attached), or other GFAP-negative cells |
| Contamination of culture[a] | Yeast (10 μm diameter with budding structures) – often as a result of breathing or coughing on the culture | Reduce talking while completing isolation, or wear surgical mask |
| | Filamentous fungi that form mats | Ensure there is no contamination in incubator. Clean incubator. |
| | Bacterial (rods/spheres, <1 μm diameter) – media will turn yellow. | Improve sterile technique.<br>Determine source of contamination: media, substrate for coverslips, dissection media<br>Replace source of contamination |

[a]We do not recommend the use of antifungal or antibacterial agents. It is better to improve sterile technique than to risk potential development of antibiotic resistance of contaminant or negative effects on the cell physiology.

An example of a neuron–astrocyte co-culture is shown in **Fig. 16.2**. We obtain populations of astrocytes with > 95% of cells positive for the astrocyte marker, glial fibrillary acidic protein (GFAP). In part, this could be due to conditions that promote GFAP+ cell growth, which are inherent to this protocol (e.g., seeding a low density, plating on laminin) (9). We suggest that this should be routinely monitored by immunocytochemistry. If a less than an ideal proportion of cells are positive for GFAP, or if a more pure population of astrocytes is desired, there are many options for attaining such a population as outlined in Chapter 11 and also reviewed in reference (9). *See* **Table 16.1** as a troubleshooting guide.

## Acknowledgments

We thank Stephanie Kaech-Petrie for her tireless advice that was instrumental in the development of this protocol.

This research is supported by the Natural Sciences and Engineering Research Council of Canada (NSERC) and the Fragile X Research Foundation of Canada.

## References

1. Bartlett WP, Banker GA An electron microscopic study of the development of axons and dendrites by hippocampal neurons in culture II. Synaptic relationships J Neurosci 1984;4:1954–1965.
2. Bi G, Poo M Synaptic modifications in cultured hippocampal neurons: dependences on spike timing, synaptic strength and post-synaptic cell type. J Neurosci 1998;18:10464–10472.
3. Yang Y, Ge W, Chen Y et al. Contribution of astrocytes to hippocampal long term potentiation through release of D-serine. Proc Natl Acad Sci (USA) 2003;11:15194–15195.
4. Kaech S, Banker G Culturing hippocampal neurons. Nat Protocols 2006;1:2406–2415.
5. Yoshida M, Saito H, Katsuki H Neurotrophic effects of conditioned media of astrocytes isolated from different brain regions on hippocampal and cortical neurons. Cell Mol Life Sci 1995;51:133–136.
6. Martin DP, Wallace TL, Johnson EM Jr. Cytosine arabinoside kills postmitotic neurons in a fashion resembling trophic factor deprivation: evidence that a deoxycytidine-dependent process may be required for nerve growth factor signal transduction. J Neurosci 1990;10:184–193.
7. Dessi F, Pollard H, Moreau J et al. Cytosine arabinoside induces apoptosis in cerebellar neurons in culture. J Neurochem 1995;64:1980–1987.
8. Chang JY, Brown S Cytosine arabinoside differentially alters survival and neurite outgrowth of neuronal PC12 cells. Biochem Biophys Res Commun. 1996;218:753–758.
9. Saura J. Microglial cells in astroglial cultures: a cautionary note. J Neuroinflammation 2007;4:26.

# Chapter 17

# Cerebellar Slice Cultures

## Josef P. Kapfhammer

## Abstract


Cerebellar slice cultures are a versatile method to study cerebellar cells in an in vitro preparation that preserves many aspects of the microenvironment of the cerebellar tissue. The setup of this slice culture model is easy, possible throughout the first two postnatal weeks in rodents, and does not require sophisticated equipment. Cerebellar slice cultures can be used for the study of electrophysiological properties of cerebellar neurons, for studying cerebellar development and various aspects of the biology of cerebellar cells. Several antibodies are available, which can be used as cell-specific markers for most cerebellar neurons and allow a good visualization of their morphology. Cerebellar slice cultures reduce the need for in vivo experiments and have developed into an important tool for the study of cerebellar biology.

**Key words:** Cerebellum, organotypic slice culture, organ culture, Purkinje cells, granule cells, deep cerebellar nuclei, dendritic development, axonal growth, postnatal, mouse.


---

## 1. Introduction

### 1.1. Slice Cultures of Nervous Tissue

The slice culture method today is a standard tool for the study of various aspects of nervous system function and development. Slice cultures are quite distinct from standard cell culture of dissociated cells. In dissociated cell cultures, individual cells grow in isolation or form an artificial monolayer on a culture substrate. Such cultures are very useful for studying individual cells, but of course they have the problem that both the microenvironment of the CNS and the neighbour relations and cell contacts of the neurons are lost. Because in slice cultures a tissue slab of considerable thickness (typically between 0.3 and 0.5 mm) is maintained


L.C. Doering (ed.), *Protocols for Neural Cell Culture*, Springer Protocols Handbooks,
DOI 10.1007/978-1-60761-292-6_17, © Humana Press, a part of Springer Science+Business Media, LLC 1992, 1997, 2001, 2010

in culture, most of the neighbour relations remain intact and the cells are retained in a virtually intact microenvironment. For neurons this means that also the local neuronal network connections are preserved, and that in CNS slice cultures local fiber projections are present. For this reason the acute slice preparation has become the standard preparation for electrophysiological studies in the CNS. The slice culture is nothing else than an acute slice preparation kept and maintained in culture for a prolonged time period. With the culture of such a thick slab of nervous tissue, several problems arise, which need to be addressed by the culture method: The thick slices show poor survival under conventional tissue culture conditions, mostly because the penetration of oxygen and nutrients is not sufficient to assure development and survival of cells making up the slice. Therefore, some special provisions have to be made in slice cultures to enhance supply of nutrients and oxygen to the slice. The most important and critical limitation in this respect is that slice cultures can only be derived from rather immature nervous tissue, typically from the early postnatal period. When slices are derived from older animals, there is poor survival of the neurons in the slice. The time window until when slice cultures can be successfully set up usually ends with the appearance of overt myelination in the tissue. For this reason, historically, the first successful attempts to culture intact slices of tissue were made in a developmental context from embryonic or neonatal animals (1–4). For the cerebellum which has a protracted development with the generation of granule cells going on for approximately three postnatal weeks in rats and mice (5), the situation is rather favorable and slice cultures can be set up from animals up to approximately 2 weeks of age (postnatal day 14, P14). The other important aspect is to provide the cultured slices with a sufficient supply of oxygen inside the slice. This has been originally achieved by the so-called roller tube method (6). In this method the tissue slice is glued onto a glass coverslip by a drop of plasma and placed in a glass tube which is slowly rotated in a roller drum maintained in an incubator at approx. 37°C. As a result, the tissue slice will be covered by culture medium for about half of the time, and exposed to air for the other half of the time. The constant agitation of the medium and the exposure to air ensure a good oxygen supply to the tissue and allow the culture of the slices for weeks and months. The critical aspect of this method is the exposure of the slice both to the air in the incubator for oxygen supply and to the culture medium for the supply of nutrients, because the oxygen supply through oxygen dissolved in the culture medium would not be sufficient for prolonged culture. More recently, a similar improved supply of oxygen to the slice has been achieved with a simplified method originally described by Yamamoto et al. (7), and now typically prepared according to the protocol of Stoppini et al. (8). In this modification, the tissue slices are placed and maintained on a permeable membrane with large pores of approx. 0.4 μm which will allow all medium components to easily penetrate the membrane but will prevent cells from migrating out of the slice. Culture medium is only supplied underneath this membrane and because of capillary forces will cover the explanted slice and even form a thin layer of medium on top of the slice (**Fig. 17.1**). The top of the slice is constantly exposed to the air interface, allowing an improved penetration of oxygen into the slice. This arrangement allows maintaining the cultures for similar periods as with the roller tube technique in regular 6-well plates without any agitation in a regular tissue culture incubator and, therefore, has now replaced the roller tube method in most laboratories. It is the method described in this protocol. It should clearly be stated that this method is in no way superior to the roller tube method and only provides a technical and practical simplification of the cultures. Where the roller tube method is available, it will certainly yield cultures of equivalent quality as obtained with the static method.

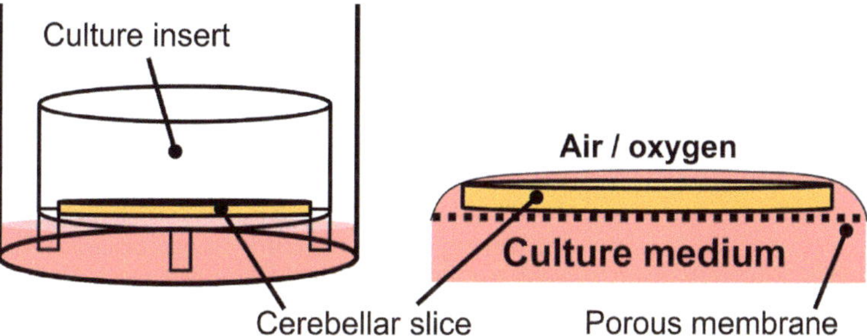


Fig. 17.1. Schematic drawing showing the arrangement of static slice cultures. The slices lie on a porous membrane in a tissue culture insert. Culture medium is provided underneath the membrane, and all medium components have free access to the slice due to large pore size (0.4 μm). Due to capillary forces the culture medium forms a thin film on top of the slice. Oxygen from the air can diffuse through this thin film of medium into the depth of the slice.

**1.2. Specific Aspects of Cerebellar Slice Cultures**

The slice culture method today is most widely used for the culture of hippocampal slices. Although the culture of cerebellar slices was part of the original publication of the roller tube method by Gähwiler (6, 9), it has been less frequently used than the hippocampal version. For setting up cerebellar slice cultures, the cerebellum is dissected out of the brain and isolated from the brainstem by cutting through the cerebellar peduncles. This results in a complete transection of all extrinsic afferents to the cerebellum, i.e., of the mossy fibers arising from the pons and other brain areas and the climbing fibers arising from the inferior olive. This will lead to the complete absence of outside afferents to the granule cells and Purkinje cells. In the same way, the efferent axons of the deep cerebellar nuclei projecting out of the cerebellum will be transected during the preparation of the culture. By co-culturing cerebellar slices with explants of the inferior olive, it has been possible to partially rebuild the climbing fiber projection to the cerebellum in slice cultures (10, 11).

Typically, cerebellar slice cultures are prepared by slicing the cerebellum in the sagittal plane. Due to the sagittal orientation of the Purkinje cells in the cerebellar folia, it is possible to preserve Purkinje cell with the majority of its processes intact within a slice. Outside the most median zone, in the parasagittal area, the deep cerebellar nuclei will be included in the slice. In such slices some Purkinje cells are maintained with their dendritic tree, and an intact axonal projection to the deep nuclei and functional connections between Purkinje cells and neurons of the DCN are present (12). The slice preparation in the sagittal plane is more traumatic for the granule cells. In vivo, their axons run perpendicular to the sagittal orientation and, therefore, become transected during slice preparation. However, because granule cells are still immature at early postnatal stages and many granule cells are only born at or shortly before the preparation of the slice, granule cell axons do regrow and a robust granule cell–Purkinje cell connection develops in cerebellar slice cultures (13). Therefore, the two-step connection granule cell–Purkinje cell–deep cerebellar nuclei is present in cerebellar slice cultures. There is also evidence for the development of most types of inhibitory interneurons in cerebellar slice cultures. After some time in culture, spontaneous electrical activity can be recorded from cerebellar slice cultures (14–16) and the cultures are a convenient system to study electrophysiological properties of cerebellar cell types (12, 17, 18). Despite the absence of extrinsic afferents, mossy fiber-like terminals are present in the cultures when analyzed by electron microscopy (19). They are likely to arise from unipolar brush cells, a glutamatergic cell type present in the granule cell layer (20). Astrocytes and Bergmann glial cells are present in slice cultures, and an extracellular matrix including the formation of perineural nets was shown to develop in slice cultures (21). One important advantage of using cerebellar slice cultures is the availability of the anti-Calbindin immunostaining procedure to label Purkinje cells. This antibody stain is Purkinje cell-specific within the cerebellum (22), and allows to brightly visualizing all Purkinje cells including dendrites, cell soma, and axon, but no other cells in the cerebellum. Due to this stain cerebellar slice cultures can be used conveniently for studies analyzing Purkinje cell morphology. Other antibodies are available to label additional cerebellar cell types, for example, anti-Parvalbumin for inhibitory interneurons (22), anti-Calretinin for unipolar brush cells (20), and NeuN for granule cells (23, 24).

<table>
<tr><td>

**1.3. Questions Addressed in Cerebellar Slice Cultures**

</td><td>

Cerebellar slice cultures have been and are used predominantly for the study of various aspects cerebellar development. I will give a few examples of how cerebellar slice cultures have been used to address such issues. One approach was to analyze the development of the slice cultures in the absence of granule cells (25) or electrical activity (26–28). Interestingly, the observed changes were rather subtle and Purkinje cell morphology developed virtually normal in such activity-deprived cultures. The dendritic development of Purkinje cells has also been studied in slice cultures, and Purkinje cell dendritic development has been shown to be affected by activation of protein kinase C (29, 30), metabotropic glutamate receptors (31), and protein tyrosine phosphatase RPTPbeta [(32), reviewed in (33)]. The possibility to transfect genes into neurons in organotypic cultures and to use tissue from mutant animals has offered additional options for the study of neuronal development. The importance of the transcription factor retinoid orphan receptor alpha (RORα) for the differentiation of Purkinje cells has been recently demonstrated by using slice cultures derived from RORα-mutant mice and by overexpressing RORα in Purkinje cells (34). By transfecting single Purkinje cells with EGFP in slice cultures, it has been possible to follow the development of the dendritic tree of these cells by two photon microscopy. These studies have shown that dendrites extend in close association with processes of Bergmann glial cells (35). The response of Purkinje cells to axonal lesions has also been studied in slice cultures (36). Because the Purkinje cell axons in the slice cultures become myelinated during the culture period, these cultures are also a valid model system for the study of myelination. Using cerebellar slice cultures it was shown that progesterone added to the cultures promoted myelination and oligodendrocyte proliferation (37, 38). Recently, cerebellar slice cultures have been reported to be an effective in vitro model for the replication of prions (39).

</td></tr>
</table>

## 2. Reagents and Equipment

**2.1. Equipment**

1. Sterile workplace, preferably a class I (EN 12469) horizontal laminar flow workbench (e.g., from BDK, Germany)
2. Stereomicroscope (e.g., Zeiss Stemi2000)
3. Tissue chopper (e.g., McIlwain)

4. Dumont No. 5 forceps for dissection, additional surgical instruments

5. Research microscope (e.g., Olympus AX70) equipped with fluorescence illumination and appropriate filters for completely separating signals of Alexa568 (red) and Alexa488 (green).

**2.2. Reagents**

Tissue-culture supplies:
1. 6-well plates (e.g., BD Falcon 353046)
2. 35 mm dishes (e.g., Greiner 627102)
3. Sterile pipettes for liquid handling, e.g., Nunc 159633
4. Millicell CM tissue culture Inserts (Millipore Cat. No. PICM 03050)
5. If viewing the insert on a non-inverted microscope is required, an alternative model with a lower rim is available (Millipore PICM ORG50)

**2.3. Media**

*2.3.1. Preparation Medium*

For 100 mL:
1. 99 mL MEM (e.g., Gibco 11012)
2. 1 mL glutamax I (Gibco 35050)
3. Adjust pH to 7.3 with HCl or NaOH

*2.3.2. Serum-Containing Incubation Medium*

For 100 mL:
1. 48.35 mL MEM (e.g., Gibco 11012)
2. 25 mL Eagle medium (e.g., Gibco Cat. No. 21400)
3. 25 mL horse serum, heat-inactivated (e.g., Gibco Cat. No. 26050)
4. 1 mL glutamax I (Gibco 35050)
5. 0.65 mL of a sterile 10% glucose solution
6. Adjust pH to 7.3 with HCl or NaOH

*2.3.3. Serum-Free Incubation Medium*

For 100 mL:
1. 97 mL neurobasal A medium (Gibco Cat. No 10888)
2. 2 mL B27 supplement (Gibco Cat. No. 17504)
3. 1 mL glutamax I (Gibco 35050)
4. Adjust pH to 7.3 with HCl or NaOH

**2.4. Fixative**

4% Paraformaldehyde (e.g., made from Merck Cat. No. 818715) in 0.1 M phosphate buffer (PB)

**2.5. Antibodies**

1. Anti-Calbindin (e.g., Swant Cat. No. CB38)
2. Anti-NeuN (e.g., Chemicon Cat. No MAB377)

3. Alexa Fluor 568 goat anti-rabbit IgG (Molecular Probes Cat. No. A11011)

4. Alexa Fluor 488 goat anti-rabbit IgG (Molecular Probes Cat. No. A11001)

***2.6. Animals***

1. Postnatal mouse pups (e.g., C57Bl6 at P8)

## 3. Protocol

### 3.1. Considerations Before Beginning

Animal model: The protocol described here is for mouse cerebellum, as the mouse has become the most commonly used animal model in neurosciences. However, this protocol can be used with slight modifications also for rat tissue.

Age of mouse pup at age of dissection: The preparation of cerebellar slice cultures is possible over a rather wide time window, certainly ranging from newborn (postnatal day 0 = P0) to P14 cerebellum. It has to be kept in mind, however, that during this time period in the mouse the cerebellum undergoes dramatic changes, and the selection of the right age for the question addressed in the experiments is a critical issue. When cultures are prepared from neonatal cerebellum (P0–P3), only few granule cells will be present at the time of culture and Purkinje cells do not yet form a strict monolayer. The culture period will thus cover the period of granule cell proliferation and migration. As for the Purkinje cells, the axons are still able to extend and dendritic development is at an early very immature stage. Cultures from that age are thus particularly suited for the study of granule cell development, early Purkinje cell dendritic differentiation, and axonal growth.

The period from P4 to P6 is not often used for making cultures because Purkinje cell survival in cultures derived from that age is reduced (36). However, a certain number of Purkinje cells will be present in these cultures, and Purkinje cell survival can be stimulated by addition of pharmacological agents (40, 41). In the period from P7 to P9 Purkinje cell survival will be good again. The arrangement of Purkinje cells in a monolayer and the polarization of Purkinje cells in cultures derived from that age will be much improved compared to cultures from younger pups. During this time period, granule cell migration from the external to the internal granular layer is at maximum and can be nicely studied in cerebellar slice cultures (13, 42). As Purkinje cells at the time of explantation only have very small dendritic arbors, such cultures are well suited for studies on dendritic extension of Purkinje cells (28, 33). In cultures derived from even older mouse pups, most Purkinje cells will be polarized and organized in a monolayer. These cultures are best suited for studies of more mature cerebellum.

**3.2. Dissection of Postnatal Mouse Cerebellum**

1. All steps need to be carried out under sterile conditions in a laminar flow workbench, using sterilized solutions and surgical instruments.

2. Decapitate mouse pups.

3. Remove brain from skull and place immediately in 35 mm dish containing 4°C cold preparation medium.
   Further steps under stereomicroscope:

4. Isolate the cerebellum by cutting through the cerebellar peduncles with dissection forceps.

5. Carefully remove visible pieces of the meninges on top of the cerebellar surface. Small rests of meninges are tolerable.

6. Transfer cerebellum to the tissue chopper, align it such that the blade will cut in the sagittal plane.

7. Cut 350 µm thick slices and transfer to a fresh dish with ice-cold preparation medium.

8. Separate slices adhering to each other with the forceps. One P8 cerebellum should yield about 10–12 slices. (Small slices from the lateral edges are not used.)

9. Transfer individual slices to the cell culture insert and place them on top of the membrane.

10. Up to 4–8 slices can be placed onto one insert.

*Selection of slices:* After cutting in the sagittal plane, the majority of the obtained slices will be derived from the cerebellar hemispheres, and a few (typically 2–4) slices from the vermis. For most applications slices from any location can be used. It should be noted, however, that vermal slices are typically larger and contain more folia compared to hemispheric slices but lack the deep cerebellar nuclei which are not present in the midline region. Purkinje cells in vermal slices thus are always disconnected from their axonal targets, whereas many Purkinje cells in hemispheric slices retain an intact axon to the deep cerebellar nuclei.

**3.3. Incubation of the Slice Cultures**

1. Place tissue culture insert in a 6-well plate with 0.75 mL incubation medium or serum-free incubation medium.

2. Incubate at 36–37°C with 5% $CO_2$.

3. Replace culture medium every 2–3 days.

4. Maintain culture for up to 4 weeks. Typical culture period is 10–14 days.

**3.4. Fixation and Immunohistochemistry**

1. Remove culture medium.

2. Carefully add 2 mL of 4°C cold fixative to each well.

3. Fix for 12–48 h at 4°C.

4. Rinse 3× with PB.

5. Cultures may be stored at 4°C for up to 2 weeks in PB prior to immunostaining.

***3.5. Immunostaining for Calbindin and NeuN***

1. Dilute rabbit anti-calbindin D-28 K (1:1000) and mouse anti-NeuN (1:500) in 0.1 M PB containing 3% goat serum and 0.3% Triton X100.

2. Incubate fixed slices overnight at 4°C under slight agitation.

3. Rinse 3× with 0.1 M PB.

4. Incubate fixed slices for 2 h with appropriate second antibodies (e.g., anti-mouse Alexa fluor 488 and anti-rabbit Alexa fluor 568) in 0.1 M PB containing 0.1% Triton X100.

5. Rinse 1× in 0.1 M PB containing 0.05% Triton X100.

6. Rinse 3× in 01.M PB.

7. For viewing slices are whole-mounted on microscopic slides and coverslipped with an aqueous mounting medium (e.g., Mowiol).

# 4. Critical Steps and Trouble Shooting

1. *Media*: should be used up within 4 weeks. pH needs to be checked and corrected if required.

2. *Dissection*: Preparation medium should be kept cold throughout (e.g., stored in refrigerator or kept on ice). Dissection should not take more than 15 min per cerebellum.

3. *Slicing*: When cerebella from older animals (e.g., P10) are used, the cerebellum might be rather large compared to the razor blade and adhere to the moving razor blade. Slicing under these conditions will not work well. The movement of the cerebellum with the blade can be prevented by keeping the plate of the tissue chopper rather dry (e.g., by sucking off surplus medium from plate). Care must be taken that the blade is well adjusted, when it is not parallel to the surface it may not cut through the cerebellum completely, and slices will be very difficult to separate.

4. *Problems with contamination*: Contaminations of the culture medium will result in bacteria or fungi growing in the medium compartment underneath the membrane.In contrast when contamination occurred during dissection and handling of the slices, bacteria or fungi will grow as colonies

on top of the membrane. This distinction helps to identify the reason for contamination.

5. *Few surviving Purkinje cells*: Check age of culture at dissection: in slices derived from P4 to P7 cerebella Purkinje cell survival is reduced.

## 5. Typical Protocol Results

The following examples are from cerebellar slice cultures derived from P8 mouse pups and cultured for 12 days. Purkinje cells are stained by anti-Calbindin (shown in red) and granule cells are stained with anti-NeuN (shown in green). **Figure 17.2** shows an example from such a culture in an area with well-preserved organization of the layers. There is an orderly arrangement of Purkinje cells in most parts of the culture. It should be noted that there is a considerable reduction of the thickness of the cerebellar slice during the culture period. Therefore, Purkinje cells from several levels of depth become compressed into one row. In the cultures, therefore, often there is no strict monolayer, but additional Purkinje cells can be located in the Purkinje cell layer (PCL) in close proximity to the original monolayer arrangement. The dendrites of the Purkinje cells extend up into the molecular layer (ML) where they overlap heavily in the plane of the culture and completely fill the molecular layer (**Fig. 17.2A, C**). The Purkinje cell axons grow across the internal granular layer (IGL) and form recurrent collaterals running back to the Purkinje cell layer. The axons form a fiber bundle in the cerebellar white matter (WM), which will run toward the deep cerebellar nuclei. Looking at the

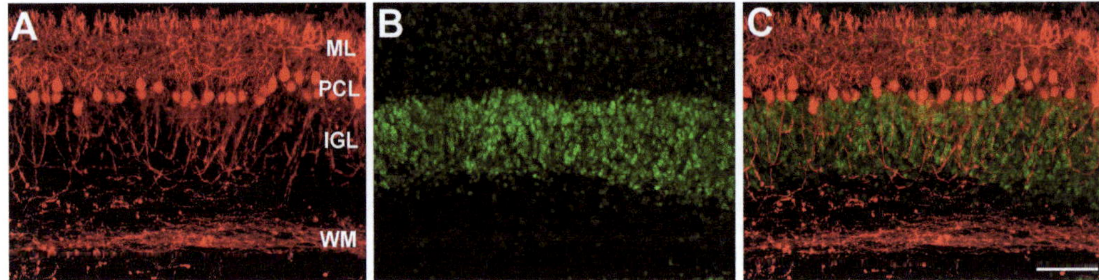


Fig. 17.2. Part of a folium in an organotypic slice culture after 12 days in vitro, photographed with the 20× lens. (**A**) Anti-calbindin staining labels Purkinje cells with their cell bodies well aligned in the Purkinje cell layer (PCL). Their dendrites extend in the molecular layer (ML) where they overlap heavily. The Purkinje cell axons grow across the internal granular layer (IGL) where they form recurrent collaterals running back to the Purkinje cell layer. The axons form a fiber bundle in the cerebellar white matter (WM), which will run toward the deep cerebellar nuclei. (**B**) Anti-NeuN staining to label postmitotoc granule cells. Granule cells are well arranged in the IGL. A substantial number of granule cells is still located in the ML probably still in the process of migration toward the IGL. (**C**) Combined image from A and B showing that the PCL is located immediately above the IGL Scale bar in C = 100 μm.

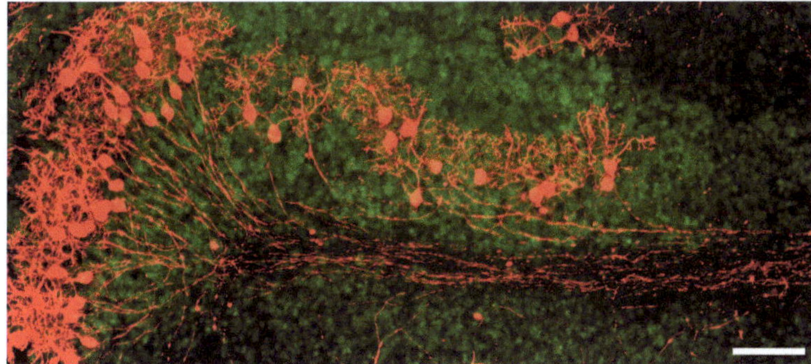

Fig. 17.3. View of folium in an organotypic slice culture after 12 days in vitro, photographed with the 10× lens. Anti-calbindin staining for Purkinje cells is shown in *red*, anti-NeuN staining for granule cells is shown in *green*. Due to the lower density of Purkinje cells in this area of the slice, the dendritic trees of individual cells can be recognized. The Purkinje cell axons assemble to form a fiber bundle in the cerebellar white matter. Scale bar = 100 μm.

granule cells (**Fig. 17.2B, C**) it becomes obvious that most granule cells are located in the IGL, directly underneath the Purkinje cells. Some granule cells are present in the ML, probably in the process of migrating toward the IGL (**Fig. 17.2B, C**).

**Figure 17.3** shows a folium in a cerebellar slice culture at a lower magnification. In this folium the density of Purkinje cells is reduced, revealing the expansion of the dendritic tree of individual Purkinje cells. The arrangement of Purkinje cells in the PCL is less strict as in the more densely populated area, but all Purkinje cells have maintained their polarization and extend their dendritic arbors into the ML. The axons of the Purkinje cells can be seen running across the IGL and assembling in the white matter to form an axon bundle running to the deep cerebellar nuclei.

In some cultures there are also areas with only few Purkinje cells present. In these areas it is possible to clearly see the entire dendritic tree of a cell without overlap with neighboring cells. An example of such an isolated Purkinje cell is shown in **Fig. 17.4**. The cell has an axon which runs to an axon bundle (*bottom*, out of focus) and has a sizeable dendritic tree with many side branches. It should be noted that the vast majority of this tree has grown during the culture period, because Purkinje cells have only a very small immature dendrite at the time the culture is set up (28).

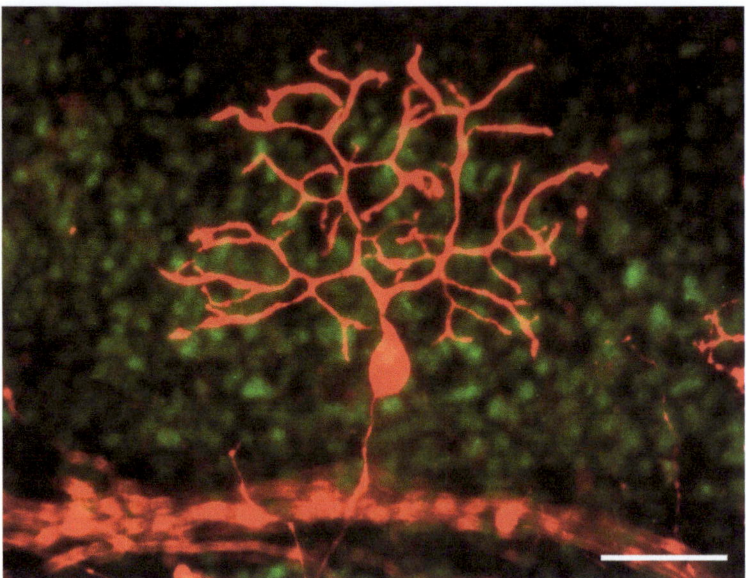

Fig. 17.4. View of a single Purkinje cell in an organotypic slice culture after 12 days in vitro photographed with the 40× lens. Anti-calbindin staining for Purkinje cells is shown in *red*, anti-NeuN staining for granule cells is shown in *green*. The large dendritic tree of this Purkinje cell can be seen. Scale bar = 50 μm.

## Acknowledgments

The author thanks Markus Saxer for technical assistance with the cerebellar slice culture, and the people who were involved in this project, in particular Alexandra Sirzen-Zelenskaya and Olivia Gugger. The work of our group is supported by the Swiss National Science foundation.

## References

1. Hild W. Myelinogenesis in cultures of mammalian central nervous tissue. Z Zellforsch Mikrosk Anat 1957; 46:71–95.
2. Bornstein MB, Murray MR. Serial observations on patterns of growth, myelin formation, maintenance and degeneration in cultures of newborn rat and kitten cerebellum. J Biophys Biochem Cytol 1958; 4:499 504.
3. Wolf MK. Differentiation of neuronal types and synapses in myelinating cultures of mouse cerebellum. J Cell Biol 1964; 22: 259–279.
4. Seil FJ. Neuronal groups and fiber patterns in cerebellar tissue cultures. Brain Res 1972; 42:33–51.
5. Altman J. Postnatal development of the cerebellar cortex in the rat. 3. Maturation of the components of the granular layer. J Comp Neurol 1972;145:465–513.
6. Gähwiler BH. Organotypic monolayer cultures of nervous tissue. J Neurosci Methods 1981; 4:329–342
7. Yamamoto Y, Kurotani, K, Toyama, K. Neural connections between the lateral geniculate nucleus and visual cortex in vitro. Science 1989; 245:192–194.
8. Stoppini L, Buchs P-A, Muller D. A simple method for organotypic cultures of nervous tissue. J Neurosci Methods 1991; 37: 173–182
9. Gähwiler BH. Slice cultures of cerebellar, hippocampal and hypothalamic tissue. Experientia 1984; 40:235–243.
10. Blank NK, Seil FJ, Leiman AL. Climbing

fibers in mouse cerebellum co-cultured with inferior olive. Brain Res 1983; 271:135–140.

11. Knöpfel T, Audinat E, Gähwiler BH. Climbing fibre responses in olivo-cerebellar slice cultures. I. Microelectrode recordings from Purkinje cells. Europ J Neurosci 1990; 3:343–348.

12. Mouginot D, Gahwiler BH. Characterization of synaptic connections between cortex and deep nuclei of the rat cerebellum in vitro. Neuroscience 1995; 64:699–712.

13. Tanaka, M, Tanaka M, Tomita A, Yoshida S, Yano M, Shimizu H. Observation of the highly organized development of granule cells in rat cerebellar organotypic cultures. Brain Res 1994; 641:319–327.

14. Leiman AL, Seil FJ. Spontaneous and evoked bioelectric activity in organized cerebellar tissue cultures. Exp Neurol 1973; 40: 748–758.

15. Seil FJ, Leiman AL. Development of spontaneous and evoked electrical activity of cerebellum in tissue culture. Exp Neurol 1979; 64:61–75.

16. Kapoor R, Jaeger CB, Llinás R. Electrophysiology of the mammalian cerebellar cortex in organ culture. Neuroscience 1988; 26: 493–507.

17. Audinat E, Knöpfel T, Gähwiler BH. Responses to excitatory amino acids of Purkinje cells' and neurones of the deep nuclei in cerebellar slice cultures. J Physiol 1990; 430:297–313.

18. Dupont JL, Fourcaudot E, Beekenkamp H, Poulain B, Bossu JL. Synaptic organization of the mouse cerebellar cortex in organotypic slice cultures. Cerebellum 2006; 5: 243–56.

19. Jaeger CB, Kapoor R, Llinás R. Cytology and organization of rat cerebellar organ cultures. Neuroscience 1988; 26:509–538.

20. Nunzi MG, Mugnaini E. Unipolar brush cell axons form a large system of intrinsic mossy fibers in the postnatal vestibulocerebellum. J Comp Neurol 2000; 422:55–65.

21. Brückner G, Grosche J. Perineuronal nets show intrinsic patterns of extracellular matrix differentiation in organotypic slice cultures. Exp Brain Res 2001; 137:83–93.

22. Celio MR. Calbindin D-28 k and parvalbumin in the rat nervous system. Neuroscience1990; 35:375–475.

23. Mullen RJ, Buck CR, Smith AM. NeuN, a neuronal specific nuclear protein in vertebrates. Development 1992; 116:201–211.

24. Weyer A, Schilling K. Developmental and cell type-specific expression of the neuronal marker NeuN in the murine cerebellum. J Neurosci Res 2003; 73:400–409.

25. Seil FJ, Drake-Baumann R, Herndon RM, Leiman AL. Cytosine arabinoside effects in mouse cerebellar cultures in the presence of astrocytes. Neuroscience 1992; 51:149–158.

26. Seil FJ, Drake-Baumann R. Reduced cortical inhibitory synaptogenesis in organotypic cerebellar cultures developing in the absence of neuronal activity. J Comp Neurol 1994; 342:366–377.

27. Seil FJ, Drake-Baumann R. Circuit reorganization in granuloprival cerebellar cultures in the absence of neuronal activity. J Comp Neurol 1995; 356:552–562.

28. Adcock KH, Metzger F, Kapfhammer JP. Purkinje cell dendritic tree development in the absence of excitatory neurotransmission and of brain-derived neurotrophic factor in organotypic slice cultures. Neuroscience 2004; 127:137–145.

29. Metzger F, Kapfhammer JP. Protein kinase C activity modulates dendritic differentiation of rat Purkinje cells in cerebellar slice cultures. Eur J Neurosci 2000; 12:1993–2005.

30. Schrenk K, Kapfhammer JP, Metzger F. Altered dendritic development of cerebellar Purkinje cells in slice cultures from protein kinase Cγ-deficient mice. Neuroscience 2002; 110:675–689.

31. Sirzen-Zelenskaya A, Zeyse J, Kapfhammer JP. Activation of class I metabotropic glutamate receptors limits dendritic growth of Purkinje cells in organotypic slice cultures. Eur J Neurosci. 2006; 24: 2978–2986.

32. Tanaka M, Tomita A, Yoshida S, Yano M, Shimizu H. Observation of the highly organized development of granule cells in rat cerebellar organotypic cultures. Brain Res 1994; 641:319–327.

33. Kapfhammer JP. Cellular and molecular control of dendritic growth and development of cerebellar Purkinje cells. Prog Histochem Cytochem 2004;39:131–182.

34. Boukhtouche F, Janmaat S, Vodjdani G, Gautheron V, Mallet J, Dusart I, Mariani J. Retinoid-related orphan receptor alpha controls the early steps of Purkinje cell dendritic differentiation. J Neurosci. 2006; 26: 1531–1538.

35. Lordkipanidze T, Dunaevsky A. Purkinje cell dendrites grow in alignment with Bergmann glia. Glia. 2005; 51:229–34.

36. Dusart I, Airaksinen MS, Sotelo C. Purkinje cell survival and axonal regeneration are age dependent: an in vitro study. J Neurosci 1997; 17:3710–3726.

37. Ghoumari AM, Ibanez C, El-Etr M, Leclerc P, Eychenne B, O'Malley BW, Baulieu EE, Schumacher M. Progesterone and its

metabolites increase myelin basic protein expression in organotypic slice cultures of rat cerebellum. J Neurochem 2003; 86:848–859.

38. Ghoumari AM, Baulieu EE, Schumacher M. Progesterone increases oligodendroglial cell proliferation in rat cerebellar slice cultures. Neuroscience. 2005;135:47–58.

39. Falsig J, Julius C, Margalith I, Schwarz P, Heppner FL, Aguzzi A. A versatile prion replication assay in organotypic brain slices. Nat Neurosci. 2008; 11:109–117.

40. Ghoumari AM, Wehrlé R, De Zeeuw CI, Sotelo C, Dusart I. Inhibition of protein kinase C prevents Purkinje cell death but does not affect axonal regeneration. J Neurosci. 2002; 22:3531–3542.

41. Ghoumari AM, Dusart I, El-Etr M, Tronche F, Sotelo C, Schumacher M, Baulieu EE. Mifepristone (RU486) protects Purkinje cells from cell death in organotypic slice cultures of postnatal rat and mouse cerebellum. Proc Natl Acad Sci U S A. 2003; 100: 7953–7958.

42. Mancini JD, Atchison WD. The NR2B subunit in NMDA receptors is functionally important during cerebellar granule cell migration. Neurosci Lett. 2007; 429:87–90.

# Chapter 18

# Hippocampal Slice Cultures

Jesse E. Hanson, Adrienne L. Orr, Silvia Fernandez-Illescas, Ricardo A. Valenzuela, and Daniel V. Madison

## Abstract

We have provided a detailed protocol for the preparation of interface hippocampal slice cultures from young mice or rats and have included modifications of the protocol necessary for culturing electrophysiologically viable slices from older animals. In addition to providing key points for successful measurements of synaptic function using electrophysiology, we have discussed approaches for studying models of neurological disease with hippocampal slice cultures. Future combination of the unique types of measurements afforded in slice cultures combined with the use of transgenic animals, exogenous reagent application, viral transfection, and the option to use slices from animals of different ages, promise to provide continued advances in understanding neural network phenotypes that may underlie neurological disease.

**Keywords:** Hippocampus, cell recording, viral transfection.

## 1. Introduction

The hippocampal slice culture preparation maintains the three-dimensional architecture of the hippocampus while at the same time permitting extensive experimental manipulation and measurement. This balance between the intactness of in vivo preparations and the tractability of dissociated neuronal cultures makes cultured slices a powerful tool for probing synaptic function in the hippocampus using techniques including electrophysiological measurement.

The traditional method of conducting electrophysiological studies on brain tissue in vitro has been to prepare brain slices from fresh tissue and then to record within a few hours of preparation (1). These "acute" slices of brain regions including the hippocampus have been used successfully for studying the electrical properties of brain for decades (2). The technical advantages

L.C. Doering (ed.), *Protocols for Neural Cell Culture*, Springer Protocols Handbooks,
DOI 10.1007/978-1-60761-292-6_18, © Humana Press, a part of Springer Science+Business Media, LLC 1992, 1997, 2001, 2010

of brain slices for conducting electrophysiological experiments are many, but mostly center around the exceptional accessibility, with landmarks such as cell body layers visible, and the ease of application of drugs and other substances. Many experiments that would be technically impossible to conduct in vivo can be accomplished relatively easily in vitro. However, acute slices also have some disadvantages. The very act of cutting brain tissue into slices severs numerous axonal connections and kills cells near the surface of the slice. In addition, while the electrical function of these slices remains intact for several hours, this decline in health renders them difficult to use for more than half a day or so, limiting the utility of acute slices for longer-duration experiments.

A solution to the short lifespan of acute slices was achieved with the advent of cultured organotypic slices using the "roller-tube" system (3). These cultured slices are prepared in a manner similar to acute slices, but are cultured in a unique roller-tube system, where they thin into monolayers that nonetheless maintain the overall organization of the source tissue. Slices can be maintained in this system for several weeks or more, allowing longer-term experiments. Although they continue to be a useful preparation, especially for applications involving imaging, (*see* for example (4)), roller-cultures never reached the popularity of acute slices, primarily because of difficulty of the preparation techniques and the difficulty in achieving consistent tissue quality. A solution to these difficulties was found with the development of the "interface" method of preparing organotypic cultures of brain tissue (5). The interface culture technique produces brain slices that maintain the three-dimensional architecture of the source tissue and can survive for many weeks, but can be produced by a technically easier and more consistent method.

There exist other earlier descriptions of this method (e.g. (5, 6)), so this chapter represents an update on the current implementation and applications of the interface type of hippocampal slice culture. The key aspect of this method is that brain slices are maintained on a porous membrane at the interface of a culture medium that provides moisture and nutrients, and the atmosphere of an incubator, allowing oxygenation at the surface of the slice. Detailed protocols for preparation and maintenance hippocampal slices from young rodents as well as modifications for making slices from older rodents are provided.

While the cultured hippocampal slice preparation is accessible to many different experimental measurements, including biochemical and imaging analyses, we will emphasize key points for successful application of electrophysiological measurements. In particular, techniques are described that are especially useful for probing the phenotypes of synaptic function in hippocampal slice culture models of neurological disease. Some specific methods of modeling disease in hippocampal slice cultures include application of pathogenic peptides during culture, transfection of slice cultures with disease-related constructs, and preparing slices from transgenic mice. The usefulness of these approaches in cultured slices has been demonstrated by advancements in understanding functional synaptic alterations in rodent models of diseases including Down Syndrome (7), Fragile X Syndrome (8, 9), X-linked mental retardation (10), HIV-associated dementia (11), and Alzheimer's disease (12, 13).

## 2. Reagents and Equipment

1. Separate 5% $CO_2$ incubators maintained at 37°C and 34°C.

2. Tissue culture hood.

3. Dissection microscope.

4. Halothane (Sigma #16730).

5. Tissue chopper (manual or automatic, e.g., Stoelting #51350).

6. "0.004" thick double-edged stainless steel razor blades, Personna (Electron Microscopy Sciences #72000).

7. Culture medium (~200 mL):
    (a) 100 mL minimum essential medium (MEM), 1× liquid with Hank's salts and glutamine (Invitrogen #11575).
    (b) 2 mL penicillin–streptomycin, liquid (Invitrogen #15140).
    (c) 2.5 ml HEPES buffer solution, 1 M liquid (Invitrogen #15630).
    (d) 50 mL Hank's balanced salt solution (HBSS), 1× liquid (Invitrogen #24020).
    (e) 50 mL horse serum, heat-inactivated (Invitrogen #26050).

8. Dissection medium (~100 mL):
    (a) 100 mL minimum essential medium (MEM), 1× liquid with Hank's salts and glutamine (Invitrogen #11575).
    (b) 1 mL penicillin–streptomycin, liquid (Invitrogen #15140).
    (c) 2.5 ml HEPES buffer solution, 1 M liquid (Invitrogen #15630).
    (d) 1 ml Tris solution, 1 M stock in sterile double-distilled water, pH 7.2 (Invitrogen #15504).

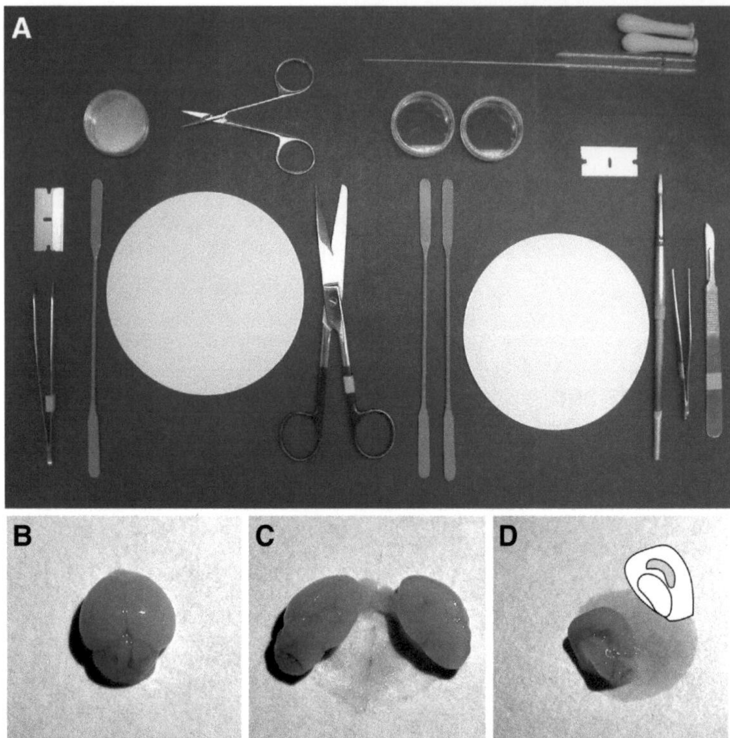

Fig. 18.1. Setup and brain dissection. (**A**) The lay out of instruments for slice culture preparation is shown. Instruments for brain removal surround the filter paper dissection surface on the left. Included are large scissors for decapitation, a razor blade for skin removal, small scissors for skull opening, forceps for skull removal, a spatula for brain removal, and a 35 × 10 mm dish for placing the brain in chilled dissection medium. Instruments for hippocampi removal surround the filter paper dissection surface on the right. Included are a razor blade for hemisphere transfection, spatulas for hemisphere opening and hippocampus extraction, a paintbrush for hippocampus transfer to the chopper, and two 35 × 10 mm dishes for collection of the intact and chopped hippocampi. Also included are tweezers and a scalpel for aiding in dissociation of individual chopped slices. Fire-polished Pasteur pipettes that are broken to accommodate individual slice transfer and unbroken for removing excess medium during plating are shown in the upper right. (**B**) The removed brain is shown. (**C**) The transfected hemispheres are shown with the medial surface oriented upward. (**D**) A hemisphere that has been spread open to expose the hippocampus is shown. The accompanying tracing illustrates the position of the hippocampus in gray. The hippocampus is lifted from the brain with a spatula and then chopped and plated as shown in **Fig. 18.2**.

9. Corning disposable sterile 250 mL filter, 0.22 μ pore size, CA membrane (Fisher #09-761-1)

10. Millipore millicell culture plate inserts, 31.5 mm, 0.4 μm pore size, hydrophobic PTFE (Fisher #PIC03050)

11. Pasteur pipette, fire-polished

12. Pasteur pipette, broken to diameter to allow collection of hippocampi, fire-polished.

13. Dissection instruments: large scissors, small scissors, forceps, two razor blades, three spatulas, paintbrush, tweezers, scalpel (*see* **Fig. 18.1A** for examples).

14. 12.5 cm filter paper sheets, Whatman grade 2 (Fisher 09-910F)

15. 35 × 10 mm BD Falcon tissue culture dish (Fisher 08-772A)

16. 60 × 15 mm BD Falcon standard dishes (Fisher 08-757-100B)

17. 150 × 15 mm BD Falcon standard dishes (Fisher 08-757-148)

18. Corning 25 cm$^2$ cell culture flasks (Fisher 10-126-30)

19. 10 mL serological pipettes (Fisher 07-200-574)

## 3. Protocol

### 3.1. Setup

All reagents and equipment are prepared in the tissue culture hood while wearing ethanol-sterilized gloves.

1. Prepare culture and dissection media by combining components listed in the reagents section in sterile filter containers and filtering with vacuum pressure. For the culture medium, the horse serum should be added last, immediately prior to filtering, to help prevent clogging of the filter membrane.

2. Prepare culture inserts by adding 1 mL of culture medium to each 35 × 10 mm culture dish using a 10 mL serological pipette. Then using fire-sterilized forceps place sterile tissue culture inserts into each culture dish. Up to seven 35 × 10 mm culture dishes are then contained in a 150 × 15 mm dish and placed in the 37°C incubator for at least 30 min prior to plating of slices to allow pH and temperature equilibration.

3. Prepare instruments for dissection in the tissue culture hood. Tools and surface for decapitation and brain removal are kept separate from tools and surface for dissection of the hippocampi (**Fig. 18.1A**). The tissue chopper is prepared by taping three sheets of filter paper, cut to size to the chopper surface, and mounting a fresh razor blade to the chopper arm. After all equipment is arranged, UV-sterilize the work area including the tools, chopper, and dissection microscope for at least 30 min.

4. Pre-chill dissection medium by placing the container from filtering in the freezer until agitating the container produces

slushy liquid. Fill three separate $60 \times 15$ mm dishes with slushy dissection medium for brain storage after removal and hippocampi storage before and after chopping. Reserve the remaining dissection medium for wetting the brain and tissue chopper surface during dissection and chopping.

### 3.2. Brain Removal

Slices are typically prepared from mice or rats on postnatal day 5–7. Modifications of the protocol for use with older animals are given in **Section 3.7**.

1. Animals are anesthetized with halothane by inhalation using a bell jar (approximately 2 MAC, 4% halothane in the air) and decapitated prior to brain removal for preparation of hippocampal slices.

2. The scalp is cut along the midline with a razor blade or scalpel and peeled to the sides.Small scissors are then used to cut the skull along the midline and forceps are used to peel the skull. Brains are removed and placed in chilled dissection medium using a spatula.

### 3.3. Dissection

Various techniques for isolating the hippocampus can be successfully used. Described here is the most common procedure used in our lab.

1. The brain is transferred to a piece of filter paper for dissection and kept moist with chilled dissection medium (**Fig. 18.1B**). The hemispheres of the brain are separated by a midline cut using a scalpel, and the medial aspect is oriented upward (**Fig. 18.1C**).

2. Using friction from two spatulas held nearly parallel to the medial brain surface, the thalamus and midbrain structures are pealed away from the cortex, exposing the hippocampus (**Fig. 18.1D**). A spatula is carefully slid underneath the hippocampus and the hippocampus is lifted out of the brain.

3. If undesired additional temporal structures remain attached to the hippocampus, the edge of a spatula may be used to trim the excess tissue.

4. Each dissected hippocampus is transferred into chilled dissection medium.

### 3.4. Slice Preparation

Manual or automatic tissue choppers may be used. The manual chopper is illustrated in this chapter.

1. The filter paper on the chopper is wet with dissection medium and excess solution is skimmed off using a Pasteur pipette. The force of the blade drop and the height of the blade stop are calibrated so that the blade leaves a mark on the top piece of filter paper but does not cause shredding of the filter paper.

2. The two hippocampi are transferred from the dissection media using a paintbrush and are oriented in parallel, longitudinally on the chopper surface. Slices are chopped at 400 micron intervals (**Fig. 18.2A**).

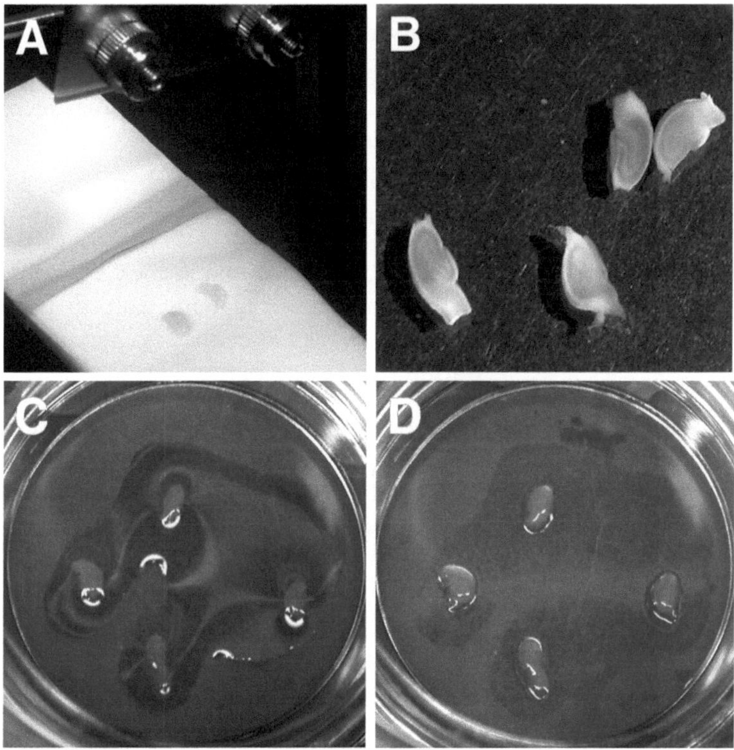

Fig. 18.2. Hippocampal slice preparation and plating. (**A**) Hippocampi are placed on the wetted filter paper, with the long axis of the hippocampus perpendicular to the chopping blade. (**B**) Floating hippocampal slices are shown prior to plating. Only slices with complete and distinct cell layers are selected for plating. (**C**) Four slices have been plated on a single culture insert. (**D**) The excess dissection medium has been removed from the surface of the culture insert membrane.

3. After chopping, the hippocampi resemble sliced loaves of bread and are transferred to chilled dissection medium in a 10 × 35 mm dish using a paintbrush or broken and fire-polished Pasteur pipette.

4. Individual slices are separated with vigorous swirling of the culture dish (with the lid on) while maintaining pressure against the surface of the culture hood to prevent overturning the dish. If adjacent slices remain in contact, small tweezers or a scalpel may be used to help dissociate the slices.

5. Good slices with intact hippocampal architecture and no damage are separated from incomplete or damaged slices, which are discarded.

### 3.5. Plating

1. Individual slices are placed on the pre-incubated culture inserts using a Pasteur pipette that has been broken to a diameter just larger than the hippocampus slice size of the species and age of animal being used (~2–7 mm) and has been fire-polished. Typically 3–4 slices are plated per culture insert.

2. An unbroken, fire-polished Pasteur pipette is used to remove the excess solution that accompanies the slice during plating.

3. Dishes of slices are then transferred to the 37°C incubator for initial storage.

### 3.6. Maintenance

Cultures are maintained in 5% CO2, at 37°C for 3 days and then at 34°C for the remaining culture period. Culture medium is replaced the day after slice preparation and then every 2–3 days during the culture period.

### 3.7. Protocol Modifications for Older Animals

While we have found that the standard protocol described above yields robust electrophysiologically viable slices when using brains from 5- to 7-day-old mice and rats, older tissue does not typically yield as healthy slices with this protocol. However, other groups have described protocols that can yield healthy slices in rats at least 40 days old as well as in mature mice (14, 15). Furthermore, such cultures could retainrobust electrophysiological properties and were maintained up to 3 months without the development of spontaneous epileptiform activity that occurs when cultures obtained from younger animals are maintained for prolonged periods. Based largely on adaptations of these previous protocols, described here are the modifications to our protocol for young animals that we have used to successful prepare slices from mature animals for electrophysiological recording.

1. Rongeurs are needed during removal of the skull.

2. After the brain hemispheres are separated, they are placed in pre-chilled (0–4°C), oxygen-saturated (95% $O_2$, 5% $CO_2$), modified Gey's balanced salt solution (mGBSS) for 20 min. The mGBSS solution is composed of (in mM) CaCl2, 1.5; KCl, 4.9; $KH_2PO_4$, 0.2; $MgCl_2$, 11.0; $MgSO_4$, 0.3; NaCl, 138.0; $NaHCO_3$, 2.7; Na2HPO4, 0.8; NaHEPES, 25; glucose 6% (w/v).

3. For the first 2 days in vitro (DIV), cultures are maintained in culture media with glucose supplemented to 28 mM and are maintained at 32°C. On DIV 3, slices are switched to a low-potassium medium consisting of 25% horse serum, 50% MEM, 25% modified Earle's balanced salt solution (mEBSS), 25 mM HEPES, and 28 mM glucose. The mEBSS is composed of (in mM) NaCl, 121.2; $NaHCO_3$, 26.2; $NaH_2PO_4$, 0.5; glucose, 5.6.

4. After 7 DIV, the cultures are moved to a 37°C and are maintained in low-potassium, low-horse serum medium consisting of 5% horse serum, 50% MEM, 45% mEBSS, 25 mM HEPES, and 28 mM glucose.

***3.8. Slice Electrophysiology***

Slices are typically recorded from after 7 to 14 days in culture. A scalpel is used to cut a piece of membrane surrounding a slice, and the slice and surrounding membrane are then carefully transferred to the electrophysiological recording chamber. All standard recording modes for analyzing synaptic input, including field potential measurements and whole cell recordings (**Fig. 18.3A**), can be readily performed in organotypic hippocampal slices. For example, spontaneous synaptic input can be measured using standard techniques of mini excitatory or inhibitory postsynaptic current recording (mEPSCs and mIPSCs; **Fig. 18.3B**). In addition, a major advantage afforded by electrophysiology in cultured rather than acute hippocampal brain slices is the ability to record from individual synaptic connections using paired whole-cell recordings (16). This is possible, because while the vast majority of axons are severed between source and target neurons in acute slices, the connections are reconstituted in organotypic slices. Of particular interest, using either measurements of individual synaptic connections or measurement of populations of synaptic connections with field potential recordings, the properties of synaptic plasticity can be readily examined in organotypic hippocampal slices (**Fig. 18.3C,D**).

***3.9. Approaches for Modeling Disease***

Using slices from transgenic mice and applying pathogenic peptides to slices during electrophysiological recording are relatively straightforward methods of modeling neurological disease in hippocampal organotypic slices. However, approaches involving transgene introduction or shRNA introduction to knock down genes of interest present challenges in cultured slices. This is because standard methods for transfecting dissociated neurons, such as calcium phosphate transfection or lipophilic reagent-mediated transfection, have proven ineffective in organotypic slices. Therefore, techniques such as viral delivery are necessary to transfect organotypic slices. While a variety of viruses may be used to transfect slice cultures including Sindbis virus, and Adenovirus, we favor lentivirus for the rapid and sustained expression of transgenes and the lack of toxicity. While undesirable submersion of the slice cultures and lack of penetration of virus through the glial cell membranes at the slice surface can hamper transfection by surface application of viral supernatant, we have found that injection of virus into the slice cultures can result in robust neuronal transfection (**Fig. 18.4**).

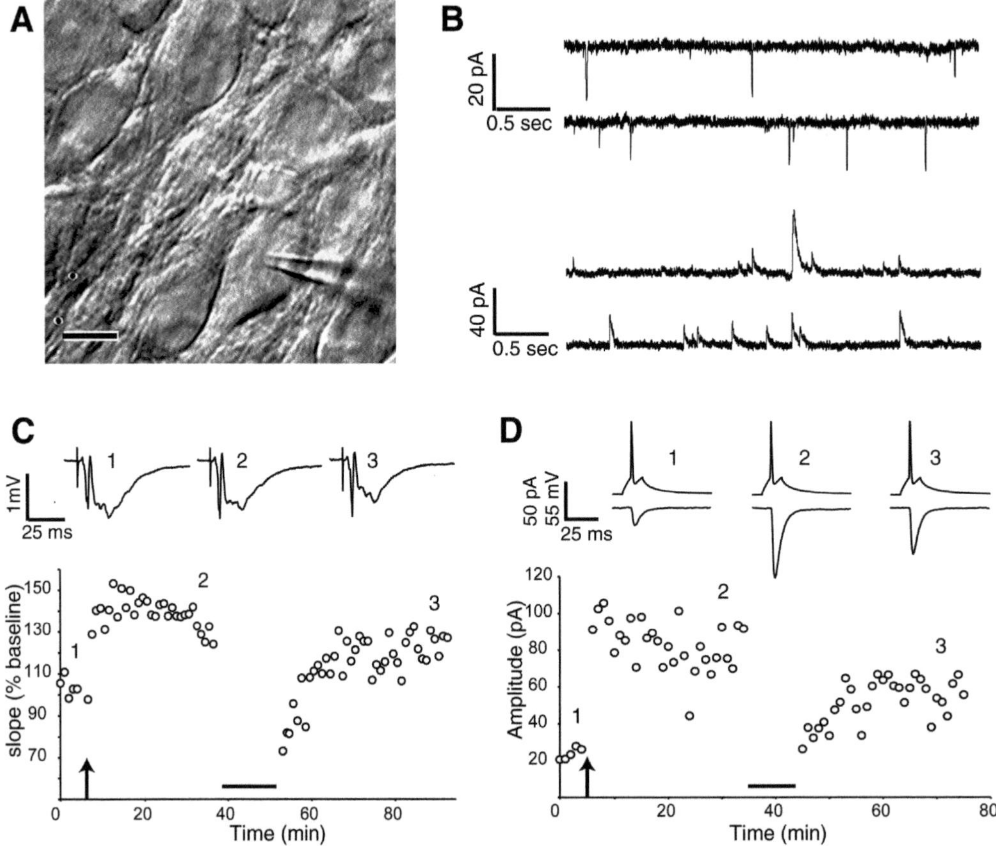

Fig. 18.3. Electrophysiological recordings from cultured hippocampal slices. (**A**) An example whole-cell patch clamp recording from a CA3 pyramidal neuron is shown. Scale bar is 10 μm. (**B**) Examples of recordings of spontaneous excitatory (*upper traces*) and inhibitory (*lower traces*) synaptic input are shown. Whole-cell recordings from pyramidal neurons in area CA3 were performed in the presence of TTX and mEPSCs were recorded in the presence of bicuculline at a −70 mV holding potential, while mIPSCs were recorded at 0 mV. (**C**) Examples of field recordings of evoked synaptic input are shown. Field recordings were made in area CA3 and field stimulation was applied in the dentate gyrus. Synaptic responses were measured as the initial slope of the response prior to occlusion by the reflected population spike. In this example experiment, synaptic response magnitudes are shown for a brief baseline period followed by induction of long-term potentiation by high-frequency stimulation (*up arrow*), followed by induction of depression (depotentiation) by 15 min of low-frequency stimulation (*horizontal bar*). Examples of synaptic responses are shown during the baseline (1), potentiated (2), and depotentiated (3) periods. (**D**) Examples of paired whole-cell recordings are illustrated. Synaptic currents were measured in one CA3 pyramidal neuron recorded in whole-cell voltage clamp mode in response to action potentials elicited by brief current injection pulses in a second CA3 pyramidal neuron recorded in whole-cell current clamp mode. Response amplitudes during a potentiation/depotentiation protocol equivalent to the one used in panel 3 are shown. Examples of presynaptic action potentials (upper) and postsynaptic EPSCs (*lower*) are shown during the three epochs of the experiment. Data shown in panels B and D are from cultured hippocampal slices prepared from young mice, while data in panel C are from cultured hippocampal slices prepared from a mature rat.

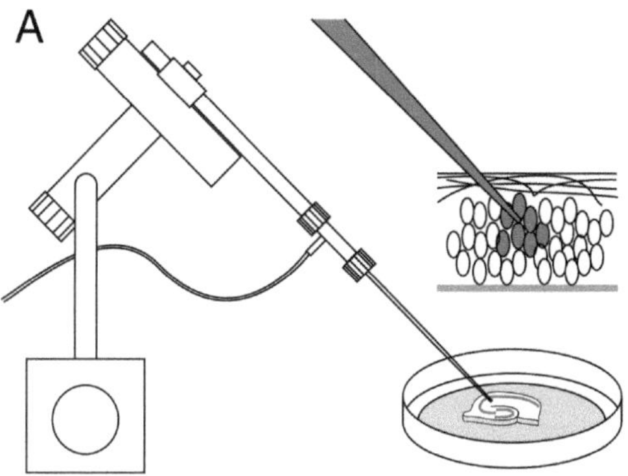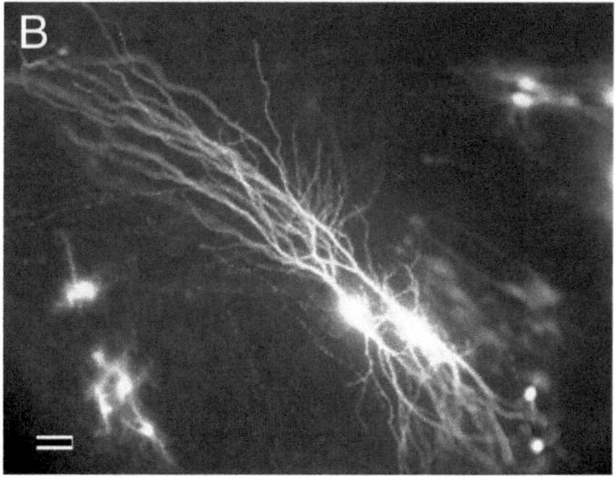

Fig. 18.4. Viral transfection of neurons in hippocampal slice cultures. (**A**) The setup for injection of virus into organotypic slices is diagramed. Virus is loaded into a micropipette that is mounted on an electrode holder, with a pressure port mounted on a micromanipulator. The micropipette is advanced through the glial membrane layer at the surface of the slice so that virus may be injected in position to infect neuronal cell bodies. Virus is injected using brief pressure pulses from a microinjector pressure pulse controller. (**B**) Example lentivirus transfected slice. Two pyramidal neurons that have been transfected with a lentiviral vector carrying an enhanced GFP transgene are clearly visible. Scale bar is 20 μm.

## 4. Critical Steps and Troubleshooting

1. One critical step for ensuring physiologically healthy slices is the selection of slices with intact hippocampal morphology during initial plating on the interface membrane. Slices without the distinct cell layers or with damage should be discarded.

2. Another critical step is the removal of excess liquid when plating slices. Initial recovery from slicing and growth in the incubator are sensitive to submersion in the dissection or culture medium. Therefore, careful removal of excess liquid from the surface of the culture insert around the perimeter of the slice should be performed using a fire-polished Pasteur pipette in order to ensure access of the surface of the slice to the atmosphere of the incubator (**Fig. 18.2C,D**).

3. For those familiar with whole-cell electrophysiology in acute slices, the presence of a layer of glial cell membranes on the surface of the cultured slices may present an unexpected obstacle. Prior to approaching the neuronal cells of interest for patch clamping, it is necessary to use relatively strong positive pressure from the tip of the pipette to pierce the glial membranes on the slice surface. Once the glial membrane has been compromised, one may proceed as normal to obtain a patch clamp recording from the underlying neuronal cell bodies.

4. One potentially significant problem in recordings aimed at measuring the monosynaptic excitatory response to

stimulation of presynaptic inputs is the presence of polysynaptic inhibitory responses that may occlude the measurement of excitation (17). These inhibitory responses are often more prevalent in slice cultures from rats compared to mice, and more pronounced in slices that have been maintained in culture longer. Unfortunately, isolation of excitatory responses cannot be achieved by application of GABA receptor antagonists alone, because loss of inhibitory function in the hippocampus leads to epileptiform bursting. This problem can be overcome in recordings of synaptic responses to the activation of populations of inputs by field stimulation, by co-applying the adenosine receptor agonist, 2-chloroadenosine (0.002 μM) along with the GABA antagonist picrotoxin (100 μM). This results in an overall quieting of synaptic activity that prevents epileptiform bursting (11). However, we have found that the suppression of excitatory input by 2-chloroadenosine precludes the robust measurement of individual synaptic responses recorded from pairs of individual neurons. Therefore, to overcome the problem of polysynaptic inhibition in paired recordings, we have used application of the mu-opioid agonist, DAMGO, which hyperpolarizes interneurons and prevents polysynaptic inhibition during the activation of single presynaptic neurons while leaving interneurons available to prevent epileptiform bursting (17).

5. Critical steps in lentiviral transfection of hippocampal slice cultures include using injection rather than surface application of viral supernatant (**Fig. 18.4A**) and minimizing epifluorescence exposure time when visualizing the fluorescent protein reporter to target neurons for whole-cell recordings. Excess fluorescence visualization can lead to phototoxicity causing neuronal disruption evident as rundown of the electrophysiological measurements. This problem can be avoided by using an automatic shutter to limit fluorescence exposure.

## 5. Typical Protocol Results

Preparation of typical organotypic slices are shown in **Figs. 18.1 and 18.2**. Visualization of the CA3 pyramidal cell layer during a typical whole-cell path clamp recording is shown in **Fig. 18.3A**. A variety of typical electrophysiological measurements from slice cultures are also shown in **Fig. 18.3**. Included are examples of spontaneous event recordings and evoked field potential recordings of synaptic plasticity. In addition, synaptic plasticity measurements are shown using a recording method uniquely possible in organotypic slices, in which pairs of neurons sharing intact synaptic connections are recorded from simultaneously. In previous studies, we have exploited this ability to record from individual connections to both gain insight into fundamental properties of synaptic plasticity and examine novel phenotypes in models of disease (7, 17, 18). Finally, example of transfection of neurons with a fluorescent reporter gene as might be used to target whole-cell recordings in transgene delivery or gene knockdown experiments is shown in **Fig. 18.4**.

## References


1. Madison, D. V. and E. B. Edson (2001). "Preparation of hippocampal brain slices." *Curr Protoc Neurosci* Chapter 6: Unit 6 4.
2. Bliss, T. V. and C. D. Richards (1971). "Some experiments with in vitro hippocampal slices." *J Physiol* 214(1): 7P–9P.
3. Gahwiler, B. H. (1981). "Organotypic monolayer cultures of nervous tissue." *J Neurosci Methods* 4(4): 329–42.
4. Bagal, A. A., J. P. Kao, et al. (2005). "Long-term potentiation of exogenous glutamate responses at single dendritic spines." *Proc Natl Acad Sci U S A* 102(40): 14434–9.
5. Stoppini, L., P. A. Buchs, et al. (1991). "A simple method for organotypic cultures of nervous tissue." *J Neurosci Methods* 37(2): 173–82.
6. Gahwiler, B. H., S. M. Thompson, et al. (2001). "Preparation and maintenance of organotypic slice cultures of CNS tissue." *Curr Protoc Neurosci* Chapter 6: Unit 6 11.
7. Hanson, J. E., M. Blank, et al. (2007). "The functional nature of synaptic circuitry is altered in area CA3 of the hippocampus in a mouse model of Down's syndrome." *J Physiol* 579(Pt 1): 53–67.
8. Pfeiffer, B. E. and K. M. Huber (2007). "Fragile X mental retardation protein induces synapse loss through acute postsynaptic translational regulation." *J Neurosci* 27(12): 3120–30.
9. Hanson, J. E. and D. V. Madison (2007). "Presynaptic FMR1 genotype influences the degree of synaptic connectivity in a mosaic mouse model of fragile X syndrome." *J Neurosci* 27(15): 4014–8.
10. Boda, B., S. Alberi, et al. (2004). "The mental retardation protein PAK3 contributes to synapse formation and plasticity in hippocampus." *J Neurosci* 24(48): 10816–25.
11. Behnisch, T., W. Francesconi, et al. (2004). "HIV secreted protein Tat prevents long-term potentiation in the hippocampal CA1 region." *Brain Res* 1012(1–2): 187–9.
12. Kamenetz, F., T. Tomita, et al. (2003). "APP processing and synaptic function." *Neuron* 37(6): 925–37.
13. Shankar, G. M., B. L. Bloodgood, et al. (2007). "Natural oligomers of the Alzheimer amyloid-beta protein induce reversible synapse loss by modulating an NMDA-type glutamate receptor-dependent signaling pathway." *J Neurosci* 27(11): 2866–75.
14. Leutgeb, J. K., J. U. Frey, et al. (2003). "LTP in cultured hippocampal-entorhinal cortex slices from young adult (P25-30) rats." *J Neurosci Methods* 130(1): 19–32.
15. Xiang, Z., S. Hrabetova, et al. (2000). "Long-term maintenance of mature hippocampal slices in vitro." *J Neurosci Methods* 98(2): 145–54.
16. Pavlidis, P. and D. V. Madison (1999). "Synaptic transmission in pair recordings from CA3 pyramidal cells in organotypic culture." *J Neurophysiol* 81(6): 2787–97.
17. Hanson, J. E., M. R. Emond, et al. (2006). "Blocking polysynaptic inhibition via opioid receptor activation isolates excitatory synaptic currents without triggering epileptiform activity in organotypic hippocampal slices." *J Neurosci Methods* 150(1): 8–15.
18. Montgomery, J. M., P. Pavlidis, et al. (2001). "Pair recordings reveal all-silent synaptic connections and the postsynaptic expression of long-term potentiation." *Neuron* 29(3): 691–701.

# Chapter 19

# Molecular Substrates for Growing Neurons in Culture

Saulius Satkauskas, Arnaud Muller, Morgane Roth,
and Dominique Bagnard

## Abstract

Culturing neurons is an effective way to analyze the basic mechanisms that govern nervous system wiring. The choice of an appropriate molecular substrate is of prime importance to the main objective of the culture. We describe the preparation of basic 2D substrates, complex 2D substrates, and 3D substrates to culture neurons to evaluate neurotrophic activity of exogenous compounds. Complex 2D substrates allow analysis of guidance signals in a mixture of molecules mimicking the in vivo environment. In this case, the spatial distribution of the factors can be controlled by forming stripes (alternating substrates) or gradients. The use of 3D substrates is in our opinion the best choice to analyze the growth of axons from explants exposed to guidance signals. This method establishes standard and highly reproducible culture conditions for a given neuronal type.

**Keywords:** Substrate, guidance signal, laminin, fibronectin, gradient, explant, axon.

## 1. Introduction

Growing dissociated neurons in culture is undoubtedly a key strategy to address the molecular and cellular mechanisms of the nervous system complexity. The intricate network of axonal connections progressively established during development requires sophisticated mechanisms ensuring appropriate neuronal differentiation and efficient growth and guidance of axons (1). The elementary but certainly not trivial task of growing an axon is the consequence of an optimal interaction between the extending process and its environment, namely the extracellular matrix (ECM). Axon elongation is primarily driven by a specialized motile structure of the distal extremity, the growth cone (2). Besides precise control of the actin cytoskeleton dynamics and myosin activity allowing efficient extension of the growth cone,

L.C. Doering (ed.), *Protocols for Neural Cell Culture*, Springer Protocols Handbooks,
DOI 10.1007/978-1-60761-292-6_19, © Humana Press, a part of Springer Science+Business Media, LLC 1992, 1997, 2001, 2010

the crucial determinant feature of axonal growth is the repeated formation and dissociation of contacts between growth cones and ECM. These contacts must be of sufficient strength to ensure growth cone adherence but sufficiently labile to avoid permanent anchoring and immobilization. This intuitive characteristic is actually the grounding element when establishing a neuronal culture. The molecular substrate has to be determined in order to provide such a subtle balance of adhesion and de-adhesion processes. The interactions between growth cones and ECM are mediated through transmembrane cell adhesion molecules, creating with their ligands a full repertoire of growth-promoting or growth-inhibiting signals. In the first part of this chapter, we describe how to prepare minimal two-dimensional substrates with standard neuronal cultures. However, depending on the neuronal type to be cultured and final objective of the culture, it may be necessary to prepare more complex substrates, mimicking in vivo-like environments. When choosing such complex substrates, it is understood that the molecular substrates will be mixtures of various compounds often produced in variable concentrations. Thus, opposed to well-defined basic substrates, complex substrates will be considered as undefined substrates. This is the case of membrane preparations (3–5) such as the one described in the second part of the chapter that provides a strong growth promoting substrate but lose the precision of exact substrate composition. Hence, we will also describe protocols allowing three-dimensional neuronal cultures. This approach is particularly suitable to study guidance signals requiring the production of molecular gradients.

## 2. Neuronal Culture on 2D Substrates

### 2.1. Basic Substrates

This part of the protocol is intended to provide a quick way to prepare molecular substrates favoring optimal axonal growth for many types of neurons (including cortical, hippocampal, and thalamic neurons among others). As discussed in the introduction, the goal is to obtain sufficient adhesion of neurons without excessive adhesion to avoid immobilization of growing axons. To this end, we routinely use poly-L-lysine at a concentration of 10 μg/mL. Nearly all types of cells adhere to this polymer of basic amino acids. While coating is possible directly on the bottom of the plastic dish, we always prefer glass coverslips coating to ensure better reproducibility of the substrate deposition and easiest handling of cultures for any subsequent immunocytochemical analysis.

***2.2. Material***

1. Poly-L-lysine (Sigma P1399; MW: 150–300 000)

2. Gey's balanced salt solution (GBSS, Gibco 24260-028)

3. 60 mm Petri dishes (BD Falcon, cat. no. 351016) or 6-well plates (BD Falcon, cat. no. 353046)

4. Round glass coverslips (Knittel Glasser, cat. no. VD 10012 Y100A) or rectangular glass coverslips (Knittel Glasser, cat. no VD1. 2250 Y100A)

5. Sterile workplace under laminar flow hood

## 3. Procedure

***Step 1: Preparation of glass coverslips***

1. Place coverslips in a glass dish and cover with xylene for 24 h

2. Remove xylene and rinse twice with acetone

3. Remove acetone and rinse twice with 100% alcohol

4. Remove alcohol and sterilize the coverslips (autoclave, generally 5 h at 180°C)

***Step 2: Preparation of stock solution and working solution***

1. Dissolve poly-L-lysine in water at a final concentration of 1 mg.mL$^{-1}$

2. Store this stock solution at –20°C in the form of 10 μL aliquots (up to 6 months)

3. Use 10 μL of this stock solution in 990 μL GBSS (keep on ice) to prepare working solution

***Step 3: Coating of round coverslips***

1. Place sterile coverslips directly in the culture dish

2. Add 70 μL of working solution per round coverslips

3. Let stand for 2 h at 37°C in a cell incubator (close the dish to avoid evaporation)

4. Remove excess of poly-L-lysine solution and rinse twice with sterile water

5. Remove water and add culture medium (generally 2 mL for a 60 mm dish)

6. Place the dishes in a $CO_2$ incubator while preparing the cells

***Step 4: Coating of rectangular coverslips***

1. Place sterile coverslips in a dish (100 mm)

2. Add 100 μL of the poly-L-lysine solution on each coverslips

3. Cover each coverslip by another one (sandwich-like structure)

4. Let stand for 2 h at 37°C in a cell incubator

5. Open the sandwich and rinse each coverslip with sterile water

6. Transfer coated coverslips to the Petri dishes (coated surface up in 60 mm or 6-well plates)

7. Place the dishes in a $CO_2$ incubator while preparing the cells

Coating should be performed directly at the bottom of the dish if the culture is devoted to protein or DNA/RNA purification. In this case, follow Step 3 after addition of the working solution of poly-L-lysine that must cover the whole surface of the dish (adjust volume accordingly). Round coverslips should be preferred in case of multiple immunocytochemical analysis of neuronal extension as a range of 3–5 coverslips per experimental condition can be added to 60 mm dishes or per well of 6-well plates. The use of rectangular coverslips has been optimized for neuronal culture grown in 60 mm dishes at a density of 200,000 cells/dish (or per well) in a volume of 2 mL culture medium. *See* **Fig. 19.1.**

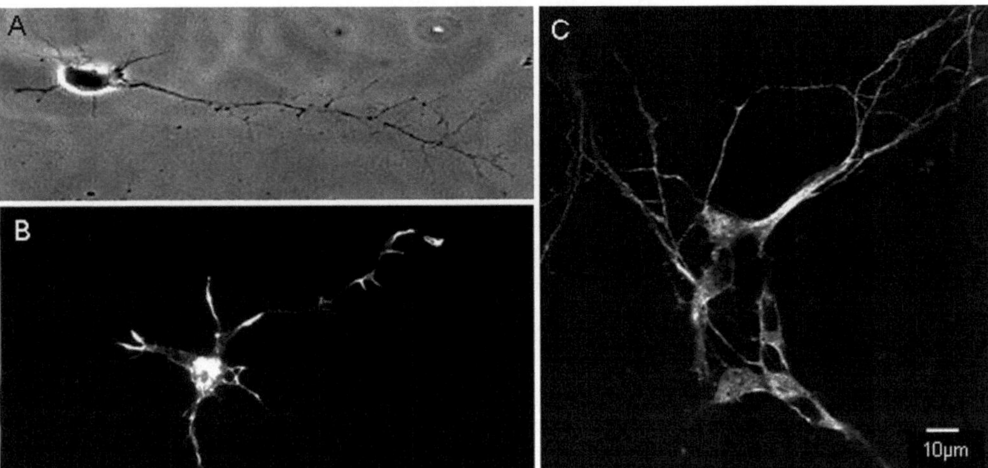

Fig. 19.1. Representative examples of neurons grown on 2D substrates composed of poly-L-lysine. **(A)** Phase-contrast microphotograph showing individual cortical axons after 24 h culture. **(B)** Visualization of the actin cytoskeleton of cortical neurons stained with Phalloïdin-Rhodamine. **(C)** Neuronal network can be obtained when increasing cell density. This microphotograph shows hippocampal neurons stained for MAP2 after 48 h culture.

## 4. Alternative Substrates

Poly-ornithine, another synthetic polyaminoacid will give similar results as that of poly-L-lysine. Virtually all growth-promoting or adhesive components of the ECM may serve as a substrate for neuronal culture. In fact, a limited number of factors are routinely used in the labs. This includes laminin (Reference: Laminin L2020, Sigma) or fibronectin (Reference: Fibronectin F1141, Sigma).

*Laminin:* Laminin is a crucial organizing element of the extracellular matrix promoting adhesion by interactions with integrins. The resulting outside-in signaling leads to the formation of focal contacts ensuring cell adhesion. A substrate composed of laminin is less adhesive compared to poly-L-Lysine, but laminin promotes rapid neuritic growth and neuronal polarity (6, 7). Additionally, laminin modulates neuronal sensitivity to trophic factors. Such influence is correlated to the concentration of laminin that may affect intracellular levels of calcium and subsequent signaling cascades. In certain cases, laminin is used to direct axon and dendrite growth (8).

A mix of laminin and poly-L-lysine (50%/50%) is frequently used in order to enhance cell survival and to trigger good axonal extension. When culturing explants, the proportion of each factor must be optimized because is it very difficult to obtain explant adhesion without poly-L-lysine while too much of poly-L-lysine will favor explant adhesion but will also favor axon fasciculation, thereby preventing individual fiber analysis. One good compromise found to be efficient for cortical explant is a mixture composed of 5 μL poly-L-lysine+15 μL laminin stock solutions in 980 μL GBSS.

*Fibronectin:* Fibronectin is involved in many cellular processes, including embryogenesis, blood clotting, tissue repair, and cell migration/adhesion. Fibronectin exists as an insoluble glycoprotein dimer that serves as a linker in the ECM (extracellular matrix) or as a soluble disulphide linked dimer found in the plasma. Many cell types including fibroblasts, chondrocytes, endothelial cells, macrophages, as well as certain epithelial cells or astrocytes produce fibronectin. Fibronectin is mainly used for the culture of neurons of the peripheral nervous system (9).

### *4.1. Complex Substrates*

Culturing neurons on basic PL or laminin substrates is particularly suitable when analyzing growth-promoting or differentiating effects of exogenous factors (neurotrophic factors, inhibitors) applied in the culture medium. These substrates provide an efficient way to produce reproducible substrates inducing easily standardized axon growth and might therefore be used for screening assays (10). On the other hand, it is often important to study molecular interactions under in vivo-like conditions. Evidently, culturing neurons will never match perfectly the in vivo environment. However, it is possible to enrich the complexity of the molecular substrate encountered by cultured neurons using various assays based on the use of membrane preparations.

### 4.2. Preparation of Uniform Complex Substrates

As previously described, membrane preparations offer a very good substrate favoring both the adhesion of neurons and explants and induce strong axonal growth. Membrane extracts can be obtained from postnatal cortical tissues (P0–P3).

### 4.3. Material

1. Homogenization buffer (10 mM Tris–HCl, 1.5 mM $CaCl_2$, 1 mM spermidine, 25 µg/mL aprotinine, 25 µg/mL leupeptine, 15 µg/mL 2,3-dehydro-2-desoxy-$N$-acetylneuraminic acid, pH = 7.4, all from Sigma)
2. PBS 1× and PBS+protease inhibitors
3. Sterile glass coverslips
4. Sterile workplace under laminar flow hood

### 4.4. Procedure

*Step 1: Membrane preparation*
1. Dissect blocks of cortical tissues in cold GBSS–glucose solution
2. Transfer blocks to the homogenization buffer
3. Homogenize the cortical tissue in ice-cold homogenization buffer first with a 1 mL pipette tip and then with a 27-gauge injection canula and incubate for 30 min at 37°C
4. Transfer homogenates in sucrose step gradients (upper phase 150 µL of 5% sucrose; lower phase 350 µL of 50% sucrose) and centrifuge (10 min, 50,000$g$)
5. Collect membrane fraction at the inter phase of the gradient (appears like a white cloudy solution)
6. Wash the fraction twice in PBS without $Ca^{2+}$ and $Mg^{2+}$ at 20,000$g$ in an Eppendorf Biofuge. PBS should be supplemented with the protease inhibitors aprotinine, leupeptine, pepstatine and 2,3-dehydro-2-deoxy-$N$-acetylneuraminic acid to prevent protein degradation.
7. After resuspension, the concentration of the purified membranes is determined by its optical density at 220 nm. Optical density of postnatal membranes solutions used for membrane carpets should be at 0.1 units. Membrane

preparations can be stored at –20°C for up to 6 months in glycerol (Vol/Vol).

### Step 2: Glass coverslips coating

1. Place a laminin-poly-L-lysine coated coverslips in a small Petri dish (50 mm diameter, Falcon) – (*see* Section 2.1)

2. Add 100 μL of membrane solution (OD: 0.1 at 220 nm)

3. Make a sandwich by placing a second coated coverslip side down onto the first coverslip

4. Let it stand for 3 h 37°C in a cell incubator

5. Separate the two coverslips and add 750 μL of culture medium onto the coated sides. Store the membrane-coated coverslips in the incubator until seeding the explants or dissociated neurons.

   **Note**: The same procedure can be applied to prepare membranes from cell lines stably expressing a factor of interest (An example is detailed in (4)). Because membrane preparation from non-neuronal cell lines (such as COS or KEK293 cells) may affect axonal growth, these extracts should be mixed with a membrane preparation from postnatal cortical tissues.

   This part of the protocol is particularly suitable to culture explants. The mixture of laminin–poly-L-lysine will provide an optimal adhesion and growth of axons while the membrane preparation carpet will provide additional growth factors and molecular partners as those found in vivo.

---

## 5. Preparation of Complex Alternating Substrates (Stripe Assay)

### 5.1. Material

1. Petri dishes filled with sterile PBS

2. Setup for stripe assay (vacuum pump, channel matrix, tubing *see* (11))

3. Silicon blocks kept in water

4. Polycarbonate membrane filters (Nuclepore or equivalent, pore size 0.1 μm)

5. Black nitrocellulose membrane filters (from Sartorius or equivalent)

6. Coated glass coverslips (*see* Section 2.1)

7. Sterile distilled water

8. Sterile PBS

9. Sterile workplace under laminar flow hood

***5.2. Procedure***

1. Sterilize the channel matrix and the silicon blocs in boiling water

2. Sterilize the nitrocellulose filters and polycarbonate filters under UV light

3. Rinse the setup for the stripe assay with 70% alcohol, water, PBS while using the vacuum pump

4. Use fresh membrane preparations or thaw frozen membrane preparations (kept in PBS/glycerol, 50/50 at –20°C)

5. Wash twice with PBS with centrifugation (10 min at 14,000 rpm)

6. Put a polycarbonate filter on the channels matrix (bright face on top) and rinsed with PBS

7. Apply pressure

8. Add 100 μL of membrane preparation

9. Maintain suction through the filter for 60–75 s

10. Remove the excess of solution

11. Transfer the filter on a Petri dish filled with PBS (coated side up)

12. Place a glass coverslips coated with laminin/poly-L-lysine in the bottom of Petri dish

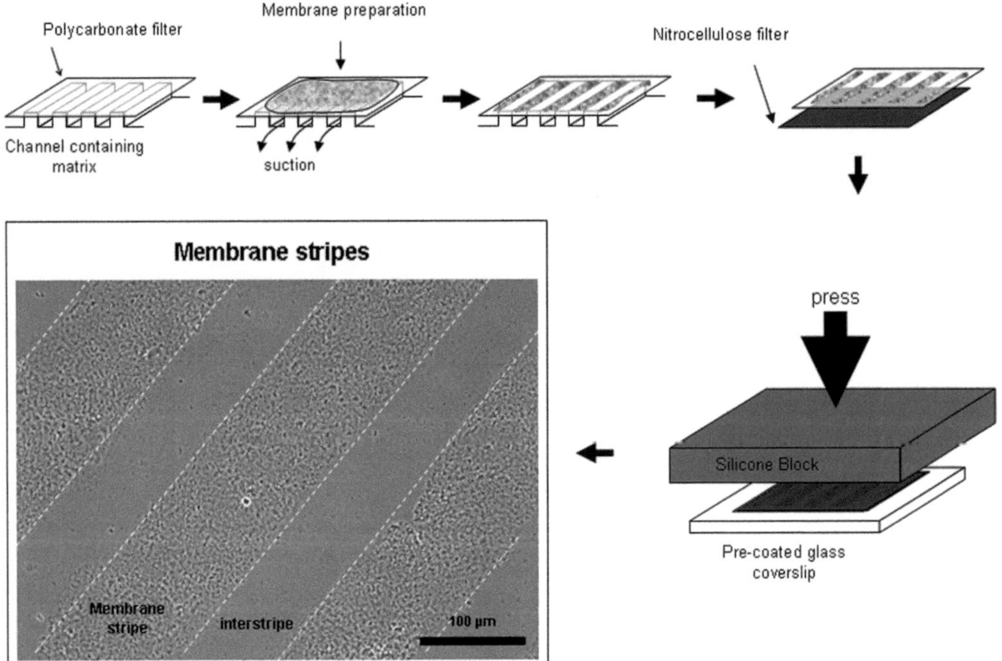


Fig. 19.2. Schematic representation of the procedure and setup used to produce alternating substrates.

13. Put the polycarbonate filter (coated side up) onto a black nitrocellulose filter. Avoid excessive drying

14. Place the filters coated face down on the glass cover slip and cover with the silicon bloc

15. Press the silicon bloc to transfer the stripes of membranes

16. Inject PBS at the junction between the silicone block and coverslips to remove the filter

17. Transfer the coverslips in a Petri dish and immediately fill with culture medium (*See* **Fig. 19.2**)
    **Note**: The initial protocol proposed by Walter and colleagues (11) allows deposition of a second membrane solution deposited onto adjacent stripes. To this end, the filter is placed on a fine nylon grid and the second membrane suspension is sucked onto the filter.

## 6. Preparation of 2D Gradients

Most of the molecular signals controlling axon growth and guidance require the formation of gradients to trigger their growth-promoting or growth-inhibiting effects. Thus, various techniques have been developed to expose cultured neurons to gradients of guidance cues. The turning assay, consisting of a microinjection of soluble factors close to growth cone in culture, is certainly a good way to study molecular gradients. However, this technique is time consuming and does not apply for membrane-associated factors and requires a complex setup ensuring continuing flow of the Petri dish to avoid saturation of the culture medium. Bayer and Boenhoffer (12) developed an alternative method to produce gradients from membrane extracts. This method has also been successfully used to study diffusible factors exhibiting strong adhesive properties toward ECM or cell membranes such as members of the class 3 Semaphorins (13). Interestingly, material and procedure are very similar to the ones used for stripe preparation.

### 6.1. Material

1. Petri dishes filled with sterile PBS

2. Setup for stripe assay (vacuum pump, channel matrix, tubing *see* (11))

3. Silicon blocks kept in water

4. Polycarbonate membrane filters (Nuclepore or equivalent, pore size 0.1 $\mu$m)

5. Black nitrocellulose membrane filters (from Sartorius or equivalent)

6. Coated glass coverslips (*see* Section 2.1)

7. Sterile distilled water

8. Sterile PBS

9. Sterile workplace under laminar flow hood

*6.2. Procedure*

1. Sterilize the channel matrix and the silicon blocs in boiling water

2. Sterilize the nitrocellulose filters and filter paper under UV light

3. Rinse the setup for stripe assay with 70% alcohol, water, PBS while using the vacuum pump

4. Use fresh membrane preparations or thaw frozen membrane preparations (kept in PBS/Glycerol, 50/50 at –20°C)

5. Wash twice with PBS with centrifugation (10 min at 20,000$g$)

6. Place polycarbonate filter on the channels matrix (bright face on top) and rinse with PBS

7. Add 100 µL of membrane preparation

8. Place a coverslips on the top of membrane solution at an angle of 45°

9. Apply pressure

10. Maintain aspiration through the filter until full aspiration of the membrane solution

11. Transfer the filter on a Petri dish filled with PBS (coated side up)

12. Place a glass coverslip coated with laminin/poly-L-Lysine in the bottom of Petri dish

13. Put the polycarbonate filter (coated side up) onto a black nitrocellulose filter. Avoid excessive drying

14. Place the filters coated face down on the glass cover slip and cover with the silicon bloc

15. Press the silicon bloc to transfer the gradient of membrane particles

16. Inject PBS at the junction between the silicone block and the coverslips to remove the filter

17. Transfer the coverslips in a Petri dish and immediately fill with culture medium
    **Note**: Two parameters define the properties of a gradient: the concentration and the slope. The concentration of the gradients can be easily modulated here by using diluted or concentrated membrane preparations (Be sure to determine OD for better evaluation of membrane concentration.). The slope of gradient can be adjusted by changing

the angle of inclination of the coverslips. Shallow gradients will be obtained with angle <45° while sharp gradient will be produced for angle >45°.

# 7. Neuronal Culture on 3D Substrates

While culturing neurons on 2D substrates is the most convenient and current way to culture neurons, it is important to keep in mind that in vivo growth is performed in a 3D environment. Thus, the whole actin dynamics and adhesive contacts are very different compared to what axons experience on 2D substrates. Moreover, as mentioned when presenting the formation of gradients of membrane particles, the spatial distribution of most of the guidance cues and neurotrophic factors is crucial to control axonal extension. In this last part of the chapter we will present a protocol allowing 3D culture of neurons to address this critical point. The proposed co-culture is intended to provide a way to analyze the responsiveness of growing axons exposed to a gradient of a define factor.

## 7.1. Reagents

1. BME (Biowest: L0046)
2. HBSS (Gibco: 24020-091)
3. MEM (Gibco: 21090-022)
4. FBS (Gibco: 10270-106)
5. Horse serum (Gibco: 26050-047)
6. Methyl cellulose (Sigma: M7027)
7. L-Glutamine 200 mM (Gibco: 25030-032)
8. Penicillin–streptomycine (Gibco: 15070-022)
9. NGF (Invitrogen: 13257-019)

The protocol described below is optimized for culture of sensory DRG neurons from E15 mice embryos. In our hands, the same protocol can be used to grow several other neuronal types including embryonic cortical or hippocampal neurons.

## 7.2. Materials

1. Two spatulas
2. Two needles
3. Cover slips (Knittel-Glaser VD1 2222 Y100A 22×22 mm)
4. Sterile workplace under laminar flow hood
5. Stereomicroscope
6. 60 mm Petri dishes (BD Falcon, cat. no. 353004)
7. Chicken plasma (Sigma P-3266, aliquots of 250 μL stored at −20°C)

8. Thrombin 10,000 units (Sigma T-4648, dissolve thrombin in 7.936 mL sterile water, centrifuge for 30 min at 2500 rpm, filter supernatant, and aliquot at 30 μL. Store at –20°C.)

9. Glucose stock solution (Dissolve glucose in sterile ddH$_2$O to obtain 50% glucose solution. Aliquot at 1 mL and store at –20°C).

10. HBSS-glucose 6.5 mg/mL (Dissolve 1.3 mL of glucose aliquot in sterile 48.7 mL ddH$_2$O. Store at 4°C for up to 2 weeks).

11. Culture medium for DRG explants (Add 95.4 mL BME, 48 mL HBSS, 50 mL FBS, 2 mL 200 mM L-glutamine final 2 mM, 2.6 mL 50% glucose final 6.5 mg/mL+0,72 mg/ml = 7.22 mg/mL and 2 mL penicillin/streptomycin).

**7.3. Procedure**

*Step 1: Cell aggregate preparation*

1. Trypsinize cells from an 80% confluent culture dish one day before co-culture experiment

2. Collect cells in 15 mL centrifuge tube and centrifuge for 5 min at 115 RCF at 4°C.

3. Resuspend the cells in 250 μL culture medium containing no selective antibiotics.
   **Caution**: To obtain stable aggregates the volume of medium added to the pellet should be adapted for each cell line. Here, the proposed volume is optimized for HEK293 cells.

4. Add 9–20 drops of 20 μL of the cell suspension on the backside of the cover of 35 mm

5. Carefully invert the cover with the drops and put it back on the Petri dish filled with 2 mL of culture media.

6. Transfer the Petri dish(es) to the incubator (37°C, 5% CO2). Aggregates are completely formed within 12–18 h.

*Step 2: Co-culture experiment in plasma clot*

1. Put dissected explants (DRGs, for example) in 35 mm Petri dish containing ice-cold glucose–HBSS.

2. Prepare cell aggregates to be co-cultured with DRGs or cortical explants (*see* above). Open the cover with aggregates and put it in inverted position on the working table. Using two spatulas collect the aggregates (one by one) and transfer them into cover-corresponding Petri dish filled with the 2 mL media. The aggregates are large enough and therefore using a scalpel can be divided in at least four pieces.

3. Thaw plasma and thrombin at 4°C and keep on ice. Dilute 30 μL thrombin in 970 μL GBSS. Keep thrombin/GBSS solution on ice.

4. Put sterile square 22×22 mm coverslips (*see* preparation of the coverslips above) in a new 60 mm Petri dish containing no media.

5. Put 20 μL of the plasma on the coverslips. Using a spatula collect one cell aggregate and transfer it into plasma. (*See* **Fig. 19.3**)

   **Caution:** Collection and transfer of the aggregate should be done with caution as some aggregates are fragile and tend to disintegrate.

6. Using a 20 μL pipette take four to six DRG explants and transfer them into the plasma containing the cell aggregate (*see* **Fig. 19.1C**).

7. Add 20 μL of the thrombin/GBSS solution close to the plasma drop. Use the needles to rapidly but carefully mix thrombin and plasma. Meanwhile, using the same needles

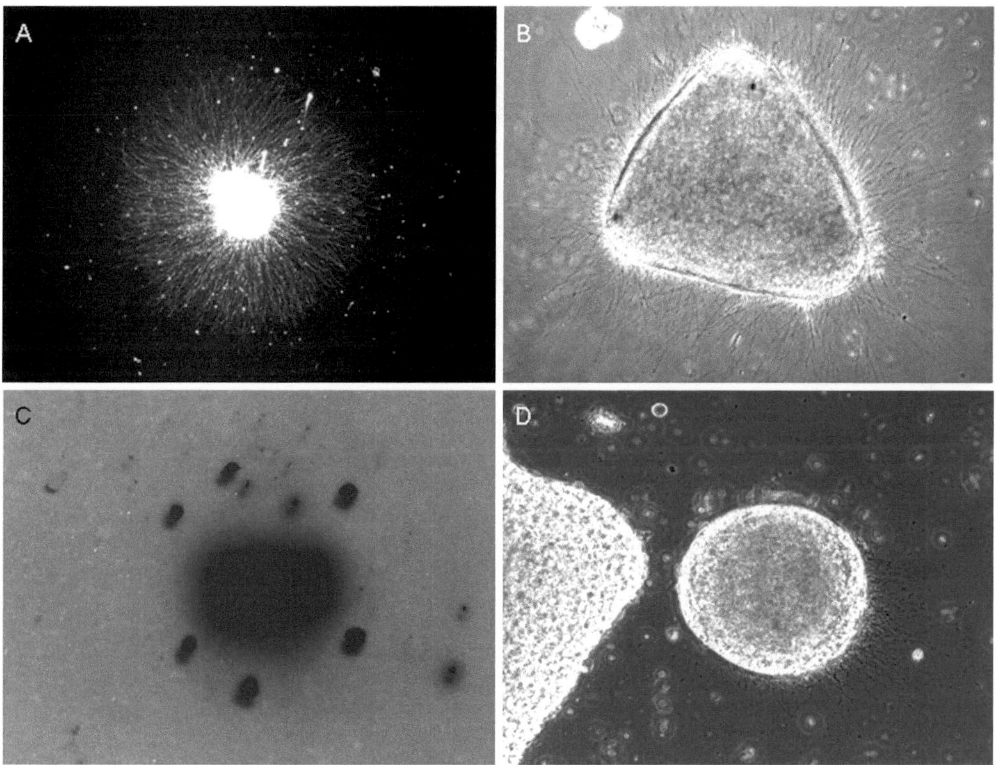


Fig. 19.3. Representative examples of neurons grown on 2D (**A**) and 3D (**B–D**) substrates. (**A**) Axons from DRG explant growing on a 2D substrate (laminin/ ploly-L-lysine). The image was taken after 24 h culture. (**B**) E15 cortical explant after 24 h culture growing on a 3D substrate (chicken plasma coagulated with thrombin). (**C**) DRG explants after 6 h culture on plasma/thrombin 3D substrate. The DRG explants were arranged around HEK293 cell aggregate expressing soluble Semaphorin3A-AP. Diffusion of Sema3A-AP was revealed with (SIGMA *FAST*$^{TM}$ BCIP/NBT, Sigma: B-5655). Note the gray clouds forming from the border of the explants. (**D**) Hippocampal explant after 24 h culture growing on plasma/thrombin 3D substrate. In this case, the gradient of Sema3A produced by the central aggregate induces strong repulsion of hippocampal axons.

quickly arrange DRG explants around the cell aggregate at a distance of 200–500 μm.

**Caution:** Thrombin immediately initiates coagulation of plasma, therefore working time to arrange DRG explants around the aggregate is between 20 and 40 s.

8. Keep coverslips (in Petri dish) for 30–40 min until plasma clot dries enough to add the media.

   **Caution:** Plasma clot must not dry out or nervous tissue will not survive.

9. Carefully add 2 mL of appropriate culture medium and transfer the Petri dishes to the cell incubator (37°C, 5% $CO_2$).

   **Caution:** Addition of media should be done with a lot of precaution, since plasma clots tends to detach. If necessary, leave plasma clot to coagulate for additional 5–10 min.

Using these culture conditions provides an efficient way to grow axons in a 3D substrate. The aggregate of cells will secrete the factor of interest that will be stabilized in the clot. Axon growth and guidance can therefore be evaluated after 24 h culture. Cell aggregates can be composed of a mixture of cells expressing various signals to generate complex substrates. A major concern could be controlling the actual concentration of the factors. However, several studies and mathematical models describing the molecular mechanisms of axon guidance showed that guidance effects are relatively independent of the absolute slope or concentration of the gradients (13, 14).


## References

1. Tessier-Lavigne, M. and Goodman, C. S. The molecular biology of axon guidance. Science, *274*: 1123–1133, 1996.
2. Bouquet, C. and Nothias, F. Molecular mechanisms of axonal growth. Adv Exp Med Biol, *621*: 1–16, 2007.
3. Gotz, M. and Bolz, J. Formation and preservation of cortical layers in slice cultures. J Neurobiol, *23*: 783–802, 1992.
4. Bagnard, D., Chounlamountri, N., Puschel, A. W., and Bolz, J. Axonal surface molecules act in combination with semaphorin 3a during the establishment of corticothalamic projections. Cereb Cortex, *11*:278–285, 2001.
5. Bagnard, D., Lohrum, M., Uziel, D., Puschel, A. W., and Bolz, J. Semaphorins act as attractive and repulsive guidance signals during the development of cortical projections. Development, *125*:5043–5053, 1998.
6. Rivas, R. J., Burmeister, D. W., and Goldberg, D. J. Rapid effects of laminin on the growth cone. Neuron, *8*: 107–115, 1992.
7. Tang, D. and Goldberg, D. J. Bundling of microtubules in the growth cone induced by laminin. Mol Cell Neurosci, *15*: 303–313, 2000.
8. Dertinger, S. K., Jiang, X., Li, Z., Murthy, V. N., and Whitesides, G. M. Gradients of substrate-bound laminin orient axonal specification of neurons. Proc Natl Acad Sci U S A, *99*: 12542–12547, 2002.
9. Smith, R. A. and Orr, D. J. The survival of adult mouse sensory neurons in vitro is enhanced by natural and synthetic substrata, particularly fibronectin. J Neurosci Res, *17*: 265–270, 1987.
10. Hanbali, M., Bernard, F., Berton, C., Gatineau, G., Perraut, M., Aunis, D., Luu, B., and Bagnard, D. Counteraction of axonal growth inhibitory properties of Semaphorin 3A and myelin-associated proteins by a

synthetic neurotrophic compound. J Neurochem, *90*: 1423–1431, 2004.

11. Walter, J., Kern-Veits, B., Huf, J., Stolze, B., and Bonhoeffer, F. Recognition of position-specific properties of tectal cell membranes by retinal axons in vitro. Development, *101*: 685–696, 1987.

12. Baier, H. and Bonhoeffer, F. Axon guidance by gradients of a target-derived component. Science, *255*: 472–475, 1992.

13. Bagnard, D., Thomasset, N., Lohrum, M., Puschel, A. W., and Bolz, J. Spatial distributions of guidance molecules regulate chemorepulsion and chemoattraction of growth cones. J Neurosci, *20*: 1030–1035, 2000.

14. Goodhill, G. J. and Baier, H. Axon guidance: stretching gradients to the limit. Neural Comput, *10*: 521–527, 1998.

# Chapter 20

# Guidance and Outgrowth Assays for Embryonic Thalamic Axons

Alexandre Bonnin

## Abstract

The assay described in this chapter allows for a relatively simple and fast screening for effects of guidance/growth promoting cues and modulators on dorsal thalamic axons growth. As mentioned in the introduction, the conditions of optimal axon growth in vitro are not identical to conditions found in vivo. Therefore, this assay like all other in vitro axon outgrowth assays does not intend to "recapitulate" in vivo situations. This is an advantage in a sense that growth conditions can be controlled and modified as desired. The results can ultimately provide a framework to establish testable hypotheses in vivo. With the emergence of less expensive and more practical real-time imaging techniques (e.g., microscope-independent and environmentally controlled culture chambers), this assay is highly adaptable to monitor the outgrowth of thalamic axons and the effect of various modulators in real time. It is possible, for instance, to visualize and measure acute effects of drugs on the response patterns of thalamic axons, on a much shorter timescale than 72 h in vitro. Furthermore, by simply modifying the dissection steps and testing different extracellular matrix substrates, the response of axons originating from any brain region of interest can be monitored. Therefore, this assay will be helpful to rapidly test the possibility that a given class of axons responds to specific guidance cues and to determine the specific pharmacological agents that may influence growth responses.

**Key words:** Dorsal thalamus, axon outgrowth, guidance cues, Netrin-1, co-cultures.

## 1. Introduction

This chapter describes a simple assay to test and quantify the effects of soluble guidance cues on the growth behaviour of embryonic dorsal thalamic axon populations. The method described here derives from original co-culture assays (1–5) in which explants or slices from embryonic mouse cortex and dorsal thalamus, or other regions, were co-cultured in order to assess their mutual influences on axon growth properties. It is impor-

L.C. Doering (ed.), *Protocols for Neural Cell Culture*, Springer Protocols Handbooks,
DOI 10.1007/978-1-60761-292-6_20, © Humana Press, a part of Springer Science+Business Media, LLC 1992, 1997, 2001, 2010

tant to note that the assays presented here aim at addressing a simple question: do dorsal thalamic axons respond (e.g., attraction, repulsion, or changes in axon outgrowth) to one specific guidance cue in a defined culture system? This assay does not intend to recapitulate the complexity of an in vivo system where the growth of thalamic axons is influenced by dynamic spatiotemporal exposure to multiple guidance and growth-promoting cues and by simultaneous changes in physical parameters of the tissue (e.g., variable surrounding cell densities, composition of the extracellular matrix, extent of axon fasciculation). Nevertheless, variations of the "single-cue guidance assay" have been used in many different applications and led to the demonstration of specific effects for several guidance cues on various types of axons [for example, *see*: (6, 7, 3)]. An advantage of this simple technique is that, once the typical response pattern of an axon population to a specific cue is established, modulation by exogenous or endogenous factors can be readily assessed. These studies can be particularly useful to identify new guidance cues or modulators, whose function should always be validated carefully in vivo. For instance, we recently demonstrated that the neurotransmitter serotonin (5-HT) changes the basal in vitro response of posterior dorsal thalamic axons to the guidance cue Netrin-1 from attraction to repulsion (*see* **Fig. 20.5** and 8). These observations led us to investigate a potential role of 5-HT on the guidance of thalamic axons in vivo. Transient gene expression manipulation in utero showed that 5-HT signalling is indeed a modulator of thalamocortical axon pathway formation in vivo.

The following protocol describes how to grow embryonic mouse dorsal thalamic explants embedded in extracellular matrix in a culture dish adjacent to a source of soluble guidance molecules, or any other soluble cue of interest. The source of guidance cues is provided by stably (or transiently) expressing live cell lines. A local source of cues is generated by aggregating these cells (using the "hanging drop" technique – *see* below and 9) and placing the aggregate in the center of an extracellular matrix drop; this disposition generates a gradient of guidance cues throughout the matrix (*see* **Fig. 20.1** and 6). Explants from the embryonic dorsal thalamus are then positioned symmetrically around, and equidistant from, the local source of guidance cues inside the matrix and therefore become evenly exposed to this gradient. Quantification can be carried out in several ways: to quickly evaluate the overall response pattern of DT axons to guidance cues, a quadrant-based system can be used to classify the outgrowth pattern of each individual explant (*see* **Fig. 20.1**). Outgrowth can be classified by an observer, blind of the conditions, as symmetrical (no preferential direction of outgrowth), away (most axons are growing away from the source of guidance cues), or toward (most axons are growing toward the source of guidance cues) – *see*

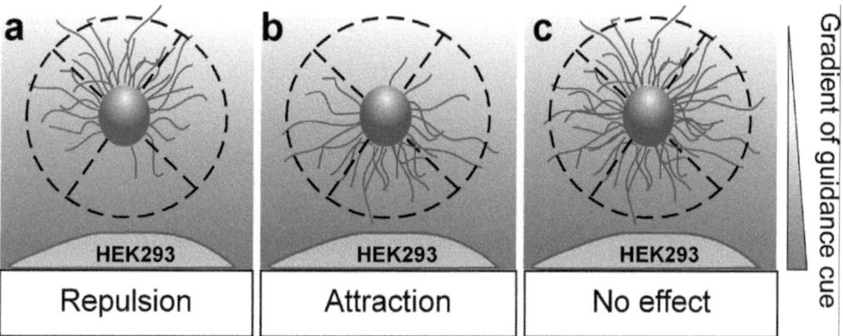



Fig. 20.1. Principle of thalamic axon outgrowth assays. Explants from the dorsal thalamus are grown in front of HEK-293 cells stably expressing the soluble cues. The release of soluble cues in the three-dimensional matrix generates a gradient influencing the direction in which thalamic axons grow. A quadrant-based system is used to quantify the response of dorsal thalamic axons: repulsion (**a**) and attraction (**b**) corresponds to more axons growing away and toward the source of guidance cues, respectively; a symmetrical outgrowth indicates that the guidance cue has no effect overall.

(8–12). This quantification can be carried out under phase-contrast illumination of an inverted microscope. A more thorough evaluation of axon outgrowth patterns can then be achieved using an endogenous transgene marker expressed by dorsal thalamic neurons, such as YFP, or immunochemical staining of cultures with an axonal marker (*see* below) and using an imaging software such as Image J (http://rsb.info.nih.gov/ij) to count and compare the number of axons growing in the proximal segment (or the area covered by stained axons, toward the source of guidance cues) to the distal segment (away from the source of cues); quantification of outgrowth pattern can then be expressed as a ratio (P/D or proximal/distal ratio) and compared across different conditions (13, 14). Whichever quantification method is used, the protocol described below can be used for a simple and relatively rapid screening of the effects of guidance cues and related modulators on thalamic axons or on axons from any other brain region of interest.

## 2. Reagents and Equipment

### 2.1. Equipment

1. HEK-293 cells growth medium: DMEM + Glutamax-1 with 4.5 g/L glucose, 25 mM HEPES and no sodium pyruvate (Invitrogen, Cat #:10564-029) with 10% FBS (Gemini Bioproducts, Cat #: 100–106).

2. Explant culture medium: Neurobasal medium w/o phenol red (Invitrogen, Cat #:12348-017), N2 (100×) supplement (Invitrogen, Cat #17502-048) and B27 (50×) supplement (Invitrogen, Cat #:17504-044).

3. Matrigel basement membrane matrix – phenol-red free (Invitrogen, Cat #: 356237)

4. TrypLE express 1× (Invitrogen, Cat #: 12605-010)

5. Trypan blue dye 0.4% (Sigma, Cat #: T8154)

6. 10 cm tissue culture dishes to grow HEK-293 cells (Corning, Cat #: 430293).

7. 10 cm sterile Petri dishes to harvest embryos (Medegen, Cat # PD1906-5005).

8. Glass bottom dishes No. 0: 35 mm dishes with a bottom glass insert to allow good confocal imaging on inverted microscopes (Mattek, Cat #:P35G-0-14C).

9. Dissection microscope.

10. Dissection tools. Forceps #5–2 pairs (FST). Small and medium scissors (F.S.T., Cat #:15005-06 and Roboz, Cat #:RS-5848). Scalpel with "Feather" scalpel blades #10 (E.M.S., Cat #: 72044-10).

11. Inverted microscope (e.g., Zeiss Axiovert 25) with digital acquisition system to capture explant pictures.

12. Phosphate-buffered saline (PBS).

13. 4% buffered paraformaldehyde (PFA), pH 7.2 in PBS:

### 2.2. Reagents

1. Paraformaldehyde powder (Sigma, Cat #P6148)

2. Sodium phosphate monobasic, anhydrous (Sigma, Cat #S6566)

3. Sodium hydroxide pellets (Sigma, Cat #S8045)
   a. Dissolve 20.0 g of paraformaldehyde powder in 300 mL distilled $H_2O$. Heat water to 60°C to aid in the process, adding a few drops of 1 N NaOH to clear solution.

   b. Cool solution to room temperature and filter using Whatman No. 1 paper.

   c. Add 9.37 g NaH2PO4 and 2.14 g NaOH to the paraformaldehyde solution.

   d. Adjust pH (if necessary) to 7–7.2 with 10 N NaOH stock.

   e. Adjust volume to 500 mL with distilled $H_2O$.

4. Mouse monoclonal (TUJ1) to neuron-specific beta-III tubulin antibody (Abcam; Cat #ab14545)

5. Donkey anti-mouse IgG Cy2 or 3 conjugated secondary antibody (Jackson Immuno; Cat #715 225-150 or 715-165-150)

## 3. Protocol

### 3.1. Generation of Hanging Drop Cell Culture

1. Allow HEK-293 cells to grow to confluence in 10 cm tissue culture-treated dishes in HEK-293 cells medium. Hanging drops preparation should start 3–4 days before the

availability of E14.5 thalamic explants because the investigator needs to allow enough time for HEK-293 cells to reach confluence (2–3 days), and hanging drops must incubate for at least 24 h prior to generation of the co-cultures in order to form solid cell clumps. Note that since repeated passaging of cells introduces increased variability in the quality of hanging drops, we usually use cells for about 2–3 weeks (4–6 passages), after which we discard them and use a new frozen stock.

2. To resuspend HEK-293 cells: aspirate media and add 2 mL of trypsin (or TrypLE express 1X).Allow trypsin 2 min to detach cells from the dish. Add 4 mL warm HEK-293 media (*see* **Section 2.2**) to dish and use a 10 mL pipette to wash cells off the bottom of dish. Transfer cells to a 15 mL conical (volume = 6 mL: 2 mL trypsin + 4 mL media) and spin for 3 min at 100$g$ in a centrifuge. Remove supernatant and resuspend cells in 4 mL warm HEK-293 media. Take 1 mL of this new cell mixture and add to 10 mL of media in a new cell culture dish (to keep the culture going for future experiments). Store in incubator (37°C, 5% CO2, 95% humidity). Split about every 3 days.

3. Determine the cell density in the remaining cell suspension.

4. Add 10 $\mu$L of cell suspension to 10 $\mu$L trypan blue dye. Add 10 $\mu$L of this mixture to a hematocytometer slide plate. Count cells and determine the concentration of the cellular suspension. To generate HEK-293 hanging drops, the optimum concentration is ~40 × 10$^5$ cells/ml. This number may vary depending on the cell line used. Optimal concentrations should be determined for each particular cell line.

5. Ensure that cells are completely in solution before making hanging drops. Using a P-20 pipette, take 20 $\mu$L quantities and place on the inversed lid of a 10 cm cell culture dish (**Fig. 20.2a, b**). Only place 12 drops on each lid, roughly 1 cm apart and avoiding the sides of the lid. Fill the bottom of the dish with 10 mL HEK-293 media. Quickly but smoothly invert the lid and place it on top of the 10 cm culture dish (**Fig. 20.2c**). Incubate for 24–48 h at 37 C, 5% CO2, 95% humidity prior to doing explant dissection. After a minimum of 24 h, clumps of cells should have formed in the bottom of the drops. These clumps should be solid enough to be carefully transferred (using gentle aspiration with a glass Pasteur pipette, or with a P-20 pipette) in the middle of a glass bottom dish and covered with a drop of Matrigel (*see* Step 10 of **Section 3.2**).

***3.2. Explant Preparation***

1. Prepare neurobasal media prior to dissecting the mice, warm to 37°C while doing dissection. For 50 mL of explant medium, prepare:

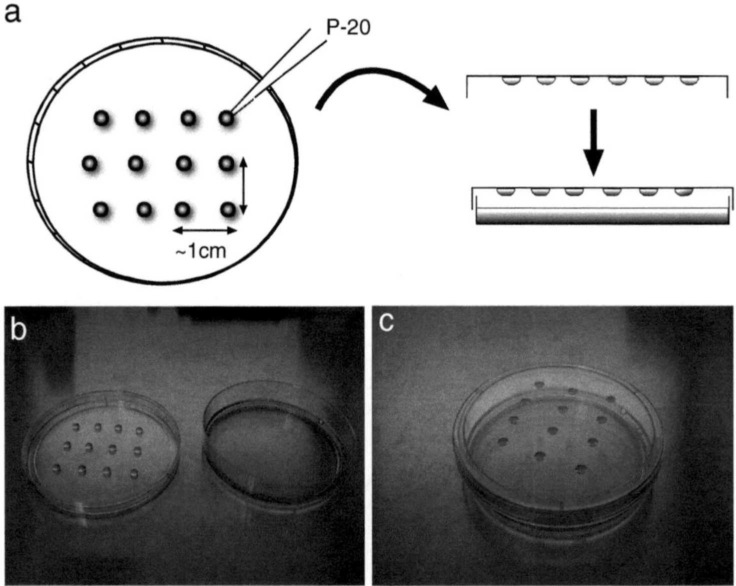

Fig. 20.2. Generation of "hanging-drop" cultures. Hanging-drops of HEK-293 cells are generated by positioning 20 µL drops of cell suspension on the inverted lid of a 10 cm tissue culture dish (**a**). The lid is positioned on top of a culture dish filled with medium and gravity allows cells to form solid clumps in the bottom of the drops. (**b, c**) Illustrate the regular arrangement of drops on a 10 cm lid, leaving enough space between drops to avoid merging.

48 mL neurobasal medium

500 µL L-glutamine

1 mL B27 supplement

500 µL N2 supplement

Total volume: 50 mL

(Volumes indicated are used straight from the stock bottle supplied – *see* **Section 2.2**.) An hour before starting the dissection, place the required amount of Matrigel stock (# of dishes required × 75 µL) on ice in order to thaw it slowly.

Once thawed, keep the Matrigel cold at all times and dispense using chilled pipette tips to avoid polymerization.

As an extracellular matrix extract, Matrigel has proven to be a good substrate for axonal growth; it does contain a variety of extracellular matrix proteins and growth factors (*see* manufacturer website). However, the most efficient substrate for a particular type of axons needs to be determined by the investigator.

*If Steps 2–9 are performed outside of the laminar flow hood, extra-caution should be used to avoid contamination (e.g., wear gloves, clean countertops, gloves, and tools with 70% ethanol)*

2. Anesthetize and sacrifice a pregnant mouse according to IACUC-accepted protocols at your institution. Spray 70% ethanol solution on the abdomen. Cut abdominal cavity open, expose uterine horns, and remove embryos quickly, placing them in ice-cold PBS as you go along. Measure the crown-rump length (CRL) of embryos to confirm expected age (e.g., E14.5 embryos: CRL = 11–12 mm).

3. Remove brains and place in ice-cold stock DMEM in 10 cm Petri dishes. Keep brains in ice-cold DMEM throughout procedure. Dissect brain region of interest under a dissecting microscope as follows:

4. Place the brains cortical surface facing down. Using a scalpel with "feather" scalpel blades, hemi-section the brains sagittaly, along the midline (*see* **Fig. 20.3a**).

5. Lay each hemi-section on its side (cortex side facing down); each half of the dorsal thalamus is visible as a circular structure, delimited on the ventral side by the zona limitens intrathalamica and on the dorso-posterior side by the axon tract of the habenula-interpedoncular pathway (**Fig. 20.3b**).

6. Using fine and sharp tweezers (#5), scoop out the dorsal thalamus following the landmarks described above – anterior

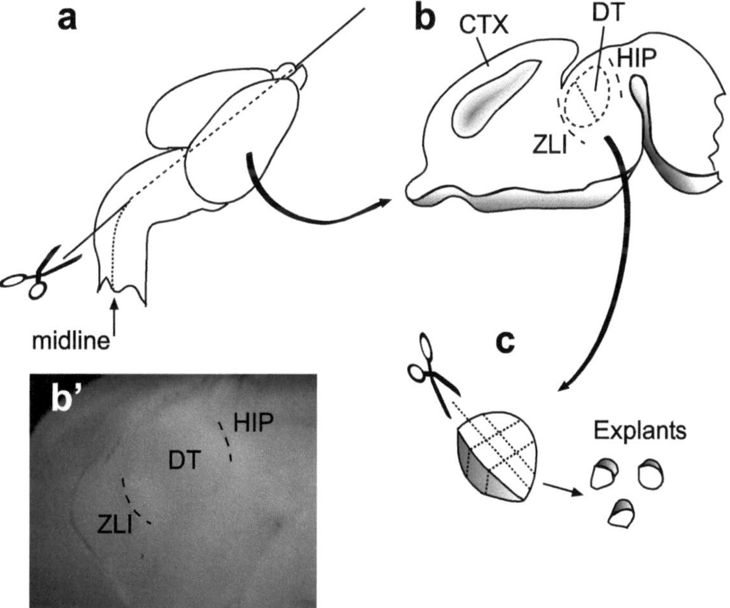


Fig. 20.3. Generation of posterior dorsal thalamic explants. Brains from E14.5 embryos are harvested and sectioned mid-sagittaly along the midline (**a**). Laying flat on the cortex (CTX), the dorsal thalamus (DT) is delineated by the zona limitens intrathalamica (ZLI) on the anterior side and the fibre tract of the habenulo-interpedoncular pathway (HIP) on the posterior side (**b, b'**). Once harvested, explants are generated by cutting the dorsal thalamus in small explants ∼200–400 μm in size (**c**).

or posterior halves of the dorsal thalamus can be harvested separately.A piece of half dorsal thalamus should be roughly 1 mm in length.

7. After isolating the brain region of interest, cut it into small explants ~200–400 μm in size using fine and sharp dissection scissors or a sharp tungsten needle (**Fig. 20.3c**). Aspirate the explants using a P-1000 pipette and transfer to a new 10 cm dish filled with ice-cold DMEM. Note that optimal explant size may vary and needs to be determined depending on the target tissue.

8. Keep the explants in ice-cold DMEM while performing the following steps.

9. Prepare the Matrigel drop containing the clump of HEK-293 cells: Steps 10–12 need to be performed one dish after another; do not prepare the Matrigel drops in advance since they will harden quickly and prevent insertion and positioning of the explants.

10. Holding the "hanging drops" Petri dish lid at an angle, hanging drops facing you, use a P-20 pipette with very gentle aspiration (so that the clump of HEK-293 cells hangs from the pipette tip) and transfer one clump of HEK-293 cells (1 hanging drop) in the center of a small Mattek glass bottom culture plate (*see* **Section 2.2**). Place 75 μL of ice-cold Matrigel (keep the stock on wet ice) on top of the HEK-293 cells in the center of the dish. Spread the Matrigel in a circular shape using pre-chilled pipette tips, avoiding disturbance of the clump of HEK-293 cells.

11. Quickly pick up four tissue explants from the cold DMEM with a P-20 pipette and, under a dissection microscope, place them in the Matrigel in a square pattern around the HEK-293 cell clump (*see* **Fig. 20.4a–c**). If necessary, use fine #5 forceps to position explants close to HEK-293 cell clump (~400–500 μm or ~1–2 explant sizes) in an equidistant square pattern. This must be done as fast as possible because the Matrigel will solidify rapidly.

12. Carefully transfer the culture dish into the tissue culture incubator (5% CO2, 37 C) for 15–30 min, without medium, to allow the Matrigel to polymerize.

13. After 15–30 min, gently add 2 mL neurobasal medium (*see* Step 1) to each culture dish, without pouring the medium directly onto the Matrigel. Allow explants to grow for 48–72 h.

14. Monitor axon outgrowth using an inverted microscope under phase-contrast illumination. Outgrowth usually becomes apparent after 24 h.

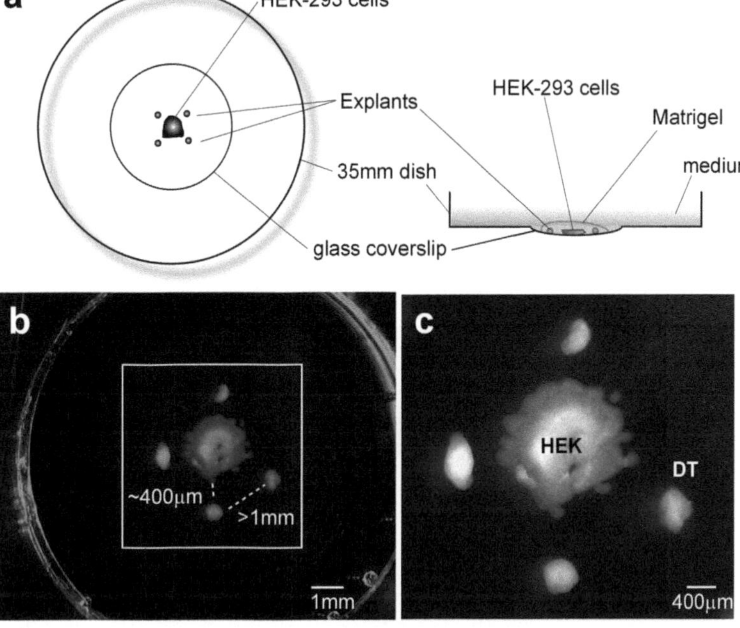



Fig. 20.4. Positioning of explants and HEK-293 in the Matrigel. (**a**) Schematic representation of a 35 mm bottom glass tissue culture dish; the clump of HEK-293 cells is positioned in the middle of the coverslip and carefully covered with Matrigel. Dorsal thalamic explants are then carefully positioned in a square pattern around the clump of HEK-293 cells. After hardening of the Matrigel, cultures are incubated in 2 mL neurobasal medium for 48–72 h. (**b**) Example of explants disposition around a clump of HEK-293 cells: explants should be as far apart as possible (~1 mm) and distant ~200–400 μm from the source of guidance cue. (**c**) Higher magnification of the boxed region in (**b**).

15. In our experience, optimal axon outgrowth length is obtained 72 h after plating. This allows enough time for thalamic axons to reach and sometimes grow over the clump of HEK-293 cells (*see* **Fig. 20.5b**).

***3.3. Immunochemical Labeling of Axons***

1. Aspirate medium and fix by adding 2 mL of 4% PFA solution to each well; incubate 2 h at room temperature or overnight at 4°C.

2. Rinse five times in 1× PBS pH 7.4 for 15 min each, with gentle agitation.

3. Block with 10% FBS/0.1% Triton X-100 in PBS pH7.4 for at least 2 h, at room temperature.

4. Incubate with primary antibody (e.g., mouse TUJ1 anti-tubulin) diluted 1:500 in blocking solution, overnight at 4°C.

5. Rinse 5 × 15 min in 1 × PBS, at room temperature with gentle agitation – an extra rinse overnight at 4°C can be added if primary antibody background is high – do not hesitate to add more rinses if background is high.

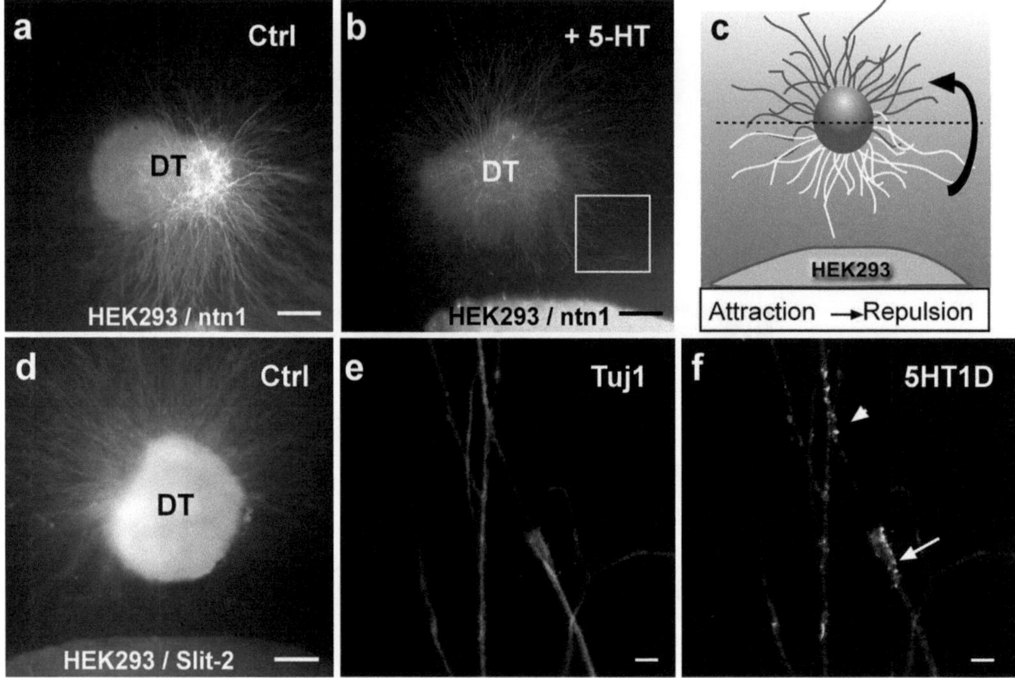



Fig. 20.5. Examples of results obtained after 72 h in culture. (**a**) In control (Ctrl) conditions, axons from a posterior dorsal thalamic explant are attracted toward a source of Netrin-1 (HEK293/ntn1). When 30 μm 5-HT (+ 5-HT) is added to the culture medium, axons are now repelled by Netrin-1 (**b**). The switch from attraction to repulsion is illustrated in (**c**) – quantification can be carried out by counting the number of axons growing proximal and distal to the source of guidance cues. (**d**) Axons from posterior thalamic explants are repelled by the guidance cue Slit-2. Because cultures are grown on bottom-glass MatTek dishes, immunofluorescence followed by confocal microscopy can be used; scale bars: 200 μm. (**e, f**) Examples of confocal imaging obtained after immunostaining of cultures for the neuron-specific beta-III tubulin (Tuj1, e) and the 5-HT receptor 5-HT1D (f). 5-HT1D receptor protein is present along (*arrow head*) and at the growth cone (*arrow*) of posterior thalamic axons in culture; the region imaged corresponds to the boxed area in (**b**); scale bars: 5 μm.

6. Incubate with anti-mouse-IgG-Cy2/3 secondary antibodies (dilute second antibody 1/1000 in block solution, incubate overnight at 4°C).

7. Rinse 5 × 15 min in 1× PBS, at room temperature with gentle agitation – an extra rinse overnight at 4°C can be added if secondary antibody background is high – do not hesitate to add more rinses if background is high.

8. Store in dark at 4°C until ready to image. The use of high-quality Mattek bottom glass culture dishes allows good fluorescence or confocal imaging on an inverted microscope (*see* **Fig. 20.5c, d**).

## 4. Critical Steps and Troubleshooting

1. The minimum incubation time of HEK-293 cells in order to form solid hanging drops is 24 h at the cell density indicated; in our experience, hanging drops can be generated

up to 48 h prior to explant cultures; increasing incubation time allows for the formation of denser, more solid, hanging drops, which may be easier to manipulate. However, care should be taken to avoid extended incubation times, because the small volume of medium in each drop will not allow cells to survive much longer past 48 h.

2. If stably expressing cell lines are not available for the factors of interest, HEK-293 can be transfected using regular techniques (e.g., calcium phosphate or commercially available kits) and hanging drops generated 24–48 h after transfection.

3. In order to achieve good survival of explants and outgrowth of axons, Steps 2–11 should be performed as fast as possible. Harvesting the brains of 10 embryos at E14.5 should take no longer than 5 min, and generating/seeding of all explants no longer than 45 min–1 h.

4. To minimize variability in explant survival/outgrowth, it is advised to not exceed 15 individual dishes per experiment (60 explants) for one operator; above this number, the time between plating of the first and last set of explants will be significant and effects on survival may affect axonal outgrowth adversely.

5. If Steps 10–12 are performed outside a tissue culture hood, surfaces and tools should be spread regularly with ethanol to minimize the risk of contamination.

6. It is critical to perform Step 10–12 rapidly because the Matrigel solidifies quickly at room temperature; once polymerization starts, it will become increasingly difficult to insert and precisely position explants around the clump of HEK-293 cells.

7. Precise positioning of explants equidistantly from the clump of HEK-293 cells is difficult to achieve; some variability will exist. For increased precision (and cost), gridded glass bottom coverslip dishes (Mattek, Cat #P35G-2-14-C-GRID) can be used.

8. Like most assays, some variability in the response of axons to a specific guidance cue will exist. The amount of variability will depend on the precision of the dissection, and of course on the heterogeneity of cell types that compose each explant. One way to minimize variability is to subdivide, when possible, the target tissue into posterior/anterior or medial/lateral segments; for instance, we observed that anterior and posterior DT axons respond in opposite ways to the same guidance cue Netrin-1. If explants were generated from the whole DT, responses to Netrin-1 would have been normally distributed between attraction, symmetrical, and repulsion. Even after dividing the DT into

anterior and posterior segments, the distribution of posterior axon responses to Netrin-1 or Slit-2 was not completely homogenous (8) in order to achieve a statistically significant measure of axon responses distribution, measures from a large number of explants, across four to five independent experiments, had to be collected.

## 5. Typical Protocol Results

**Figure 20.5** illustrates typical results obtained from explant co-cultures. In these experiments, explants from the posterior half of the DT at E14.5 were grown in front of clumps of HEK-293 cells expressing soluble forms of the guidance cues Netrin-1 or Slit-2. After 72 h, cultures were stained for neuron-specific beta-III tubulin and images captured using a stereomicroscope equipped for fluorescence imaging. Posterior DT axons are mainly attracted by Netrin-1 (**Fig. 20.5a**), but as recently demonstrated the attractive effect of Netrin-1 can be switched to repulsion by exposing co-cultures to 30 μM 5-HT (**Fig. 20.5b, c** and 8). **Figure 20.5d** illustrates the repulsive effect of Slit-2 on posterior DT axons, which are mainly growing away from the source of Slit-2. There is some variability in the response of posterior DT axons to guidance cues; therefore, for each condition, quantification was carried out from at least four to five independent experiments (at least 30 explants; *see* 8 for examples of quantifications). Because of inter-experiment variance, analysis of a large number of explants is required to reach the statistical power required for the comparison of DT axons response distributions to different cues, or to different modulators (8). We demonstrated that 5-HT effect is mediated by two different receptors, 5-HT1B and 5-HT1D. Immunocytochemical staining demonstrated that the 5-HT1D receptor protein is indeed expressed along and at the growth cones of posterior DT axons growing in vitro in the co-culture assay (**Fig. 20.5e, f** and 8). The use of glass bottom MatTek culture dishes enabled us to perform very high-resolution confocal imaging of growing DT axons embedded in the Matrigel (**Fig. 20.5e, 20.5f**).

## Acknowledgments

This work was supported by a Conte Center grant P50 MH078028; A.B. is recipient of a NARSAD Young Investigator Award. The author would like to thank Dr Pat Levitt for his outstanding support and for his help with this manuscript, and Dr Jane Wu for providing Netrin-1 and Slit-2-expressing HEK 293 cells.

## References


1. Bolz J, Novak N, Staiger V. Formation of specific afferent connections in organotypic slice cultures from rat visual cortex cocultured with lateral geniculate nucleus. J Neurosci 1992;12:3054–70.

2. Rennie S, Lotto RB, Price DJ. Growth-promoting interactions between the murine neocortex and thalamus in organotypic co-cultures. Neuroscience 1994;61:547–64.

3. Richards LJ, Koester SE, Tuttle R, O'Leary DD. Directed growth of early cortical axons is influenced by a chemoattractant released from an intermediate target. J Neurosci 1997;17:2445–58.

4. Tuttle R, Braisted JE, Richards LJ, O'Leary DD. Retinal axon guidance by region-specific cues in diencephalon. Development 1998;125:791–801.

5. Molnar Z, Blakemore C. Development of signals influencing the growth and termination of thalamocortical axons in organotypic culture. Exp Neurol 1999;156:363–93.

6. Wu W, Wong K, Chen J, et al. Directional guidance of neuronal migration in the olfactory system by the protein Slit. Nature 1999;400:331–6.

7. Poe BH, Brunso-Bechtold JK. Directed outgrowth from a subset of cochlear nucleus fibers in a collagen-gel matrix. Brain Res Dev Brain Res 1998;105:153–7.

8. Bonnin A, Torii M, Wang L, Rakic P, Levitt P. Serotonin modulates the response of embryonic thalamocortical axons to netrin-1. Nat Neurosci 2007;10:588–97.

9. Del Duca D, Werbowetski T, Del Maestro RF. Spheroid preparation from hanging drops: characterization of a model of brain tumor invasion. J Neurooncol 2004;67:295–303.

10. Braisted JE, Tuttle R, O'Leary DD. Thalamocortical axons are influenced by chemorepellent and chemoattractant activities localized to decision points along their path. Dev Biol 1999;208:430–40.

11. Braisted JE, Catalano SM, Stimac R, et al. Netrin-1 promotes thalamic axon growth and is required for proper development of the thalamocortical projection. J Neurosci 2000;20:5792–801.

12. Li HS, Chen JH, Wu W, et al. Vertebrate slit, a secreted ligand for the transmembrane protein roundabout, is a repellent for olfactory bulb axons. Cell 1999;96:807–18.

13. Lin L, Rao Y, Isacson O. Netrin-1 and slit-2 regulate and direct neurite growth of ventral midbrain dopaminergic neurons. Mol Cell Neurosci 2005;28:547–55.

14. Alcantara S, Ruiz M, De Castro F, Soriano E, Sotelo C. Netrin 1 acts as an attractive or as a repulsive cue for distinct migrating neurons during the development of the cerebellar system. Development 2000;127:1359–72.

# Chapter 21

# Detection of Cell Death in Neuronal Cultures

**Sean P. Cregan**

## Abstract

In this chapter we describe a technically simple and cost-effective method for quantifying apoptotic and non-apoptotic cell death in primary neuronal cultures. The method consists of three assays: cell viability assay, nuclear morphology assay, and caspase-3 activity assay (Cregan et al. J Cell Biol 2002; 158(3):507–517). A significant advantage of these assays is that they do not require fixation or washing steps that cause dead cells to detach from the tissue culture dish, resulting in an underestimation of the extent of cell death.

**Key words:** Apoptosis, nuclear assay, caspase-3, cerebellar neuron.

## 1. Introduction

Neuronal cell death is known to play an important role in brain development and to contribute to the loss of neurological function in acute and chronic neurodegenerative conditions. Thus extensive research is being undertaken to elucidate the signalling pathways that regulate neuronal cell death in these settings.

Apoptotic and non-apoptotic modes of cell death (e.g., excitotoxicity, necrosis, autophagy) have been implicated in neurodegenerative processes. The role of apoptosis has received particular attention as it is known to be a genetically programmed cell death pathway amenable to therapeutic intervention. Neurons undergoing apoptosis exhibit distinct morphological features including plasma membrane blebbing, chromatin condensation, and shrinkage or fragmentation of the nucleus. These morphological alterations are mediated primarily by the caspase family of proteases, which cleave specific regulatory and structural elements in the cell.

L.C. Doering (ed.), *Protocols for Neural Cell Culture*, Springer Protocols Handbooks,
DOI 10.1007/978-1-60761-292-6_21, © Humana Press, a part of Springer Science+Business Media, LLC 1992, 1997, 2001, 2010

Model systems developed in primary neuronal cell cultures are a valuable tool for investigating signalling pathways involved in the regulation of neuronal survival and cell death. The method outlined here has been used to identify several key molecular processes involved in the regulation of neuronal cell death, including the death stimulus specific role of different Bcl-2 family proteins and the involvement of apoptosis-inducing factor in apoptotic and non-apoptotic cell death pathways (1–5).

### 1.1. Cell Viability Assay

In this assay neuronal viability is determined using two probes that measure different survival endpoints: intracellular esterase activity and plasma membrane integrity. Importantly, this assay detects both apoptotic and non-apoptotic forms of cell death. Procedurally, neurons are incubated with the cell-permeable substrate Calcein AM and the cell-impermeable dye ethidium homodimer-1 (EthD-1). Live cells contain endogenous esterases and are identified by their ability to convert calcein AM to the green fluorescent product calcein. Ethidium homodimer-1 (EthD-1) is excluded from live cells, but enters dead cells due their compromised plasma membrane. EthD-1 binds DNA with high affinity and produces a bright orange-red fluorescence in the nucleus of dead cells. Neuronal death/survival is then determined by scoring the fraction of green fluorescent (live) and red fluorescent (dead) neurons using a fluorescence microscope.

### 1.2. Nuclear Morphology Assay

Although a number of techniques have been developed to identify apoptotic cells, nuclear morphology remains the gold standard. In this assay neurons are incubated with the cell-permeable dye Hoechst 33342, which intercalates into A-T-rich regions of DNA and produces a blue fluorescence under UV illumination. Hoechst-stained neurons are examined on a fluorescence microscope and the fraction of apoptotic nuclei is counted.

In contrast to the diffuse heterogeneous fluorescence pattern observed in healthy neurons, the nuclei of apoptotic neurons are markedly shrunken (karyopyknosis) and/or fragmented (karyorrhexis) and exhibit a bright uniform fluorescence due to chromatin condensation.

### 1.3. Caspase-3 Activity Assay

Caspase activation is a hallmark of apoptotic cell death. These proteases cleave cellular substrates at aspartatic acid residues proximal to tetrapeptide sequences specific to each caspase (6). Pharmacological and genetic evidence indicate that caspase-3 is activated in the vast majority of apoptotic cell death paradigms (7, 8). In this assay caspase-3 activity is determined in neuronal extracts by measuring the cleavage of a synthetic fluorescent substrate consisting of the caspase-3-specific peptide motif DEVD linked to the fluorochrome 7-amino-4-trifluoromethyl coumarin (AFC). Cleavage of the DEVD-AFC substrate results in a shift in fluorescence emission of AFC (from blue to green), which can be measured on a fluorescence microplate reader.

## 2. Reagents and Equipment

1. Neurobasal media (Invitrogen, cat # 21103-049) supplemented with B27 (Invitrogen cat # 17504-044) and 25mM KCL

2. Phosphate-buffered saline (PBS), pH 7.4

3. Calcein AM (Invitrogen, cat # C3099)

4. Ethidium homodimer-1 (Invitrogen, cat # E1169)

5. Hoechst 33342, bizBenzimide trihydrochloride (Sigma, B2261)

6. Ac-DEVD-AFC (Biomol cat # P-409) dissolved in DMSO (10 mM)

7. Boc-aspartyl (OMe)-fluoromethylketone (Sigma cat # B2682)

8. Caspase lysis buffer: 10 mM HEPES, 1 mM KCl, 1.5 mM $MgCl_2$, 0.1% $NP_4O$, 10% glycerol; supplement fresh with protease inhibitors: 5 µg/mL aprotinin, 2 µg/mL leupeptin, 200 µg/mL PMSF, and 1 mM DTT

9. Caspase reaction buffer: 25 mM HEPES (pH 7.5), 10% sucrose, 10 mM DTT, and 0.1% CHAPS

10. Multiwell tissue culture dishes, 1.9 $cm^2$/well (Nunc cat# 176740)

11. Tissue culture dishes, 8.8 $cm^2$/well (Nunc cat # 150318)

12. 96-well black microplates (Nunc cat # 237108)

13. Microfuge tubes (1.5 mL)

14. Cell scrapers

15. Refrigerated microcentrifuge

16. Fluorescence microplate reader (e.g., SpectraMax M5, Molecular Devices)

17. Inverted fluorescence microscope
    a. Filter set for calcein fluorescence (e.g., Semrock FITC-3540B, Omega XF23, or Chroma 31001)

    b. Filter set for EthD-1 fluorescence (e.g., Semrock TRITC-A, Omega XF35, or Chroma 31005)

    c. Filter set for Hoechst fluorescence (e.g., Semrock DAPI-5060B, Omega XF06, or Chroma 31000)

    d. (Recommended) CCD camera (e.g., QiCam, Qimaging)

    e. (Recommended) Image capture and analysis system (e.g., Northern Eclipse, EMPIX imaging Inc.)

## 3. Protocol

The following protocol has been optimized for the quantification of cell death and caspase-3 activity in primary cultures of mouse cerebellar granule neurons. For cell viability and nuclear morphology assays, neurons are plated in 1.9 cm$^2$ multiwell tissue culture dishes at $0.4 \times 10^6$ cells/500 µL neurobasal media and for caspase-3 activity assay neurons are plated in 8.8 cm$^2$ dishes at $2 \times 10^6$ cells/2 mL neurobasal media. Neuronal cultures are exposed to a death stimulus and at selected times are assayed for cell death as follows.

### 3.1. Cell Viability Assay

1. Prepare a 10X working strength dilution of Calcein AM (5 µM)/EthD-1 (30 µM) in neurobasal media (*see* **Note 1**). The volume of labelling solution required will depend on the number of samples to be analyzed (50 µL required per well).

2. Add 50 µL of 10X Calcein AM/EthD-1 staining solution to each well and incubate neurons for 15 min at room temperature. The effective concentration will be 0.5 µM Calcein AM and 3 µM EthD-1 (*see* **Notes 2–4**).

3. Visualize neurons on an inverted fluorescence microscope.Calcein (green) fluorescence can be viewed using a filter set compatible with excitation at ~495 nm and emission at ~515 nm (e.g., Semrcok FITC-3540B, Omega XF23 or Chroma 31001) and EthD-1 (orange/red) fluorescence can be visualized using a standard compatible with excitation at ~530 nm and emission at ~620 nm (e.g., Semrock TRITC-A, Omega XF35 or Chroma 31005) (*see* **Note 5**).

4. Count the number of live (green fluorescent) and dead (orange-red fluorescent) neurons. A minimum of five random fields and at least 500 cells should be counted for each sample (*see* **Note 6**).

5. Calculate percent survival as # calcein positive cells (green)/# total cells (green + orange) × 100%. Alternatively, calculate percent cell death as # EthD-1 positive cells (orange)/# total cells (green + orange) × 100%.

**3.2. Nuclear Morphology Assay**

1. Prepare a 20 mM stock solution of Hoechst 33342 in water and store aliquots at –20°C.

2. Prepare a 10X working dilution of Hoechst 33342 (10 μM) in Neurobasal media.

3. Add 50 μL of 10X Hoechst 33342 staining solution to each well and incubate for 15 min at room temperature.

4. Visualize neurons on an inverted fluorescence microscope. Hoechst (blue) fluorescence can be examined using a compatible with excitation at ~350 nm and emission at ~460 nm (e.g., Semrock DAPI-5060B, Omega XF06, or Chroma 31000).

5. Count the fraction of nuclei that exhibit distinct apoptotic morphology (*see* **Note 7**). A minimum of five random fields and at least 500 cells should be counted for each sample (*see* **Note 8**).

**3.3. Caspase-3 Activity Assay**

*3.3.1. Preparation of Neuronal Extracts*

1. Remove culture media by gentle aspiration and add 500 μL of ice-cold PBS to each 8.8 cm$^2$ culture dish.

2. Gently remove cells from the surface of the dish using a cell scraper and transfer to a 1.5 mL microfuge tube.

3. Centrifuge at 400$g$ for 5 min at 4°C.

4. Remove supernatant and resuspend cell pellet in 100 μL of ice-cold caspase lysis buffer and extract for 20 min on ice.

5. Centrifuge at 10,000$g$ for 10 min at 4°C.

6. Transfer supernatant (soluble cell extract) to a new microfuge tube.

7. Determine the protein concentration of cell extracts (*see* **Note 9**).

*3.3.2. Determination of Caspase-3 Activity*

1. Prepare a volume of caspase reaction buffer sufficient for the number of samples to be analyzed and warm to 37°C in a water bath. To determine the volume required, consider that each sample will be analyzed in triplicate and that 200 μL is required for each replicate.

2. Split the caspase reaction buffer into two equal volumes and label the tubes "A" and "B". Add Ac-DEVD-AFC to the tube labelled "B" to a final concentration of 30 μM (this is a 2X solution).

3. Add 100 μL of caspase reaction buffer "A" to wells of a 96-well fluorescence assay plate. Set up triplicate wells for each sample to be analyzed and one well as a reagent blank (i.e., reaction buffer but no cell extract).

4. To each well add a volume of cell extract equivalent to 5 μg of protein (*see* **Note 10**).

5. To each well add 100 µL of caspase reaction buffer "B".

6. Place the microplate in a fluorescence microplate reader pre-heated to 37°C and measure the fluorescence output using filter sets compatible with excitation at 400 nm and emission at 505 nm. Record the fluorescence signal every 15 min over a 2 h period.

7. To calculate caspase activity select a 60 min interval in which the increase in fluorescence signal is most linear, and determine the change in fluorescence signal over this 60 min interval for each sample.Caspase activity can be reported as the change in fluorescence signal over the 60 min period or as a ratio relative to the change in fluorescence signal of an untreated control sample (i.e., fold increase over control) (*see* **Note 11**).

## 4. Critical Steps and Troubleshooting Notes

1. Calcein AM is susceptible to hydrolysis; therefore, aqueous solutions should be prepared fresh and used within 12 h.

2. Optimization of dye concentration. We have found concentrations of 0.5 µM Calcein AM and 3 µM EthD-1 to be optimal for mouse cerebellar granule neuron as well as mouse cortical neuron cultures. However, the concentrations may have to be optimized for other types of neuronal cultures or mixed neuron-glia cultures. The optimal concentration of Calcein AM will depend on the level of esterase activity in different cell types. To determine the optimal conditions, test a range of Calcein AM concentrations (0.1–5 µM) and EthD-1 concentrations (0.5–5 µM) in neuronal cultures that have been treated with a known death agonist for 24 h. Select an EthD-1 concentration that produces a bright fluorescence signal in the nucleus but not the cytoplasm of pyknotic, phase bright cells (dead cells). Select a Calcein AM concentration that does not produce a green fluorescence signal in dead cells, but generates significant signal in the cell body and neuritis of live cells.

3. The Calcein AM/EthD-1 staining solution is added directly to the culture media of neurons cultured in N2/B27-supplemented neurobasal media (i.e., serum-free media). However, if neurons are cultured in serum-based media, it may be necessary to replace the culture media with a balanced salt solution (e.g., DPBS) prior to adding

the staining solution as extracellular esterases present in serum may lead to the extracellular hydrolysis of Calcein AM and consequently elevated background fluorescence. If a media change is necessary, then this should be done carefully to minimize the detachment of dead cells from the surface of the culture dish.

4. The duration of the dye incubation may also have to be optimized for different types of neuronal cultures. In any case it is advisable to analyze all samples within a similar time frame as the calcein signal will continue to develop with time. Furthermore, if staining is done in a balanced salt solution, then the incubation period should be kept to a minimum as neurons tend to undergo excitotoxic cell death during prolonged incubation in BSS. Therefore, if analyzing many samples, we recommend arranging them in batches and staggering the start of the dye incubations for the different batches.

5. If the fluorescence microscope is equipped with a fluorescein long pass and dual-emission filters (e.g., Omega XF25 or Chroma 41012), the calcein and EthD-1 fluorescence can be viewed simultaneously.

6. We highly recommended that microscope images be acquired using a CCD camera and image processing system (e.g., Northern Eclipse, EMPIX) so that cell counting can be performed at a later time.

7. Apoptotic cells are defined as those exhibiting a shrunken nucleus with bright – uniform fluorescence (karyopyknosis) or a fragmented nucleus with bright – uniform fluorescent bodies (karyorrhexis).Importantly, cells exhibiting a shrunken nucleus, but not uniform fluorescence are not considered to be apoptotic (*see* examples in **Fig. 21.1 B**).

8. Hoechst fluorescence is rapidly bleached under UV illumination, so exposure time should be minimized. Therefore, we recommended that microscope images be acquired using a CCD camera and image processing system so that cell counting can be performed at a later time.

9. Caspase lysis buffer is compatible with most standard protein quantification assays, including Pierce BCA$^{TM}$ Protein Assay (ThermoFisher) and Biorad DC Protein Assay (Biorad).

10. We have found that the fluorescence signal produced increases in a linear fashion over a protein range of 1–20 $\mu$g for both cerebellar granule and cortical neuron cultures exposed to an apoptotic death stimulus. However, the amount of protein required may need to be optimized

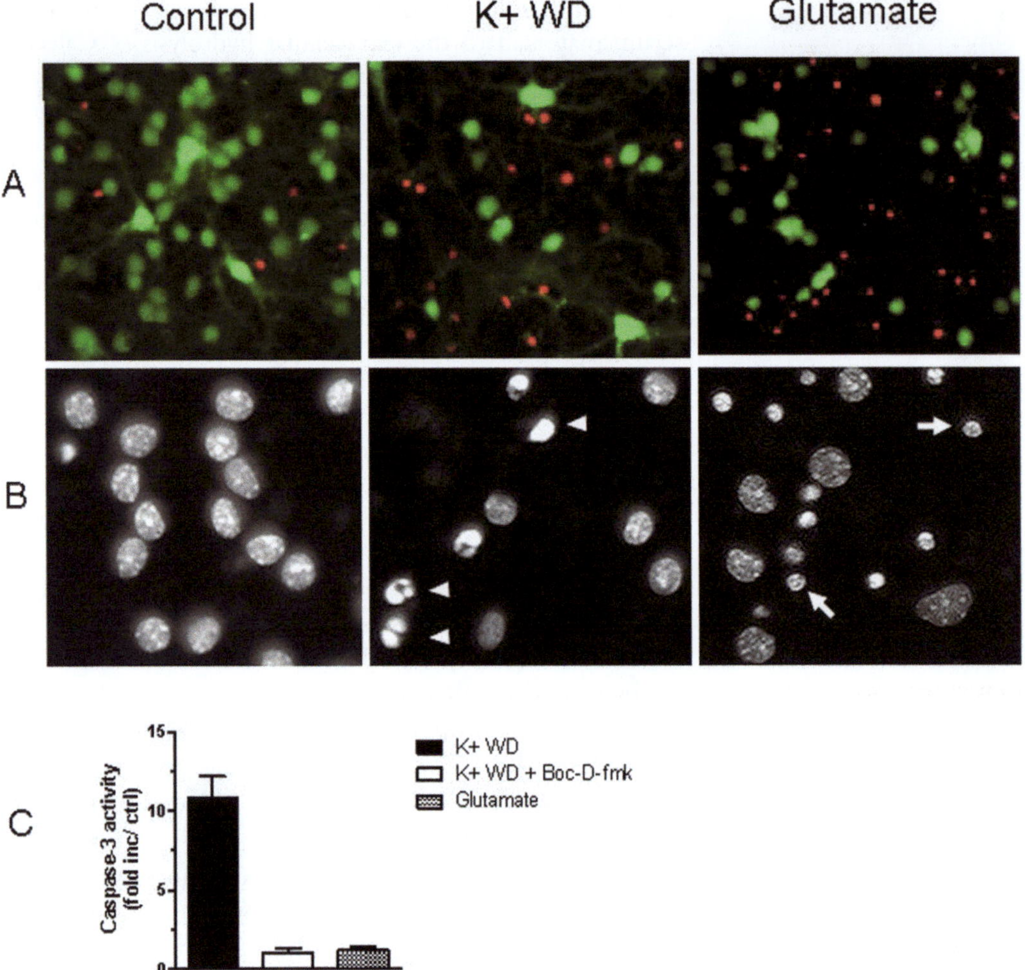

Fig. 21.1. CGNs cultured for 7 days in vitro were either treated with glutamate (100 $\mu$M) to induce excitotoxic (non-apoptotic) cell death or switched to media containing non-depolarizing concentrations of potassium (5 mM, K+ WD) to induce apoptosis. (**A**) Representative images of cell viability staining performed 24 h after treatment. Note the increase in EthD-1-positive (red/dead) neurons and decrease in Calcein-positive (green/live) neurons following both treatments. (**B**) Representative images of Hoechst 33342-stained neurons 24 h post-treatment. Neurons exhibiting classical apoptotic morphology are indicated by *arrowheads* whereas shrunken nuclei of neurons undergoing non-apoptotic cell death are indicated by *arrows*. (**C**) Neuronal extracts isolated 16 h following treatment were analyzed by caspase-3 activity assay. Activity is reported as fold increase over untreated control cells. Where indicated, the pan-caspase inhibitor Boc-D-fmk (50 $\mu$M) was added to cultures at the time of treatment.

for other types of neuronal cultures. This can be done by examining the change in fluorescence signal produced over a 1 h interval by increasing amounts of protein. Select a protein concentration that produces a fluorescence signal that lies within the linear range.

11. To confirm that an increase in Ac-DEVD-AFC cleavage is due to caspase activation, verify that addition of the pan-

caspase inhibitor Boc-D-FMK (50 μM) to the neuronal cultures blocks the cleavage activity.

## 5. Typical Protocol Results

CGNs cultured for 7 days in vitro were either treated with glutamate to induce excitotoxic (non-apoptotic) cell death or switched to media containing non-depolarizing concentrations of potassium to induce apoptosis. Both of these treatments resulted in a significant increase in the fraction of EthD-1-stained (dead) neurons and a corresponding decrease in Calcein-stained (live) neurons (**Fig. 21.1A**). CGNs deprived of depolarizing levels of potassium exhibit classical apoptotic nuclear morphology and high levels of caspase-3 activity (**Fig. 21.1B and C**). In contrast, the nuclei of CGNs treated with glutamate appear shrunken, but they do not exhibit chromatin condensation and cell death is not associated with caspase-3 activity.

## References


1. Cheung EC, Melanson-Drapeau L, Cregan SP et al. Apoptosis-inducing factor is a key factor in neuronal cell death propagated by BAX-dependent and BAX-independent mechanisms. J Neurosci 2005; 25(6): 1324–1334.
2. Cregan SP, Fortin A, Maclaurin JG et al. Apoptosis-inducing factor is involved in the regulation of caspase-independent neuronal cell death. J Cell Biol 2002; 158(3):507–517.
3. Miller TM, Moulder KL, Knudson CM et al. Bax deletion further orders the cell death pathway in cerebellar granule cells and suggests a caspase-independent pathway to cell death. J Cell Biol 1997; 139(1): 205–217.
4. Putcha GV, Moulder KL, Golden JP et al. Induction of BIM, a proapoptotic BH3-only BCL-2 family member, is critical for neuronal apoptosis. Neuron 2001; 29(3):615–628.
5. Steckley D, Karajgikar M, Dale LB et al. Puma is a dominant regulator of oxidative stress induced Bax activation and neuronal apoptosis. J Neurosci 2007; 27(47): 12989–12999.
6. Thornberry NA. Caspases: a decade of death research. Cell Death Differ 1999; 6(11): 1023–1027.
7. McStay GP, Salvesen GS, Green DR. Overlapping cleavage motif selectivity of caspases: implications for analysis of apoptotic pathways. Cell Death Differ 2008; 15(2): 322–331.
8. Woo M, Hakem R, Soengas MS et al. Essential contribution of caspase 3/CPP32 to apoptosis and its associated nuclear changes. Genes Dev 1998; 12(6):806–819.

# Chapter 22

# Live Imaging of Neural Cell Functions

Sabine Bavamian, Eliana Scemes, and Paolo Meda

## Abstract

To determine the cell autonomous and environmental factors that control the differentiation of neurons, astrocytes, and oligodendrocytes, we have used neurospheres made of primary neural progenitor cells. These organoids are amenable to the live cell imaging of several parameters which are central to the proper control of neuron and glial cell differentiation, as well as to the function of the resulting fully differentiated neural cells. Here we report on the methods to study in living cells the connexin-dependent cell-to-cell coupling, the oscillations in intracellular $Ca^{2+}$, and specifically the intercellular synchronization of such events, and the ATP release by either exocytosis of vesicles or through specialized membrane channels. The methods rely on the combination of a variety of state-of-the art microscopy and biophysical methods.

**Keywords:** Neurospheres, neurons, astrocytes, oligodendrocytes, cell coupling, $Ca^{2+}$, ATP, exocytosis, fluorescence microscopy, confocal microscopy, TIRF microscopy, mouse.

## 1. Introduction

Among the various models of neural cell culture that are available for functional studies, we have focused on primary cultures of neural progenitor cells which, in the presence of epidermal and basic growth factors, have the ability to self-renew and to grow into neurospheres (1–4). After removal of growth factors and adhesion to a coated support, these spheric cell aggregates give rise to progenitor cells that migrate out of the neurosphere and form layers of differentiating neurons, astrocytes, and oligodendrocytes (**Fig. 22.1**). Neurospheres are therefore regarded as assemblies of multipotent precursor cells that represent simplified in vitro models of CNS differentiation (1–4).

L.C. Doering (ed.), *Protocols for Neural Cell Culture*, Springer Protocols Handbooks,
DOI 10.1007/978-1-60761-292-6_22, © Humana Press, a part of Springer Science+Business Media, LLC 1992, 1997, 2001, 2010

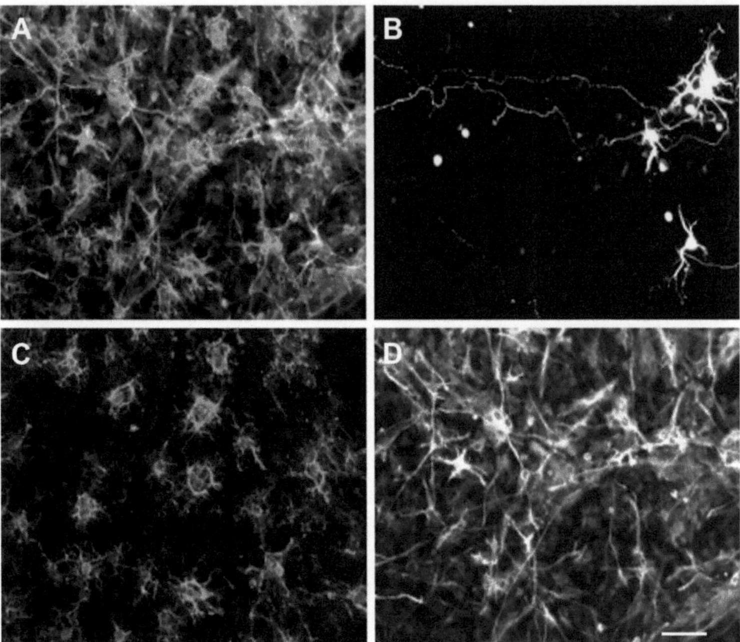

Fig. 22.1. After adhesion to a support and withdrawal of growth factors, neurosphere progenitors differentiate into different types of neural cells. Immunolabeling for GFAP, βIII tubulin, and O4 and revealed the presence of differentiated astrocytes (**A**), neurons (**B**) and oligodendrocytes (**C**) in the outgrowth of neurospheres. Panel **D** (merge) shows the intricate arrangement of the three cell types within neurospheres. Bar, 100 μm.

In order to determine the cell autonomous and environmental factors that control this differentiation, we have tested neural progenitor cells from mouse embryo CNS for the extent, spatial organization, and cell-specific connexin-dependent coupling, intracellular $Ca^{2+}$ and ATP release. These somewhat interdependent parameters are central to the proper control of neuron and glial cell differentiation, as well as to the function of the resulting fully differentiated neural cells (2–4). Connexin-dependent coupling is an almost obligatory form of cell-to-cell communication in animal tissues (5, 6). It is mediated by the gap junctional channels made of different combinations of the 20 proteins that form the connexin family, and allows for direct cell-to-cell exchanges of current-carrying ions and a number of other cytosolic metabolites (5, 6). Such exchanges, also referred to as intercellular coupling, have been shown to be implicated in the control of multiple cell functions (5, 6), including the differentiation and function of CNS networks (7). The relevance of this implication is stressed by the increasing number of genetic and acquired human diseases shown to be associated to connexin and/or coupling alterations (8), including in the CNS (9). Oscillations of free cytosolic $Ca^{2+}$ also play important roles in the physiology and pathology of the CNS (4, 10–12), whether they are restricted within one cell or transmitted to adjoining cells as intercellular waves. Depending on the cell type, the propagation of $Ca^{2+}$ waves across a tissue depends on the diffusion of second messengers through connexin gap junction channels and/or on ATP released from some cells and that results in the activation of purinergic receptors, of both ionotropic (ligand-gated) and metabotropic type (G-protein-coupled) on neighboring cells (13–15). Some of these receptors generate cAMP via stimulation of adenylyl cyclase, and all lead to $Ca^{2+}$ mobilization via the production of phospholipase C. Spontaneous calcium oscillations in neural progenitors are also dependent on the activation of purinergic receptors, likely because cells release ATP that, in turn, can activate these receptors and increase cytosolic $Ca^{2+}$ in the same (autocrine signaling) or the nearby cells (paracrine signaling) (13–15). Release of ATP in the extracellular medium can occur by exocytosis of vesicles or secretory granules and/or through specialized membrane channels, made of pannexins, and in some cases also possibly of connexin proteins (6, 15).

Given the limited amount of tissue that can be prepared from primary cultures of neural precursors, the structural and the functional heterogeneity of these preparations, the different patterns of expression of connexins, purinergic receptors, and pannexins in different types of neural cell types, and the rapidity of the events these proteins mediate, we used a variety of state-of-the-art microscopy methods that allow for live and cell-specific imaging, as well as for the capture of fast neural and glial cell functions. Here, we review these approaches.

## 2. Monitoring Cell Coupling in Neural Cell Cultures

Among the various methods that can be used to evaluate cell-to-cell coupling (16), we choose the microinjection of fluorescent tracers that are impermeant to the cell membrane, due to the tridimensional organization of neurospheres (2, 3). Furthermore, the microinjection method allows to compare the coupling of different types of cells in their native shape, and provides direct evidence for the cell-to-cell exchange of signal molecules, which is expected to occur alike the cell-to-cell exchanges of $Ca^{2+}$-mobilizing second messengers, nucleotides, and metabolites (5, 16, 17).

### 2.1. Reagents

#### 2.1.1. Gap Junction Tracers

For evaluation of cell-to-cell coupling, tracers should be nontoxic, small enough to cross gap junction channels (cutoff size is about 900 Da in vertebrate cells), and should not be able to leak across the non-junctional areas of normal cell membranes

(2, 3, 5, 16, 17). The choice of the most adequate tracer is dependent on several factors, including the scope of the experiment, the conditions under which the study is conducted, and the pattern of connexins in the system investigated. Most connexins are permeant to several tracers and, as yet, no molecule has been shown to permeate only one type of gap junction channel, i.e., is specific for any connexin isoform. However, different connexins form channels with distinct permeabilities to molecules, and these specific properties allow for subtle discrimination between molecules which are only slightly different in size and/or electrical charge (16–18). This is particularly relevant in neural cell cultures, in as much as the channels expressed in neurons (which are mostly made of Cx36) are quite different from those connecting progenitor cells and differentiated astrocytes (which are mostly made of Cx43) and oligodendrocytes (which are mostly made of Cx37 and Cx42 (7). We have used the following tracers, which are commercially available from several companies, including Molecular Probes Inc. (http://www.probes.com), Sigma Chemical Co. (http://www.sial.com), Eastman Organic Chemicals (http://www.eastman.com), and Vector Laboratories (http://www.vectorlabs.com):

1. Lucifer Yellow CH (net charge −2; mol.wt 457; size 12.6 × 14 × 5.5 Å): this tracer has a high fluorescence efficiency (**Fig. 22.2**), which ensures its detection in minute levels.

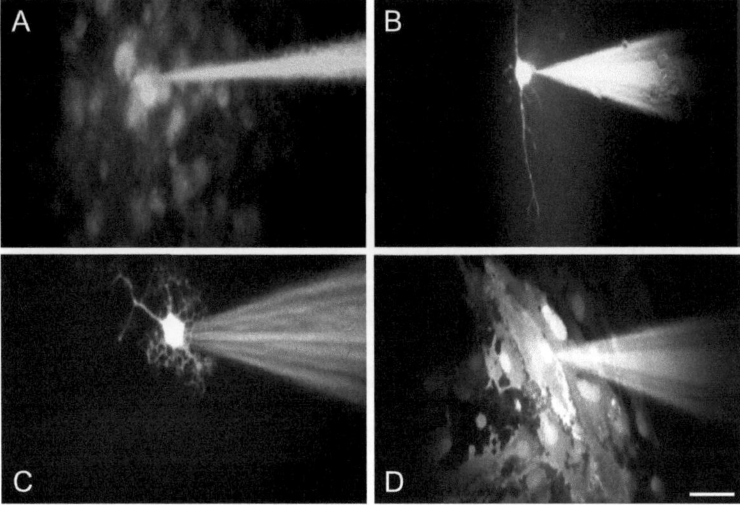



Fig. 22.2. Microinjection of Lucifer Yellow reveals distinct coupling patterns of neuron and glial cells within differentiating neurospheres. Lucifer Yellow was microinjected by iontophoresis into single cells of adherent neurospheres and fluorescence photographs taken 5 min later. Dye coupling was strong between the progenitor cells located in the neurosphere center (**D**) and between the astrocytes that grew out of these structures (**A**). In contrast, no diffusion of the tracer to adjacent cells was observed when oligodendrocytes (**C**) and neurons (**B**) were microinjected. Bar, 15 μm. Modified from (3). Reproduced with permission of the Company of Biologists.

It also binds to cell components after aldehyde fixation, a property exploited to identify the cell(s) containing the tracer after histological processing at both light and electron microscopy levels (16, 19). The tracer is usually injected as a 2–4% solution in either distilled water or 150 mM LiCl, and can be stored for weeks in the dark at 4°C. The original dye does not dissolve in $K^+$-containing solutions. However, a potassium salt is now commercially available if there is such a need. Lucifer Yellow is observed using a barrier filter with an emission between 520 and 560 nm and an excitation filter with a range between 430 and 435 nm.

2. 6-carboxyfluorescein (net charge −2; mol.wt 376; size 12.6 × 12.7 × 8.5 Å): this tracer has a lower fluorescence efficiency than Lucifer Yellow CH and does not resist histological processing. After recrystallization (20), it can be stored in the dark at 4°C as a 20% solution in distilled water, pH 8–9. It is used as a 4% solution in either distilled water or 3 M KCl. Carboxyfluorescein is observed with the same barrier and excitation filters used for Lucifer Yellow.

3. Neurobiotin (MW 287, 1 positive charge, size 12.7 × 5.4 Å): has molecular dimensions that are less than those of Lucifer Yellow and, hence, is expected to permeate more easily connexin channels, notably of neurons (21). The tracer (Vector Laboratories, Burlingame, CA) is typically used at a concentration of 1–4% in 0.25 M lithium acetate (pH 7.3) and resists histological processing after fixation in 4% paraformaldehyde. To be detected, neurobiotin should be visualized with either streptavidine coupled to a fluorochrome (cyanine dyes, fluorescein, or rhodamine) or to horseradish peroxidase. If peroxidase-conjugated streptavidin is used, the activity of the enzyme may be revealed after blockade of the endogenous peroxidase activity (by two successive 30 min incubations in a 0.1 M Tris–HCl buffer (pH 7.6) supplemented with 0.5 % $H_2O_2$), and membrane permeabilization (by a 10 min incubation in a 0.1 M Tris–HCl buffer (pH 7.6) supplemented with 0.25 % Triton-X). To this end, cells are incubated 1–10 min in a 0.1 M phosphate buffer supplemented with 1 mg/mL diaminobenzidine, 1.2 mM cobalt chloride, 0.8 mM nickel chloride, and 0.9 mM $H_2O_2$. The reaction is stopped by replacing the developing mixture with distilled water.

4. Ethidium bromide (net charge +1; mol.wt 394; size 11.6 × 9.3 × 4.3 Å) or propidium iodide (PI; net charge +1; mol. wt 661; size 12.9 × 9.3 × 4.5 Å). These two molecules are most useful to evaluate coupling of cells connected by Cx36 channels, such as neurons (16, 18), as well as to screen for pannexin/connexin "hemi-channels" (15, 16).

5. 2-[*N*-(7-nitrobenz-2-oxa-1,3-diazol-4-yl)amino]-2-deoxy-D-glucose (2-NBDG; MW = 342).This molecule is a fluorescent derivative of glucose that has been used to evaluate. glucose uptake by living cells (22). When used at a concentration of 5 mg/mL, it may be transferred through some junctional connexin channels, thereby providing evidence for metabolic coupling of the cells.

6. Other gap junction-permeant tracers that have been used less frequently and may be useful in particular cell systems or under specific conditions are DAPI (MW 279, 1 positive charge), biocytin (MW 373, no electrical charge) and biotine-X cadaverin (MW 442, 1 positive charge), 2′,7′-dichlorofluorescein (MW 401, 1 negative charge). The latter molecule has a permeability 4–6 times higher than 6-carboxyfluorescein, presumably because of a smaller hydration radius which is due to its minor electrical charge. It has little affinitiy for binding to cytoplasmic proteins and, thus, does not resist histological processing. Most of these tracers can be dissolved in LiCl and used at a final concentration of 0.5–1% (16).

7. Dextrans have molecular weights and sizes which largely exceed the maximum cutoff limit of gap junction connexin channels. As such, these macromolecules are useful to ensure that the intercellular transfer of any of the above-listed tracers is actually dependent on gap junctions, and is not accounted for by alternative pathways, such as the persistence of cytoplasmic bridges at the end of mitosis (16). To this end, we have used 10 kDa dextrans coupled to rhodamine, which, at the concentration of 100 mg/mL, may be introduced into cells as the other tracers.

8. It is obviously feasible to use the above-listed tracers in combination. We have often combined Lucifer yellow with either dextran-rhodamine, neurobiotin, or propidium iodide (18, 23). In the latter case, the two tracers were each used at a 2% concentration and injected into the same cell by electrical pulses of alternated polarity (negative pulses preferentially inject Lucifer Yellow, whereas positive pulses preferentially inject neurobiotin and propidium iodide).

## 2.2. Equipment

### 2.2.1. Electrodes

Electrodes are pulled from filament containing borosilicate glass capillary tubing (F. Haer & Co., Brunswick, ME) using a micropipette puller (we use either the BB-CH puller from Mecanex, Geneva, Switzerland, or the PC-10 puller from Narishige Co., Tokyo, Japan) that is adjusted to give the desired electrode resistance. Electrodes of 50–60 MΩ when filled with 3 *M* KCl are adequate for most uses (when filled with solutions of tracers, these electrodes show resistances 4–20 times higher, depending on the solvent used). The electrodes are pulled before the experiment, bent on a small gas flame and filled (tip only) with a tracer, using a 34-gauge nonmetallic syringe needle (Micro-Fil; World Precision Instruments, Stevenage, UK) fitted on a 0.22 μm-pore filter. The filled electrodes are stored in a clean, humidified container and used within the next 12 h.

*2.2.2. Setup*

As a function of the scope of the study, of the cell type investigated, and of the available funding, different setups may be found suitable for the intracellular microinjection and recording of gap junction tracers. Basically, a general purpose setup should include a microscope, one or two micromanipulators, an electronic rack for performing and controlling the cellular microinjection, and a camera system to record the experiment. Most of the required equipment is commercially available from several companies, including Zeiss (http://www.zeiss.com), Nikon (http://www.nikon.com), Axon Instruments Inc. (http://www.axonet.com), World Precision Instruments (Aston, UK; http://www.wpiinc.com) and Tektronix, Inc. (Wilsonville, OR; http://www. tek.com). Other necessary pieces of equipment (electrode puller, etc.) and materials may be found through the BioSupplyNet Source Book (http://www. biosupplynet.com) and the Axon Guide for Electrophysiology & Biophysics Laboratory Techniques (http://www.axonet.com).

For injection of the tridimensional neurospheres, we use an inverted Zeiss IM35 microscope equipped for both fluorescence (XBO 75 W Osram bulb, excitation and barrier filters for fluorescein and rhodamine detection) and phase-contrast views. Under both illuminations, a ×25 objective was used to visualize the entire neurosphere and the differentiating cells that migrated out of it. The microscope is placed on an antivibration VX95 table; (Micro-Controle; http://www.newport.com) and has a homemade stage that can be held at different temperatures through the water circulation provided by a Minisitat pump (Huber, Germany).

For holding and positioning the electrodes, we used electronic micromanipulators (PatchMan, http://www. eppendorf.com).

We usually microinject tracers by iontophoresis (2, 3, 18, 23, 24), using a setup that includes a programmable pulse generator (Master-8, A.M.P. Instruments, http://www.ampi.co.il), and an amplifier (AxoClamp2B, Axon Instruments Inc. http://www.moleculardevices.com). Current pulses and recorded voltages are displayed on separate channels of a digital storage oscilloscope (2214, Tektronix Inc.), whose outputs are traced on a four-channel TA550 paper recorder (Gould; http://gould.co.uk). The system is connected by Ag/AgCl wires to the tracer-filled electrode. When large cells (e.g., astrocytes) are tested, pressure injection, as provided by a Picospritzer II pulse generator (General Valve Co., Fairfield, NJ) may also be adequate.

***2.3. Protocol***

1. A preliminary condition for a successful microinjection is that the neurospheres under study do not move during impalement. To this end, standard culture Petri dishes were successively coated with poly-D-lysine (10 μg/mL in water) and fibronectin (10 μg/mL in PBS).

2. Ten floating neurospheres were plated in 20 μL standard culture medium within each dish.

3. Neurospheres were incubated for 2 h to initiate cell adhesion, and culture medium without EGF was gently added.

4. Adherent neurospheres were cultured for 8 days, during which time half of the volume medium was changed twice.

5. At the time of microinjection, the culture medium is replaced with a Krebs-Ringer-bicarbonate buffer, supplemented with 10–30 mM Hepes and 5 mg/mL bovine serum albumin.

6. Dishes with the specimen to be injected are covered with a thin layer of light liquid paraffin (BDH Laboratory Supplies, Poole, UK) to avoid medium evaporation, and transferred on the heated (37°C) stage of the microscope.

7. An electrode, previously filled in the tip with the selected tracer, is completely filled with 3 M Hepes-buffered (pH 7.2) lithium chloride and mounted on the injection setup.

8. Under phase-contrast illumination, the electrode is positioned over the area of the neurosphere to be injected.

9. The electrode is then slowly lowered against the membrane of an individual cell.

10. The cell is penetrated by briefly "vibrating" the electrode, using the negative capacitance control of an amplifier. Correct penetration may be judged by the rapid filling of the injected cell (if the tracer selected is fluorescent) and by a drop of membrane potential (for all tracers dissolved in a salt solution).

11. As soon as the adequacy of the penetration has been ascertained, all illumination is turned off to prevent any photoabsorption, which could damage the cell (if a video recording is made, the illumination should be decreased as much as possible with inert absorption filters, and the signal amplified with an adequate high sensitivity camera), and to

minimize the binding of tracers to cytoplasmic and nuclear components.

12. The iontophoretic injection of the tracer is then pursued in the dark, using hyperpolarizing or depolarizing current pulses (or a combination of these), depending on the charge of the selected tracer. The duration, amplitude, and frequency of these pulses can be varied depending on the type of electrode used (the amount of current that can be injected decreases per unit time with increasing electrode resistance), the amount of tracer to be injected, and the cell type. We routinely use square, 0.1 nA pulses of 900 ms duration and 0.5-Hz frequency for 3–10 min.

13. Maintenance of adequate penetration should be evaluated by the persistence of a stable resting membrane potential, indicating a good sealing between the electrode and the cell membrane, and, in the case of fluorescent tracers, by the persistence of a high cell fluorescence after interruption of the injection, indicating lack of diffusion of the tracer in the surrounding medium.

14. At the end of the experiment, the current injection is interrupted before the electrode is gently pulled out of the cell.

***2.4. Analysis of the Microinjected Cells***

If a fluorescent tracer was chosen, the injected cell and its coupled neighbours may be immediately visualized at the end of each injection (2, 3, 18, 23, 24). We usually record each injection photographically, using either Kodak Ektachrome 400 daylight film for color slides and a constant exposure time of about 20 s or a digital camera (AxioCam MRm, Leitz; C5810 3CCD camera, Hamamatsu Photonics), and excitation and barrier filters appropriate for the specific fluorescence of the choosen tracer.

Scoring and identification of the coupled and uncoupled cells, at either light or electron microscopy levels, is facilitated using tracers that bind to cell components. To permanently fix the tracers for light microscopy, Zamboni's fluid or a 4% solution of paraformaldehyde in PBS are the preferred fixatives since they do not contribute to background fluorescence and ensure the covalent binding of the injected tracers to cytoplasmic structures. After a 30–90 min fixation at room temperature, the specimen is rinsed three times 10 min in PBS, dehydrated by 10 min passages in a sequence of 30, 50, 70, and 95 % ethanol, followed by two 20 min passages in absolute ethanol. The preparation is then soaked twice 15 min in propylene oxide (absolute ethanol should be used if the preparation was grown in a plastic dish), 2 h in a 1:1 (v/v) mixture of propylene oxide (or absolute ethanol) and Epon 812, and infiltrated 12 h with pure Epon. The specimen is then embedded in freshly prepared Epon 812 and cured according to standard procedures. In sections, the injected cell can be recognized by its fluorescence labeling, and is amenable to additional immunolabeling using specific antibodies (16, 17, 23, 24).

***2.5. Critical Steps and Troubleshooting***

The criteria for the proper selection of the tracers are discussed above (**Section 2.1.1**). The main problems of the microinjection procedure are the killing of the cells during electrode penetration, current injection and visualization (that should both be minimal), and the difficulty to avoid a double penetration that would result in artifactual coupling. When cells are intricated as in neurospheres, this is a major concern that can be controlled by parallel electrophysiological recording. The quality of the electrodes is also essential, not only for both fast and non-traumatic penetration of the cell but also for its sealing on the electrode tip that prevents tracer leakage. Whatever the theory, the microinjection procedure remains to a large extent investigator-dependent, and problems decrease with errors, trials, and time (16).

## 3. Monitoring Cytosolic $Ca^{2+}$ in Neural Cell Culture

Differentiation of neural precursors and proper function of the differentiated neurons and glial cells are modulated by the levels and transient oscillations of cytosolic $Ca^{2+}$, as well as by the intercellular synchronization of these transients and the intercellular wave diffusion of $Ca^{2+}$, whether via connexin gap junction channels or via some paracrine signalling (10–15). We have used fluorimetric means to monitor these $Ca^{2+}$ changes in cultures of neural cells and other cell types that, alike neurons, are coupled by Cx36 channels (4, 25, 26).

### 3.1. Reagents

#### 3.1.1. $Ca^{2+}$ Tracers

Out of the many $Ca^{2+}$ indicators (27–29) that are commercially available, we have tested the following:

1. Fura-2 and indo-1: at low concentration, these ratiometric dyes (excitation: 340/380 nm emission, 512 nm for fura-2; excitation 351 nm, emission 405/485 nm for indo-1) allow for accurate measurement of intracellular $Ca^{2+}$ concentrations, by correcting for uneven dye loading, leakage and photobleaching, and unequal cytoplasmic thickness (30). Therefore, these probes are preferred for evaluation of actual $Ca^{2+}$ levels, after proper callibration.

2. Fluo-3 and Fluo-4/AM: these non ratiometric tracers obtained by combining BAPTA and a fluorescein-like structure (31) have absorption and emission peaks of 506 and 526 nm, respectively. Their absorption spectrum is therefore compatible with excitation at 488 nm by argon-ion laser sources, which permits detection of 40- to 200-fold

changes in fluorescence intensity once $Ca^{2+}$ binds to the tracer (31–34). Fluo-4 is an analog of Fluo-3, which features increased fluorescence excitation at 488 nm and, thus, provides larger signal under confocal microscopy. Both tracers are excited at a single wavelength. An advantage is that the required equipment is much simpler than that required for ratiometric tracers, and that the dynamic of $Ca^{2+}$ binding is fast. Thus, Fluo3 or 4 are preferred to monitor rapid changes in $Ca^{2+}$ transients, such as the intercellular synchronization of these oscillations or the fast diffusion of $Ca^{2+}$ waves, as evaluated by changes in the relative fluorescence intensity. In contrast, these tracers should not be selected to obtain absolute evaluations of $Ca^{2+}$ levels.

3. Highly localized $Ca^{2+}$ signals may escape detection with the above-mentioned tracers, also because the loading of these tracers is often heterogenous in the cytoplasm, presumably due to compartimentalization of the membrane permeant AM-ester form of the tracers and/or to incomplete AM ester hydrolysis or leakage. In this case, genetically encoded $Ca^{2+}$ probes that are targeted to specific cell compartments ("cameleons") offer distinct advantages (35). The yellow cameleon 3.6 is a fusion protein comprising: 2 modified green fluorescent proteins (ECFP and EYFP) linked by a calmodulin domain and a CaM-binding peptide. When the calmodulin domain binds to $Ca^{2+}$, it associates with the CaM-binding peptide, causing a conformational change that brings the modified GFP domains closer to the EYFP domain. Part of the excitation energy from ECFP is then transferred to EYFP, resulting in fluorescence resonance energy transfer (FRET). Thus, increased $[Ca^{2+}]_i$ results in a proportional increase of the 535/480 emission ratio, which can be easily monitored using a fast confocal microscope (36). In cells transfected as per standard methods with the cameleon protein, a homogenous loading of distinct cell organelles is expected, which should facilitate the detection of $Ca^{2+}$ oscillations. A limitation of the approach is the frequent splitting of the indicator fusion protein with cell passage.

*3.1.2. $Ca^{2+}$ Buffer*

1. Make a 1 M stock solution of following salts: NaCl (29.22 g/500 mL $H_2O$), KCl (37.28 g/500 mL $H_2O$), $MgCl_2$ 2.033 g/10 mL $H_2O$, Hepes free acid (23.83 g/100 mL $H_2O$, CaCl2 (1.47 g/10 mL $H_2O$). Sterilize by autoclave and store at 4°C.

2. Make the final solution: by mixing 125 mL stock NaCl (125 mM final), 5.7 mL stock KCl (5.7 mM final), 1.2 mL stock $MgCl_2$ (1.2 mM final), 10 mL stock Hepes (10 mM

final), 1.2 mL stock $CaCl_2$ (1.2 mM final). Add $H_2O$ to 1000 mL. Warm solution to 37°C and adjust pH to 7.4 with 10 N NaOH.

*3.1.3. Fluo-3*

Make a 5 mM stock solution of Fluo-3/AM (Molecular Probes, Eugene, OR) in DMSO. Store in the dark on ice.

*3.1.4. Pluronic Acid*

The solubilisation of the nonpolar AM ester forms of the tracers, which allows for their entry into cells, is enhanced by addition of the non-ionic detergent Pluronic® F-127 (Molecular Probes P-3000MPl). Prepare a stock solution of 20% (w/v) pluronic acid into the loading buffer.

*3.1.5. Loading Buffer*

To the $Ca^{2+}$ buffer described above (**Section 3.1.2**) add 2 mM glucose and 0.1% bovine serum albumin (fraction V; Sigma Chemical).

**3.2. Equipment**

To monitor changes in cytosolic $Ca^{2+}$ with sufficient spatial and temporal resolution, we used an epifluorescence microscope (Eclipse TE2000-S; Nikon, Japan), equipped with a CCD camera (Orca-ER; Hamamatsu, Japan). Fluo-3 fluorescence intensity emitted at one excitation wavelength (488 nm) was continuously acquired at a rate of 1.0 Hz using combined systems of filters and shutter (Lambda DG-4 Diaphot, Sutter Instruments, Burlingame, CA) driven by a computer through Metafluor software (Universal Imaging Systems, Downingtown, PA). Other experiments were run on a laser confocal microscope, whose scanning speed is increased by multiple beams scanning, as provided by a spinning disk. This disk contains multiple sets of spirally arranged pinholes placed in the image plane of the objective lens. Each illuminated pinhole is imaged by the objective to a diffraction-limited spot on the specimen. Light reflected from the specimen can be observed in the camera system after it has passed back through a conjugate pinhole in the spinning disk. In this way, several thousand points are simultaneously illuminated on the disk, mimicking the effect of several thousand confocal microscopes running in parallel. The rapidly spinning disk compensates for the spaces between the holes to create a real-time confocal image, hence providing for a high-resolution and rapid monitoring of fast cell events. In our setup, the incident laser beam (488 nm) passes through a spinning disk (QLC 100, Visitech International) confocal scanner head, mounted on an inverted Axiovert 200 M microscope (Zeiss). A 40×, 1 NA oil immersion objective (Zeiss) is used to acquire fluorescence images, using a 520-nm long-pass filter and a Cascade II 16 bits cooled EMCCD frame transfer camera (Photometrics-RopperScientific).

For high-throughput $Ca^{2+}$ measurement, we have used a multiwell plate reader (ImageXpress Micro, Molecular Devices) that is fitted on an inverted and fully automated epifluorescent micro-scope, equipped with a $10\times$ SFluor, NA 0.45 objective (Nikon). The system is equipped with a Photometrics CoolSNAP HQ 12-bits digital interline CCD camera.

**3.3. Protocol**

*3.3.1. Confocal Microscopy*

1. Glass cover slips with a diameter suitable for the record-ing chamber are cleaned 30 min in acetone and 30 min in ethanol before insertion into sterile 35-mm culture Petri dishes (Falcon).

2. Plate 10 neurospheres or $3\times10^5$ cells per coverslip.

3. Grow the cultures for 3 days in an air/7% $CO_2$ incubator to obtain neurosphere outgrowth or almost confluent mono-layer clusters.

4. Just before the experiment, dilute the stock solutions of Fluo-3/AM and pluronic acid in the $Ca^{2+}$ buffer to obtain a final concentration of 5 $\mu$M and 0.02%, respectively.

5. Incubate the cells with the AM ester form of the selected $Ca^{2+}$ tracer. We typically used a 60 min incubation with Fluo-3 at 37°C, within an air/7% $CO_2$ incubator (4, 25, 26). However, both the concentration of the tracer and the loading conditions have to be adapted as a function of the cell type, to obtain a sufficiently high, still homogeneous loading of most of the cells with the lowest possible dye concentration.

6. Wash the cells in the indicator-free $Ca^{2+}$ buffer to remove any dye nonspecifically bound to the cell surface.

7. Incubate the cells again for 20 min to allow for complete de-esterification of the AM esters.

8. Place the cover slip with the tracer-loaded cells in the recording chamber, which is continuously perfused with the recording solution.

9. Simultaneously measure the fluorescence due to Fluo-3 exci-tation in 5–20 cells per neurosphere/cluster at 37°C.

*3.3.2. High-Throughput Microscopy*

1. Plate about $2\times10^4$ cells/well in 96-well plates.

2. Place the plate on a rocking shaker for 30 min to homoge-nously distribute the cells in the wells.

3. Culture as described above for confocal micrsoscopy.

4. Insert the plate in the recording chamber of the reader.

5. Record short movies (75 time points with an interval of 1 s) sequentially in each well, using the autofocusing ability of the equipment.

**3.4. Analysis**

To measure the intercellular synchronization of $Ca^{2+}$ spikes, the recordings of regions of interest are analyzed with either the Metamorph (confocal studies) or the ImageXpress software (high-throughput studies) as a function of time. Fluo-3 fluorescence intensity ($F$) obtained from regions of interest were normalized to initial values ($F0$) and expressed as relative changes ($F/F0$). The curves of calcium activity recorded in parallel in several individual cells (**Fig. 22.3**) are compared with a dedicated program, which indicates significant delays and computes a synchrony index.

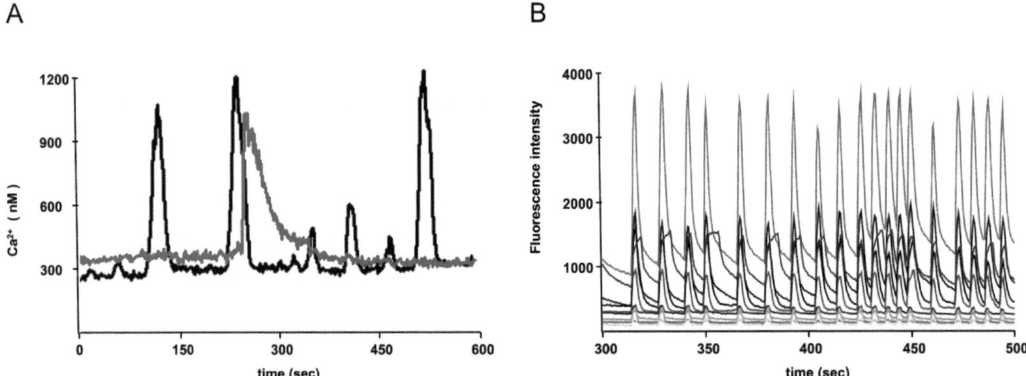



Fig. 22.3. Spontaneous Ca$^{2+}$ transients take place in progenitor cells derived from neurospheres and are synchronized in Cx36-expressing cells (**A**) Spontaneous Ca$^{2+}$ transients were recorded in Fura2-AM-loaded astrocyte precursors that were identified by immunolabeling for the CD44S marker. Modified from (4). Reproduced with permission of John Wiley & Sons, Inc. (**B**) In cultures of the insulin-producing MIN6 cells, comparable transients are induced during stimulation by 20 mM glucose plus 15 mM tetraethylammonium, and are fully synchronized by Cx36 channels in most cells of a cluster. In both panels, each trace corresponds to a different cell.

### 3.5. Critical Steps and Troubleshooting

The criteria for the selection of the most adequate tracer are discussed above (**Section 3.1.1**). The quality of the equipment is essential. Another critical step is a proper loading of the cells with the tracers, which can be judged both by the percentage of cells that actually retained the tracer and by its uniform distribution throughout the cell. Subcellular compartmentalization is a frequent problem with the AM esters, which is usually lessened by lowering the incubation temperature. Thus, the conditions summarized above need to be empirically adapted to specific cell types (28, 33).

## 4. Monitoring ATP Release in Neural Cell Cultures

ATP is an important paracrine signal in the nervous system that contributes to the intercellular propagation of Ca$^{2+}$ waves (13, 15). The nucleotide can be directly exchanged by adjacent cells via gap junction channels, ensuring metabolic cell-to-cell coupling. It can also be released across cell membranes by either specific channels or by exocytosis of ATP-containing vesicles (4, 13–15). In the latter cases, ATP then activate cells carrying the cognate purinergic receptors, whether these are the very same cells that released the nucleotide (autocrine signaling) or nearby cells (paracrine signaling).

***4.1. Reagents***

Several companies provide kits for ATP determination by bioluminescence (37) and the choice will depend on the researcher's need. We prefer kits that provide separated (instead of pre-mixed) reagents (4), since this allows flexibility to optimize the proportion of luciferin and luciferase according to a particular system, sample requirements, and instrumentation. To measure the ATP released from astrocyte progenitors, we have used a kit from Invitrogen-Molecular Probes (http://probes.invitrogen.com).

The reaction solution is prepared under a cell culture hood by mixing 8.9 mL $dH_2O$, 0.5 mL 20× buffer (500 mM Tricine buffer, 100 mM MgSO4, 2 mM EDTA, 2 mM sodium azide, pH 7.8), 0.1 mL DTT (100 mM), and 0.5 mL luciferin (10 mM in 1× buffer). This reaction solution can be stored as 1 mL aliquots for several months at –20°C, if protected from light.

***4.2. Equipment***

The luciferin–luciferase bioluminescence assay measures ATP based on the production of light (575 nm), which is generated when firefly luciferase catalyzes the oxidation of luciferin (37–39), as per the reaction: luciferin + ATP + $O_2$ $\xrightarrow{\text{luciferase}}$ oxiluciferin + AMP + pyrophosphate + $CO_2$ + light. It can detect as little as 0.1 pmol ATP. You need therefore a sensitive luminometer. We have used a Sirius Luminometer (Berthold Detection Systems, Oak Ridge, TN, USA), operated with a 10 s integration time. Depending on the luminometer type and experimental conditions, adjustment of the integration time (1 s or above) may be needed.

***4.3. Protocol***

1. Culture adherent neurospheres as described above.

2. Expose the cells to the experimental condition to be tested for 2 min at 37°C.

3. Sample the medium and centrifuge it 3 min at 1000 rpm and 4°C.

4. Freeze the media aliquots at –20°C until measurements.

5. Lyse the cells in a Krebs-Ringer-bicarbonate buffer containing 0.1% triton-X100 (for evaluation of total ATP content).

6. Freeze the cell extracts at –20°C until measurements.

7. Thaw an aliquot of the reaction solution containing luciferin and add to it 0.25 μL luciferase.

8. Gently mix this reaction solution (do not vortex to avoid enzyme denaturation).

9. Add 50 μL of the reaction solution containing luciferase in each of three wells of a sterile 96-well (flat bottom) Microlite TCT plate (http://www.thermo.com/) or three tubes, depending on the type of available luminometer (to obtain triplicate measurements).

10. Perform background luminescence measurements in these three wells/tubes.

11. Prepare serial dilutions of ATP (50 μM, 5 μM, 500 nM, 50 nM) in the same medium/buffer, which was used during the 2 min test incubation of the cells.

12. Perform luminescence measurements of these samples to establish a standard curve (luminescence by ATP concentration) to transform the sample luminescence values into ATP concentration.

13. Thaw the first experimental sample (media or cell extract).

14. Add 5 μL of this sample to 50 μL of the reaction solution (**Step 9**) and read luminescence values. Use triplicates.

15. Repeat **Step 14** for all other samples.

16. For analysis, subtract background luminescence from the values obtained for the samples and standard curve, and express the amount of released ATP relative to total ATP (or protein) levels.

### *4.4. Critical Steps and Troubleshooting*

The following considerations are important in planning the experiments: (1) the amount of non-lytic release of ATP typically represents less than 1% of total intracellular ATP. Thus, the volume of the experimental solution should not be too large for the total number of cells analyzed. Preferably use confluent cultures and the minimal amount of solution to cover the cells; (2) due to the spontaneous hydrolysis of ATP in solution and the presence of ectonucleotidases at the cell surface, minimize the time between the exposure of cells to the experimental solution and the sample collection; (3) to avoid the presence of floating cells on samples (which would dramatically alter the amount of ATP in solution), spin down the collected samples and retain only the supernatant for analysis; (4) if multiple ATP measurements are planned on the same cells, an interval, which should be empirically determined for each cell type, is required between each test, presumably to allow for refilling the ATP stores or for recovery of other components of the release mechanisms; (5) all solutions and materials should be retained free of biological contaminants (notably skin cells and bacteria by using gloves), given that the assays used to detect ATP are very sensitive and can be affected by even a minimal bacterial contamination; (6) protect luciferin from light and perform the measurements at or close to 28°C, which gives the optimal ezymatic activity; (7) the ATP standards, needed to establish the reference curve, should be made in a solution with the very same composition used, since ionic composition may alter luciferase activity. It is important to mention that luciferase consumes not only ATP but also ADP and their synthetic analogs, but does not degrade pyrimidine nucleotides. Thus, stimulation of cells with purine di-or trinucleotides is not recommended (37–39).

## 5. Imaging Exocytosis from Astrocyte Progenitor Cells

We have used total internal reflection fluorescence (TIRF) microscopy (40) to visualize the release of small ATP-containing vesicles, labeled by quinacrine, at the cell membrane–glass interface (4, 40, 41). The method detects optical changes when light crosses media of different refractive indexes. TIRF uses an objective (with a NA 1.45 or higher) to send a beam of laser or arc lamp light to the coverglass-cell interface. When the angle of a light beam obliquely incident upon a substrate–liquid interface is greater than the critical angle of refraction, the light is totally reflected at the interface. At the point of total reflection, a light electromagnetic field (evanescent field) penetrates into the liquid and propagates parallel to the surface, with an intensity that decays exponentially with the z-axis. For most cells adherent to a substrate, the energy of the evanescent field is usually sufficient to excite fluorophores located at and near (up to 100–200 nm) the cell surface (40, 41).

### 5.1. Reagents

1. A stock solution of 5 mM quinacrine (Sigma), a weak base that accumulates in acidic compartments such as secretory vesicles (42–44) and ATP-containing vesicles (4, 45), was made in DMSO.

2. A stock solution of 0.5 mM fluorescent ATP analog, Mant-ATP [2(3)-O-(N-methylanthraniloyl)-adenosine-5-O-triphosphate (BioLog Life Science), was made in DMSO (4).

### 5.2. Equipment

Images were acquired by either laser or white light TIRF. The laser system consisted of an inverted microscope (Zeiss Axiovert 100 M) modified for through-the-objective evanescent field illumination, equipped with an APO100× oil immersion objective (1.45 NA; 0.17 WD), filter set (488 nm), and a CCD camera (ORCA-ER, Hamamatsu, Japan), driven by OpenLab software on a Macintosh computer. The white-light system (Nikon) consisted of a special arc shape diaphragm placed in the path of a lens-collimated light generated by an arc lamp and mounted on an inverted TE-2000 Nikon microscope equipped with filter sets (488 nm and 594 nm), a 100× (1.45 NA) oil immersion objective, and a CCD camera (Orca-ER, Japan)

### 5.3. Protocol

1. Glass bottom dishes (http://www.mattek.com) are coated with 10 μg/mL fibronectin (Sigma) and 10 μg/mL poly-D-lysine (Sigma).

2. Floating neurospheres (about 100 μm diameter) are plated on these dishes and cultured as described above.

3. When progenitor cells grow out the neurospheres, the cultures are exposed to the various experimental conditions to be tested.

4. Cultures are then either exposed 5 min at 37°C to 5 μM quinacrine (to load all vesicles) or 5 h at 37°C to 50 μM Mant-ATP (to detect the ATP content).

5. The cells are washed several times in culture medium.

6. Cultures loaded with quinacrine are immediately transferred to the stage of the TIRF microscopes. Cultures loaded with Mant-ATP are screened 3.5 h after washout of the fluorescent ATP analog.

7. Fluorescence reflecting incorporated quinacrine (**Fig. 22.4**) or Mant-ATP was visualized at 512 nm when excited at 488 nm, and images acquired at 0.6 s intervals for 1–2 min.

### 5.4. Analysis

For evaluation of quinacrine or Mant-ATP fluorescence changes during exocytosis, the acquired images are played back to define regions of interest within the evanescent field, using the Metamorph software (Universal Imaging). Changes in fluorescence intensity within these regions are displayed either as absolute fluorescence intensity values or as fraction of initial values (**Fig. 22.4**), to determine the time course and the number of spontaneous and evoked exocytic events, which are defined as the quasi-instantaneous disappearance of the fluorescent vesicles within the evanescent field.

### 5.5. Identification of the Cells Monitored

For in vivo detection of the astrocyte precursors (46–48) to be monitored by TIRF, we have used extracellular mouse monoclonal antibodies (http://www.chemicom.net) directed against the cell adhesion glycoprotein CD44S (48). To this end, adherent progenitors are incubated with anti-CD44S (1:200; Chemicon) in DMEM-F12 for 30 min at 37°C. After several washes with culture medium, the cells are incubated with Alexa-Fluor (Molecular Probes)-tagged secondary antibodies (1:1000 in DMEM-F12) for 30 min at 37°C. Depending on the study to be performed, cells can be loaded with quinacrine during the last 5 min of the immunostaining protocol.

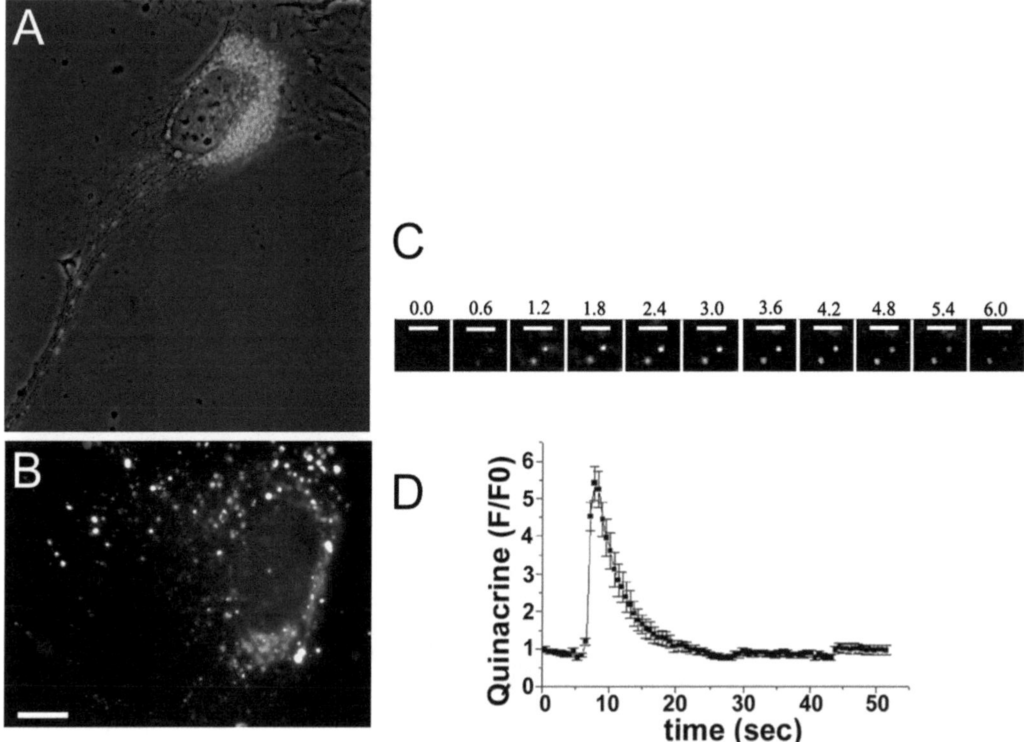

Fig. 22.4. Total internal reflection fluorescence microscopy detects exocytosis of individual ATP-containing vesicles. **(A)** Live fluorescence view of a neural cell progenitor loaded with quinacrine. The base accumulated in small, perinuclear vesicles. **(B)** White light TIRF image shows the quinacrine-loaded vesicles at the membrane–glass interface of unstimulated cells. Bar: 10 $\mu$m. **(C)** Sequential TIRF images showing the time course (s) of the appearance and disappearance of quinacrine-loaded vesicles in the evanescent field, after stimulation of exocytosis by 10 $\mu$M the calcium ionophore A23187. Bars: 4 $\mu$m. **(D)** The ionophore induced a rapid recruitment of vesicles into the evanescenet field, as reflected by the immediate upstroke of the relative fluorescence intensity. Fluorescence then decreased in the evanescent field as quinacrine-loaded vesicles released their content in the extracellular medium, as a function of exocytosis. Modified from (4). Reproduced with permission of John Wiley & Sons, Inc.

## Acknowledgments

Work of PM is supported by grants from the Swiss National Science Foundation (310000-122430), the Juvenile Diabetes Foundation International (1-2007-158), and the European Union (Betaimage FP7-222980; IMIDIA, IMI C-2008.T7). ES is funded by the NIH (NS052245).

## References

1. Reynolds, BA, Weiss, S. Generation of neurons and astrocytes from isolated cells of the adult mammalian central nervous system. Science 1992; 255: 1707–10.

2. Duval N, Gomes D, Calaora V et al. Cell coupling and Cx43 expression in embryonic mouse neural progenitor cells. J Cell Sci 2002; 115: 3241–51.

3. Scemes E, Duval N, Meda P. Reduced expression of P2Y$_1$ receptors in connexin43-null mice alters calcium signaling and migration of neural progenitor cells. J Neurosci. 2003; 23:11444–52.

4. Striedinger K, Meda P, Scemes E. Exocytosis of ATP from astrocyte progenitors modulates spontaneous Ca$^{2+}$ oscillations and cell migration. Glia. 2007; 55:652–62.

5. Bittar EE, Hertzberg EL, (eds). Gap Junctions, Adv Mol Cell Biol 2000:1–407

6. Shestopalov VI, Panchin Y. Pannexins and gap junction protein diversity. Cell Mol Life Sci. 2008; 65:376–94.

7. Söhl G, Maxeiner S, Willecke K. Expression and functions of neuronal gap junctions. Nat Rev Neurosci. 2005; 6:191–200.

8. Lai-Cheong JE, Arita K, McGrath JA. Genetic diseases of junctions. J Invest Dermatol. 2007; 127:2713–25.

9. Mas C, Taske N, Deutsch S et al. Association of the connexin36 gene with juvenile myoclonic epilepsy. J Med Genet. 2004; 41:e93–98.

10. Parri HR, Crunelli V. The role of $Ca^{2+}$ in the generation of spontaneous astrocytic $Ca^{2+}$ oscillations. Neuroscience. 2003; 120:979–92.

11. Beck H, Yaari Y. Plasticity of intrinsic neuronal properties in CNS disorders. Nat Rev Neurosci. 2008; 9:357–69.

12. Iino M. Identification of new functions of $Ca^{2+}$ release from intracellular stores in central nervous system. Biochem Biophys Res Commun. 2008; 369:220–4.

13. Scemes E, Giaume C. Astrocyte calcium waves: what they are and what they do.Glia. 2006; 54:716–25.

14. Scemes E. Components of astrocytic intercellular calcium signaling. Mol Neurobiol. 2000; 22:167–79.

15. Spray DC, Ye ZC, Ransom BR. Functional connexin "hemichannels": a critical appraisal.Glia. 2006; 54:758–73.

16. Bruzzone R and Giaume C (eds). Connexin methods and Protocols. Humana Press, Totowa, USA. 2001:1–491.

17. Harris AW. Emerging issues of connexin channels: biophysics fills the gap. Quart Rev Biophys 2001; 34: 325–472.

18. Charpantier E, Cancela J, Meda P. Beta cells preferentially exchange cationic molecules via connexin36 gap junction channels. Diabetologia. 2007; 50:2332–41.

19. Stewart WW. Functional connections between cells as revealed by dye-coupling with a highly fluorescent naphthalimide tracer. Cell 1978; 14:741–59.

20. Socolar SJ, Loewenstein WR. Methods for studying transmission through permeable cell-to-cell junctions. Methods Membr. Biol. 1979; 10:123–79.

21. Peinado A, Yuste R, Katz LC Extensive dye coupling between rat neocortical neurons during the period of circuit formation. Neuron 1993; 10:103–14.

22. Blomstrand F, Giaume C. Kinetics of endothelin-induced inhibition and glucose permeability of astrocyte gap junctions.J Neurosci Res. 2006; 83:996–1003.

23. Caton D, Calabrese A, Mas C et al. Lentivirus-mediated transduction of connexin cDNAs shows level – and isoform-specific alterations in insulin secretion of primary pancreatic beta-cells. J Cell Sci. 2003; 116:2285–94.

24. Cohen-Salmon M, Regnault B, Cayet N et al. Connexin30 deficiency causes instrastrial fluid-blood barrier disruption within the cochlear stria vascularis. Proc Natl Acad Sci U S A. 2007; 104:6229–34.

25. Ravier MA, Güldenagel M, Charollais A et al. Loss of connexin36 channels alters beta-cell coupling, islet synchronization of glucose-induced $Ca^{2+}$ and insulin oscillations, and basal insulin release. Diabetes. 2005; 54:1798–807.

26. Calabrese A, Zhang M, Serre-Beinier V et al., Connexin 36 controls synchronization of $Ca^{2+}$ oscillations and insulin secretion in MIN6 cells. Diabetes. 2003; 52:417–24.

27. Kotlikoff MI. Genetically encoded $Ca^{2+}$ indicators: using genetics and molecular design to understand complex physiology. J Physiol. 2007; 578:55–67.

28. Simpson AW. Fluorescent measurement of $[Ca^{2+}]c$: basic practical considerations. Methods Mol Biol. 2006; 312:3–36.

29. Meldolesi J. The development of $Ca^{2+}$ indicators: a breakthrough in pharmacological research. Trends Pharmacol Sci. 2004; 25:172–4.

30. Grynkiewicz G, Poenie M, Tsien RY. A new generation of $Ca^{2+}$ indicators with greatly improved fluorescence properties. J Biol Chem. 1985; 260:3440–50.

31. Minta A, Kao JP, Tsien RY. Fluorescent indicators for cytosolic calcium based on rhodamine and fluorescein chromophores. J Biol Chem. 1989; 264:8171–8.

32. Harkins AB, Kurebayashi N, Baylor SM. Resting myoplasmic free calcium in frog skeletal muscle fibers estimated with fluo-3. Biophys J. 1993; 65:865–81.

33. Kao JP. Practical aspects of measuring $[Ca^{2+}]$ with fluorescent indicators. Methods Cell Biol. 1994; 40:155–81.

34. Lambert, D. Calcium Signaling Protocols. Methods in Molecular Biology, Humana Press 1999; 114:125–33

35. Miyawaki, A, Llopis, J, Heim, R. et al. Fluorescent indicators for $Ca^{2+}$ based on green fluorescent proteins and calmodulin. Nature 1997; 388:882–87.

36. Nagai T, Yamada S, Tominaga T et al. Expanded dynamic range of fluorescent indicators for $Ca(^{2+})$ by circularly permuted

yellow fluorescent proteins. Proc Natl Acad Sci U S A 2004; 101:10554–59.

37. Wood KV, Lam YA, McElroy WD. Introduction to beetle luciferases and their applications. J Biolumin Chemilumin. 1989; 4: 289–301.

38. Gould SJ, Subramani S. Firefly luciferase as a tool in molecular and cell biology. Anal Biochem. 1988; 175:5–13.

39. Nichols WW, Curtis GD, Johnston HH. Choice of buffer anion for the assay of adenosine $5'$-triphosphate using firefly luciferase. Anal Biochem. 1981; 114:396–7.

40. Axelrod D. Cell-substrate contacts illuminated by total internal reflection fluorescence. J. Cell Biol. 1981; 89:141–45.

41. Smyth JW, Shaw RM. Visualizing ion channel dynamics at the plasma membrane. Heart Rhythm. 2008; 5(6 Suppl):S7–11

42. Breckenridge LJ, Almers W. Final steps in exocytosis observed in a cell with giant secretory granules. Proc. Natl. Acad. Sci. USA 1987; 84: 1945–49.

43. Pralong WF, Bartley C, Wollheim CB. Single islet beta-cell stimulation by nutrients: relationship between pyridine nucleotides, cytosolic $Ca^{2+}$ and secretion. EMBO J. 1990; 9:53–60.

44. Coco S, Calegari F, Pravettoni E et al. Storage and release of ATP from astrocytes in culture. J. Biol. Chem. 2003; 278: 1354–62.

45. Zimmerberg J, Curran M, Cohen FS et al. Simultenous electrical and optical measurements show that membrane fusion precedes secretory granule swelling during exocytosis of beige mouse mast cells. Proc. Natl. Acad. Sci. USA 1987; 84: 1585–89.

46. Liu Y, Rao M. Glial progenitors in the CNS and possible lineage relationship among them. Biol. Cell 2004; 96: 279–90.

47. Liu Y, Wu Y, Lee JC. et al. Oligodendrocyte and astrocyte development in rodents: an in situ and immunohistological analysis during embryonic development. Glia 2002; 40: 25–43.

48. Rao MS, Mayer-Proschel M. Glial-restricted precursors are derived from multipotent neuroepithelial stem cells. Dev. Biol. 1977; 188: 48–63.

# Chapter 23

# Tissue Culture Procedures and Tips

Arleen Richardson and Sergey Fedoroff

## Abstract

This chapter covers the basic and essential aspects of "good practice" tissue culture. Core information on autoclaving and the sterilization procedures for tissue culture media and reagents are supplied. Descriptions of the classical methods of chemical and mechanical separation used for subculturing cells are presented. We provide important directions on how to care for incubators, prevent and control bacterial, fungal, and mycoplasm contamination. Finally, tips on water purification and the care and cleaning of dissection instruments are outlined.

**Key words:** Subculture, EDTA, trypsin, sterilization, biological contamination, water purification, incubator, dissection instruments.

## 1. Subculturing Cells

Cells attach to the culture substratum with varying degrees of adherence (1). Cells that must attach to the substratum are referred to as adhering cells (anchorage-dependent cells) and those that do not adhere and grow as cell suspensions in the medium are referred to as non-adhering cells (non-anchorage-dependent cells).

When cells proliferate in culture, they eventually completely populate the culture vessel and must be subcultured (passaged). To keep cells growing logarithmically, it is important to know how the culture should be split, i.e., how many new culture flasks can be initiated from a single culture that has become confluent. Slow-growing cell cultures are split twice, i.e., cells from one confluent culture are passaged into two new culture vessels. A rapidly growing cell culture may require passaging into as many as six

L.C. Doering (ed.), *Protocols for Neural Cell Culture*, Springer Protocols Handbooks,
DOI 10.1007/978-1-60761-292-6_23, © Humana Press, a part of Springer Science+Business Media, LLC 1992, 1997, 2001, 2010

or eight culture vessels. Cell cultures that are extremely sensitive to subculturing methods may require combining several cultures together in order to ensure that the new culture begins with a sufficient number of cells to make it viable.

It is convenient to arrange cell subculturing for every 7–10 days and to feed cultures every 3–4 days.

It is easy to subculture cells of non-adhering cell types, because they grow in suspension. An aliquot containing a defined number of viable cells may simply be used to inoculate new flasks containing medium.

Subculture of adherent cells requires removal of cells from the substratum and preparation of a cell suspension containing a defined number of viable cells. There are different methods to remove cells from substratum for different adherent cell types. The method chosen should be one that results in the least amount of damage to the cells. Two general approaches are used: mechanical and chemical.

***1.1. Mechanical Separation***

1. *Shaking*. When the culture contains only loosely adherent cells, a portion of the medium is removed and the cells in the small volume of remaining medium are dislodged from the substratum by sharply rapping the culture vessel against the hand or by using a mechanical shaker. Creation of air bubbles or frothing of the medium should be avoided (2).

2. *Washing*. When the cells are not firmly attached to the substratum, they may be removed by repeated washing of the culture with medium flushed from a Pasteur pipette. Washing should be started at the top of the flask proceeding toward the base, holding the flask at a slant. The creation of air bubbles or frothing of the medium should be avoided.

3. *Scraping*. A soft rubber policeman may be used to remove more adherent cells by gentle scraping. The most efficient type of rubber policeman has the end sealed flat and cut on an angle (Lab Scientific, Livingston, NJ). The wing-type rubber policeman can be used, but it is difficult to insert into the neck of a small flask. A rubber policemen can also be made by using a short piece of tissue culture-grade tubing (approximately 1 1/2–2 1/2 cm long), with half of the tubing removed lengthwise. This piece is then inserted into the short bend of a glass rod bent at a 45° angle. Such a device is particularly useful for scraping the cells off the flask's surface; it may be autoclaved and can be made to any length and degree of bend.

A number of disposable cell scrapers and lifters are available commercially. Cell scrapers are designed for use in flasks and lifters are designed for use in dishes. The cell scrapers usually have blades that swivel, are variable in length, and are angled for easy access into flasks (BD Falcon cell scrapers, BD Biosciences, Bedford, MA; Costar cell scrapers and lifters, Corning Incorporated Life Sciences, Acton, MA).

***1.2. Chemical Separation***

1. *EDTA (Versene)*. The disodium salt of ethylenediaminetetra-acetate (EDTA) or versene, in a $Ca^{+0+}$- and $Mg^{++}$-free medium, may be used to disaggregate cells by chelating calcium ions. This procedure is often used in conjunction with enzymatic digestion. It should not be used with enzymes that require calcium for activation such as collagenase, dispase, pronase, and DNase.

   Cultures should be washed with $Ca^{++}$- and $Mg^{++}$-free BSS, such as Puck's BSS (*see* Chapter 14) before use of EDTA. Otherwise, EDTA in the medium will be quickly neutralized by the high concentration of divalent ions present in the growth medium. The lowest possible effective concentration of EDTA should be used.

   When used together with trypsin, EDTA reduces the trypsin activity somewhat (to about 80%). However, the combination of EDTA and trypsin reduces the aggressive nature of trypsin and may improve the viability of the released cells.

2. *Enzymes.*
   Enzymes are used to digest the extracellular matrix proteins and the proteins of cell junctions responsible for cell adhesion to each other and to the substratum.

   Trypsin is the most commonly used enzyme because it is active within the physiological pH range (optimum pH 8.0) and can easily be inactivated. Trypsin is available as crude, crystalline, or purified.

   Crude trypsin contains a number of other enzymes: proteases, polysaccharidases, nucleases, and lipases, as well as impurities. It is very effective in releasing cells from the intercellular matrix of the tissues and is less expensive than crystalline trypsin. However, batches of crude trypsin vary greatly in activity and amount and type of impurities and may be cytotoxic. Therefore, each batch must be pre-tested before it is acceptable for use in a laboratory. In addition, crude trypsin may contain viruses, mycoplasma, and various bacteria and therefore should be sterilized by filtration through 0.2 μm pore size membranes (Supor membrane disc filter, Pall Life sciences, VWR International, Mississauga, ON).

   **Note:** Even though the trypsin may be sterile, it may contain variable amounts of endotoxins.

Crystalline trypsin is less variable and is very effective for cell harvesting. It should be noted that crystalline trypsin is not as efficient in disaggregating tissues as crude trypsin.

Generally, enzymes are supplied as a lyophilized powder or in solution and should be stored at 2–8°C and protected from moisture. The vials should be brought to room temperature before opening. Open vials must be stored in desiccators because the powder is very hydrophilic. Proteolytic enzymes diluted in medium or PBS are unstable and should be used as soon as possible or frozen at –20°C. Trypsin does not dissolve well in PBS. It should first be reconstituted in 0.001 $N$ HCl and then added to the medium or PBS. To minimize damage to the cells, the pH of the trypsin solution must be adjusted to physiological range by gassing with $CO_2$/air or $O_2$ mixture. Damage to the cells is also minimized by trypsinizing at 4°C. At that temperature, trypsin does not enter the cell and the viability of cells is improved.

To prepare cultures for trypsinization, cells grown in a serum-containing medium should first be rinsed with a $Ca^{++}$- and $Mg^{++}$-free BSS, such as Puck's BSS (*see* Chapter 14) to remove all traces of serum before the addition of the enzyme. After rinsing the culture, crude trypsin at concentrations of 0.025–0.05 (w/v) or crystalline trypsin at concentrations of 0.01–0.05 (w/v) in Puck's BSS is added to the culture vessel to just cover all the cells. Trypsin solution can be standardized by using the activity (units per milligram protein or units per milliliter) of the enzyme.

**Note:** In general, the smallest possible volume of enzyme solution, the lowest concentration, and the shortest possible exposure time of the enzyme solution should be used to release cells from the substrate and each other.

To further disaggregate cell clumps following incubation with proteolytic enzyme, the cell suspension is triturated. This is a very crucial step in cell disaggregation. It has been recommended that a 10 mL pipette be filled and emptied at a rate of about 3.0 mL/s without causing bubbles to form in the cell suspension.

**Note:** To avoid bubbles when triturating, do not pull the tip of the pipette out of the medium. Fill and empty the pipette while it is submerged in the cell suspension.

The most effective way to determine the endpoint of trypsinization is by observing the culture under the inverted phase-contrast microscope. When the cells round up and begin to come off the substrate, enzymatic activity should be inhibited. Trypsin activity can be inhibited by adding an equal volume of medium containing 5–10% serum to the cell–trypsin solution. In serum-free medium and medium low in serum, soybean trypsin inhibitor (1 mg/mL) may be added to the medium to neutralize the trypsin. Excess of the antitryptic agent should be avoided, because it is somewhat inhibiting to cell growth. Crude trypsin contains a variety of enzymes and the soybean trypsin inhibitor does not neutralize all the enzymes. Therefore, either crystalline trypsin should be used in this procedure, or, after trypsinization with crude trypsin, the cells should be centrifuged, washed, and resuspended in new medium.

In some cases, other enzymes such as pronase, dispase, and collagenase are more effective by themselves or in combination. These enzymes require the presence of calcium for their activity, and should be rinsed with a BSS containing calcium and magnesium, e.g., Hanks' BSS (*see* **Section 3.2, 2c**). In addition, they are not, or are only partially, neutralized by serum. Therefore, these enzymes should be removed from the cells by centrifugation.

## 2. Sterility Control

### 2.1. Laboratory Environment

The tissue culture laboratory may be a single room designated for cell culture or may simply be space in a larger multipurpose room. It is usually defined by the presence of a laminar flow cabinet and an incubator. A laminar flow hood is advantageous for maintaining a sterile environment for preparation of cultures and provides personnel or product protection, or both.

It is important to be familiar with the environment of the room in terms of air flow, traffic patterns, and possible sources of contamination. The placement of bacteriological plates throughout the room and at different times of day and night can aid in determining the changes needed to optimize the area for cell culture.

The plates should be exposed for 30 min to 1 h and then incubated. Bacterial (sheep blood agar, tryptic soy agar, etc.) and fungal (Saboraud dextrose agar and others) plates are placed in front of all air vents and around doors entering the room. Plates are also placed on and around major equipment in the room (centrifuges, refrigerators, ice machines, and the like), on and around the laminar flow hood and incubator, and areas of traffic (including the floor) that are near the working space. The growth of bacteria or fungi on the plates will pinpoint problem areas and appropriate steps can then be taken to eliminate or minimize the sources of contamination.

### 2.2. Autoclaving

Autoclaves are effective sterilizers only if materials to be autoclaved are properly wrapped and placed in such a way that the steam has access to them. The autoclave should not be overloaded. The autoclave must be operating properly, maintaining both temperature and pressure (121°C, 15 psi).

To ensure that sterilization has occurred, commercial indicators can be used. The best indicators of sterilization take into account the parameters of time, temperature, and steam penetration. Indicators, such as Comply Chemical Indicators, 3 M; Indicator-Plus Strips (VWR International, Mississauga, ON, Canada) can be placed in the center of a pack, inside capped tubes, etc., and only when the temperature is reached (121°C), and 12–16 min have elapsed, do the indicators change color. It should be noted that the endpoint for most indicators is variable and that the minimum autoclaving time is 15 min.

Autoclave tapes are available to attach to the outside of packs. They change color during autoclaving; most do not indicate that the product is sterilized, but simply indicate that items have been exposed to autoclaving.

Nontoxic materials should be used to wrap items to be autoclaved. There are many commercially available wrappers (Dualpeel® Tubing, VWR CanLab Mississauga ON, Canada; Chemitest Autoclave Indicator bags/pouches, Thomas Scientific, Swedesboro, NJ) designed for use in autoclaves. In addition, unbleached cotton muslin may be used for bagging large glassware and envelopes of various sizes can be constructed from 27 lb uncoated vegetable parchment paper (available from restaurant supply companies).

Steam used for sterilization in autoclaves can contain many impurities, including the contaminants present in the feedwater and additives that prevent scale formation and corrosion. These contaminants are deposited on the sterilized material, which then may be dissolved in solutions used in cell culture. The addition of a steam filter (Balston, MA), placed as close to the steam inlet as possible, eliminates impurities in the steam before it enters the autoclave. The steam filters should be changed on a regular basis.

### 2.3. Medium and Other Cell Reagents

As an integral part of a quality control program, it is important to monitor the sterility of media and other cell reagents. Media may be tested for bacteria and fungi by inoculating fluid thioglycolate broth, trypticase soy broth, Sabouraud dextrose broth, and blood agar plates.

To ensure sterility of large volumes of media, it is important that adequate sampling is done. In our laboratory, we remove a sample of the medium at the beginning and end of filtration and, in addition, a 25 mL sample is removed after filtering every liter of medium. All bottles of filtered medium and samples are numbered in sequence so that problems that occurred during filtration can be readily identified. These samples are incubated at room temperature until the entire batch of medium is used. Any samples that show signs of microbial contamination are matched to the numbers on medium bottles. Those bottles of medium are discarded.

Media are also tested for growth-promoting qualities on primary cultures or cell lines that are routinely used in the laboratory.

**2.4. Incubators**

1. *Carbon dioxide.* For control of pH during culturing, incubators in which the amount of carbon dioxide in the atmosphere can be controlled are used. For proper pH control, the carbon dioxide concentration in the incubator should correspond to the concentration of the bicarbonate buffer in the medium. Purchased media with specified concentrations of bicarbonate buffer require specified concentrations of carbon dioxide. For example, Eagle's minimum essential medium (MEM) with Hanks' BSS does not require additional carbon dioxide in air. Such a medium is designed for a regular atmospheric environment. Eagle's MEM with Earle's salts requires an atmosphere of 5% carbon dioxide in air and Dulbecco's modified Eagle's medium requires an atmosphere containing 10% carbon dioxide.

   The gas mixtures fed into the incubator from pressurized tanks may contain impurities that may be cytotoxic. Therefore, an in-line hydrophobic filter should always be inserted between the gas cylinder and the incubator.

2. *Cleaning incubators.* Incubators that have high humidity and operate at 37°C provide an excellent environment for growth of molds and yeasts. Therefore, incubators should be cleaned on a regular basis (every 2–3 months), or at the first sign of mold or yeast contamination. The following procedure is for incubators that are used for non-biohazardous cultures:

   a. The incubator is turned off and unplugged.

   b. All removable parts (shelves, humidity pan, etc.) are taken out, washed, rinsed with good-quality water, and autoclaved.

   c. The incubator is dried with lint-free towels and carbon dioxide sensors are covered with plastic. The incubator is then sprayed with a disinfectant appropriate for the application (e.g., 70% alcohol; Incubator Clean, AppiChem, Inc, Cheshire, CT), paying particular attention to corners, seam lines, etc. The incubator is closed and left for approx 2 h.

   d. The incubator is rinsed with distilled water until all traces of the disinfectant are removed. It is then dried with lint-free towels.

   e. The incubator is sprayed with 70% ethanol, again paying particular attention to corners and seam lines. A pan of 70% ethanol is placed in the incubator. The incubator is closed and left for approx 24 h.

   f. After 24 h, the 70% ethanol is removed, all surfaces are dried and the incubator is reassembled. Fill the humidity tray with sterile distilled water and replace all dirty air and $CO_2$ filters as needed.

## 3. Biological Contamination

### 3.1. Cell Lines

Cell lines should be obtained only from reputable cell banks such as the ATCC (Manassas, VA). Cell lines from other sources should be verified by agencies specializing in authentication of cells by DNA finger printing techniques (ATCC, Manassas, VA; BioReliance, Rockville, MD). The profiles that are generated provide a baseline for future comparison.

In laboratories that carry several cell lines, the risk of cross-contamination among cell lines is very high. "Thus, only one cell line should be permitted inside a safety cabinet at any one time, while separate supplies of culture medium, etc., should be dedicated to each cell line" (3). When any change in the cell line behaviour is observed, authentication of the cell line is mandatory.

### 3.2. Bacterial Contamination

Bacterial infection in cultures is manifested by cell debris, usually a decrease in pH of the medium, and/or turbidity of the medium. Bacteria in a culture can be seen under the microscope at high power. A low level of bacterial contamination or slow-growing bacteria may go undetected; however, the cells in the culture will exhibit slowed and uncharacteristic growth.

Cultures containing bacteria, as well as any reagents used with those cultures, must be autoclaved and discarded.

In exceptional cases, an attempt can be made to save irreplaceable cultures by extensive washing, treatment with high concentrations of antibiotics, and subculturing the cells into as many cultures as possible to enhance the probability that some will be free of contamination. If the use of antibiotics is unacceptable, extensive washing (ten times or more) of the culture and subculturing into 96-well plates may produce some cultures free from bacterial contamination.

1. Procedure for decontamination

   a. Prepare a solution of antibiotics in Hanks' balanced salt solution (HBSS). We use a highly concentrated penicillin and streptomycin mixture, which has a broad bactericidal spectrum against gram-positive and gram-negative bacteria. The solution is made up at $50\times$ the recommended working concentration (*see* below).

   b. Wash the culture with Hanks' BSS at least five times, or more, if heavily infected.

   c. Treat with the antibiotic solution for 1–3 h. Examine the cultures about every 20 min. As soon as the morphology of the cells begins to change (round up), remove the antibiotic solution and replace with fresh medium.

   **Note:** The antibiotics at $50\times$ the recommended concentration can be cytotoxic to cells; therefore, it is important to remove the solution as soon as the cells show any sign of change in normal morphology.

d. Incubate cultures (with no antibiotics) for 24 h at 37°C.

e. Examine the cultures for the presence of bacteria. If bacteria are present, the cultures may either be treated with Gentamycin (50 μg/mL) or be washed, and split into many small cultures using 96-well plates.

2. Solutions.

a. *Penicillin and streptomycin solution.* The recommended concentrations of antibiotics for prophylactic use in cell culture are 100 U/mL penicillin G and 100 μg/mL streptomycin. However, the concentration of antibiotics tolerated by cells will vary according to the cell type. The high concentrations of antibiotics in Hanks' BSS that we use for decontamination are: penicillin G, 5000 U/mL; streptomycin, 5000 μg/mL.

b. *Gentamycin sulfate.* Gentamycin is a broad-spectrum antibiotic widely used to treat persistent contaminations. It is bactericidal in vitro and is effective against gram-negative and some gram-positive bacteria, some penicillin-resistant strains of bacteria, and some mycoplasmas. The effective prophylactic concentration is 50 μg/mL.

c. *Hanks' BSS.*

| Component | g/L |
| --- | --- |
| NaCl | 8.0 |
| KCl | 0.4 |
| $Na_2HPO_4 2H_2O$ | 0.06 |
| $KH_2PO_4$ | 0.06 |
| $CaCl_2$ | 0.14 |
| $MgSO_4 7H_2O$ | 0.2 |
| $NaHCO_3$ | 0.35 |
| Phenol red | 0.02 |
| Glucose | 1.0 g/L |

**3.3. Fungal Contamination**

Common fungi in cultures grow as single cells, such as yeast, or as multicellular mycelium (molds). Mold infections of cultures may be unnoticed until they have grown large enough to be readily apparent (many long intertwined branching filaments in one section of the culture dish). A change in pH and turbidity of the medium may indicate the presence of yeast. Yeast is usually noticed first when the culture is examined under the microscope.

Yeast appears as refractile ovals in chains or branches, or shows evidence of budding, and is usually found throughout the culture. The common sources of fungal contamination in the tissue culture laboratory are humidified incubators, cardboard, and the tissue culture personnel.

Cultures containing fungi must be autoclaved and discarded. In addition, the incubators in which the cultures were grown must be thoroughly cleaned and disinfected, and all reagents used with the contaminated cultures discarded.

1. Antifungal agents commonly used in tissue culture include amphotericin B (Fungizone®) and nystatin (Mycostatin®). The prophylactic concentration of amphotericin B is 2.5 µg/mL and that of nystatin is 50 µg/mL or 100 U/mL.

2. Amphotericin B, depending on the concentration used and the susceptibility of the fungal contamination, is either fungistatic or fungicidal. Nystatin is fungistatic. Therefore, if these antibiotics are used prophylactically, they should be used continuously throughout the experiment, since removal of the antibiotic may allow any fungi present to proliferate.

### 3.4. Mycoplasmal Contamination

1. Mycoplasmas have many effects on cell cultures (e.g., changes in growth rate and metabolism of cells, induction of chromosomal aberrations, and depletion of arginine in culture medium) that are difficult to detect. The major sources of mycoplasma in the tissue culture laboratory are cross-contamination from infected cell cultures, serum and non-autoclavable medium additives, primary tissue (gastrointestinal and respiratory tract), and laboratory personnel.

2. Since contamination of cell cultures by mycoplasma is difficult to detect, periodic screening of cultures is recommended. Although many screening methods are available (4), it is usually more accurate, less time consuming, and therefore cheaper to send samples to a laboratory that specializes in testing for mycoplasma contamination (ATCC, Manassas, VA; Bioreliance, Rockville, MD).

3. Cultures infected with mycoplasma and all reagents used with them must be autoclaved and discarded.
**Note:** Mycoplasmas are not removed by conventional 0.22 µm porosity filters.

### 3.5. Endotoxin Contamination

Endotoxins (lipopolysaccharides) from the membranes of gram-negative bacteria are common contaminants in tissue culture. They have many and varied effects on cells, even when present in nanogram quantities. Endotoxins are released during the growth of bacteria when the membranes are sloughed off, or when the bacteria die and lyse. Bacterial contamination in the water purification system or during preparation of culture media will result in contamination with endotoxins, even if the bacteria are removed by filtration. Water is the major source of endotoxin contamination, but endotoxins are also present in sera and other biological materials, such as soybean trypsin inhibitor, bovine serum albumin, growth factors, and so on.

1. The *Limulus* Amebocyte Lysate (LAL) test method (Pyrotell, Gel-clot Formulation, Associates of Cape Cod Inc.) is recommended for the quantification of endotoxin in water and for monitoring levels of endotoxin in reagents and in tissue culture media and additives.

2. It is difficult to remove endotoxins from serum or other complex cell culture solutions. Use of methods such as ultrafiltration, charcoal and solvent extraction results in protein denaturation and the loss of serum quality. Solid-phase endotoxin-adsorbing reagents (Acticlean Etox, Sterogene Bioseparations, Inc. Carlsbad, CA) that have high specific affinity for endotoxins have been developed and can be used to remove endotoxin from serum and other biological solutions.

3. Serum used in tissue culture medium should be selected for its low endotoxin level.

**3.6. Use of Antibiotics**

Good aseptic technique is usually superior to the use of antibiotics in tissue culture, but in certain cases the prophylactic use of antibiotics is necessary. Antibiotics should be used when the starting material for primary cultures is not sterile, e.g., tissue obtained from the abattoir, human tissue obtained during surgery or postmortem, and tissue from skin, nasal passages, gastrointestinal, and urogenital tracts. In such cases, the tissue should be bathed in a solution containing a high concentration of antibiotics before initiating cultures. Prophylactic dosage of antibiotics may be used for as short a time as possible (1–2 d) or continuously, if these primary cultures are to be terminated at the end of an experiment.

**Note:** As a general rule, avoid prophylactic use of antibiotics in cell cultures. In exceptional cases, antibiotics may be used to decontaminate irreplaceable cultures (*see* **Section 3.2**).

Cell lines should be free of biological contamination.

## 4. Water

The importance of high-quality water in cell culture cannot be overemphasized. High-quality water is essential not only as a solvent for culture medium and reagents but also as the final rinsing step used in the preparation of glassware for tissue culture. Water quality is especially critical when cells are grown in serum-free medium.

### 4.1. Water Purification

Water can be purified by distillation, ion exchange (deionization), carbon adsorption, microporous membrane filters, ultrafiltration, and reverse osmosis. Each of these procedures has certain advantages and disadvantages, but no one procedure by itself is satisfactory for obtaining water of a quality high enough for tissue culture purposes. The solution is to combine a number of the procedures in sequence. Systems are available that use a number of replaceable cartridges for carbon adsorption, ion exchange, ultrafiltration, and finally a 0.22 μm porosity filter. Such a system can remove, by ultrafiltration, particles with a molecular weight of more than 10 kDa. Most endotoxins (lipopolysaccharides) are larger than 20 kDa and therefore are removed by ultrafiltration. Sometimes it is necessary to pre-purify the water before it is fed into the system, to extend the life of the cartridges.

Using equipment composed of several purification steps, it is possible to obtain large amounts of ultrapure, type I reagent grade water as defined by the College of American Pathologists (CAP), the American Society for Testing and Materials (ASTM), and the National Committee for Clinical Laboratory Standards (NCCLS). Water acceptable for tissue culture use must be free of endotoxins and organic contaminants, be essentially ion-free, and have a resistivity close to 18 M$\Omega$/cm at 25°C (resistivity of chemically pure water has been calculated to be 18.3 M$\Omega$/cm).

The water content of inorganic and ionizable solids, organic contaminants, and particulate matter varies from one geographical region to another and, therefore, the requirements for purification of water may vary. The water purification system (Barnstead/Thermolyne, IA, Millipore, MA) to be used depends greatly on the amount of water required and the geographical location.

1. Ultrapure water should be used within a day of its production and should not be stored longer, since the highly purified water has the capability of leaching out impurities from the storage container (Pyrex$^{TM}$ or plastic). These contaminants then become part of the medium and tissue culture solutions or are deposited on glassware.

2. All water purification systems require monitoring, proper maintenance, and cleaning to ensure continued water quality. All tubing in contact with the water purification system should be changed frequently to prevent microbial and algae growth. Bacterial growth can occur in resin beds, storage vessels, distillation apparatus, and so on, and therefore maintenance should be carried out on a regular basis.

## 5. Visible Fluorescent Light

Visible fluorescent light is detrimental to media, sera, and cells in culture because of the generation of free radicals in the media. Light affects tryptophan and tyrosine, with riboflavin acting as a photosensitizer. Wavelengths up to 450 nm are responsible for the deleterious effects and to eliminate them certain precautions must be taken.

1. Fluorescent tubes in the light fixtures in the laminar flow hoods can be replaced by yellow fluorescent tubes, which will reduce the formation of free radicals in media.

2. Sera, especially fetal bovine and calf serum, which have low levels of catalase activity, should be stored in the dark and protected from light by the use of yellow plastic bags.

3. HEPES-containing media is particularly sensitive to the effects of visible fluorescent light.Bottles containing media with HEPES as a buffering agent should always be wrapped in aluminum foil.

4. As a rule, all media should be stored and cultures grown in the dark.Incubators with glass doors should have the glass covered to prevent any light from entering the incubator.

## 6. Dissecting Instruments

Once you have determined what instruments are needed for dissection, one should consider size and styles. For example, forceps may be straight, curved, or sharply angled. The tips range in size from 0.05 mm × 0.01 mm to much larger and may be serrated or smooth. The tips can be examined under 40× magnification and should close flatly. The greater the distance the tips are closed flat, the better the forceps.

In addition, one should consider that instruments are made of a variety of materials such as tungsten carbide, steel alloys, ceramics, titanium, and titanium alloys. Each of these materials has different characteristics that are important for your specific application. In general, dissecting instruments should be corrosion-resistant, stainless, non-magnetic, heat resistant (e.g., titanium resists the temperature up to 440°C and Dumostar alloy up to 550°C) and of adequate hardness.

**Note:** Dissecting instruments are precision-made, expensive, and require special care. They should be the responsibility of the individual and should not be given to others to use. Instruments should be used only for the purposes for which they are designed.

Care and cleaning of instruments are also important. Instruments used for tissue culture are not to be oiled, because this may create problems when the oil mixes with the dissected tissues or becomes burned to the steel. Scissors with spring handles should be stored in the open position to prevent tiring of the spring. Particular care must be given to the tips of fine forceps and scissors. Short lengths of flexible tubing may be used as tip protectors during storage.

Instruments should be routinely examined under a dissecting microscope. The ends of forceps and scissors should be of equal length, the tips should meet perfectly, and the shearing edges should be sharp. Stored instruments should be firmly secured within the instrument box, so that they do not move when the box is handled. Gauze may be placed between layers of instruments.

After the instruments have been used, they should be put in a tray lined with soft material and soaked in distilled water. Manual cleaning of the instruments should be done carefully, one at a time, with gauze, clean cloth, or a soft brush to remove any tissue remnants or blood. A very dilute nontoxic soap solution may be used for cleaning followed by careful rinsing, first with warm tap water and then with distilled water. Instruments should be thoroughly dry before sterilization.

Ultrasonic cleaners may also be used to clean instruments, although they are not recommended for delicate instruments. Manufacturers provide recommendations for the cycle time. The instruments, in the open position, are placed into the cleaner and completely submerged. After ultrasonic cleaning, instruments are rinsed carefully in warm tap water followed by distilled water to completely remove cleaning solution. It is advisable that instruments of similar metal content should be cleaned together.

The instruments can be sterilized by alcohols, autoclaving, or hot bead sterilizers. Methanol, ethanol, and isopropanol are used as bactericidal agents at concentrations of 70–80% alcohol in water. Alcohols leave no residues on surfaces and are effective agents for sterilization and disinfection. Instruments can be immersed in 70% alcohol and allowed to dry in air inside the laminar flow hood, repeating the procedure three times. Another method is to dip the instrument in alcohol and then pass it through the flame of a gas burner, repeating the procedure three times. Scissors should be in the open position while flaming.

**Note 1:** Make sure the flame is completely out before re-immersing the instrument in the alcohol. If the alcohol does catch on fire, immediately cover with a metal lid.

**Note 2:** Instruments may be damaged by high heat, depending on the alloy used and the degree of hardness. Do not keep the instrument in the gas flame while the alcohol is burning, just pass it through the flame to ignite the alcohol. Do not heat the instruments to red-hot heat.

Instruments may also be sterilized by autoclaving. Steam sterilization may damage instruments that are of inferior quality and these should not be mixed with those of higher quality. After autoclaving, the instruments should be placed immediately into a drying oven until they are completely dry.

Tips and blades of instruments can be rapidly and conveniently sterilized in hot bead or dry sterilizers (Inotech Biosystems International, Lansing, MI; Fine Science Tools, Inc., North Vancouver, British Columbia, Canada). These sterilizers can be maintained at the work station, at a temperature of 250°C. Dry, clean instruments are inserted into the hot beads for at least 20 s for complete sterilization. The instruments are then removed and placed in a sterile glass Petri dish in the biological safety cabinet and allowed to cool before use.

## References


1. Letourneau, P.C (2001), Preparation of substrata for in vitro culture of neurons, in: *Protocols for Neural Cell Culture*, Fedoroff, S. and Richardson, A., eds., Humana Press, Totowa, New Jersey, pp. 245–254.
2. Cole, R and de Vellis, J (2001), Preparation of astrocyte, oligodendrocyte and microglia cultures from primary rat cerebral cultures, in: *Protocols for Neural Cell Culture*, Fedoroff, S. and Richardson, A., eds., Humana Press, Totowa, New Jersey, pp. 245–254.
3. MacLeod, R. A. F., Dirks,W. G., Matsuo, Y., Kaufmann, M., Milch, H. and Drexler, H. G. (1999), Widespread intraspecies cross-contamination of human tumor cell lines arising at source. Int. J. Cancer **83,** 555–563.
4. Hay, R. J., Caputo, J. and Macy, M. L., eds. (1992), *ATCC Quality Control Methods for Cell Lines*, American Type Culture Collection, Rockville, MD.


## Further Reading


Bottenstein, J., Hayashi, I., Hutchings, S., Masui, J., Mather, J., McClure, D. B., Ohasa, S., Rizzino, A., Sato, G., Serrero, G., Wolfe, R. and Wu, R. (1979), The growth of cells in serum-free hormone-supplemented media. in: *Methods in Enzymology Cell Culture*, Jakoby, W. B. and Pastan, I. H., eds., Academic Press, New York, pp. 94–109.

Fogh, J., ed. (1973), *Contamination in Tissue Culture*. Academic, New York, pp. 1–288.

Gabler, R., Hegde, R. and Hughes, D. (1983), Degradation of high purity water on storage. *J. Liquid Chromatog.* **6,** 2565–2570.

Goule, M. C. (1984), Endotoxin in vertebrate cell culture: its measurement and significance. in: *Uses and Standardization of Vertebrate Cell Cultures*, Tissue Culture Association, MD **5,** 125–136.

Mather, J., Kaczarowski, F., Gabler, R. and Wilkins, F. (1986), Effects of water purity and addition of common water contaminants on the growth of cells in serum-free media. *BioTechniques* **4,** 56–63.

McKeehan, W. L. (1977), The effect of temperature during trypsin treatment on viability and multiplication of single normal human and chicken fibroblasts. *Cell Biol. Int. Reports* **1,** 335–343.

Melnick, J. L. and Wallis, C. (1977), Problems related to the use of serum and trypsin in the growth of monkey kidney cells. in: *Developments in Biological Standardization*, Perkins, F. T. and Regamey, R. H., eds., S. Karger, Basel, pp. 77–82.

Perkins, J. J. (1969), *Principles and Methods of Sterilization in Health Sciences*. Charles C. Thomas, Springfield, IL, pp. 1–549.

Perlman, D. (1979), Use of Antibiotics in cell culture media. in: *Methods in Enzymology: Cell Culture,* vol. 58, Jacoby, W. B. and Pasten, I. H., eds., Academic, New York, pp. 110–116.

Rottem, S. and Barile, M. F. (1993), Beware of mycoplasmas. *Trends Biotechnol.* **11,** 143–150.

Ryan, J. A. (1994), Understanding and managing cell culture contamination. Corning, Inc. Technical Publication TC-CI-559, Corning, NY.

Taylor, W. G. (1984), Toxicity and hazards to successful culture: cellular responses to damage induced by light, oxygen or heavy metals. in: *Uses and Standardization of Vertebrate Cell Cultures,* Patterson, M. K., ed., Tissue Culture Association, MD, 58–70.

Wang, R. J. and Nixon, T. (1978), Identification of hydrogen peroxide as a photoproduct toxic to human cells in tissue-culture medium irradiated with "daylight" fluorescent light. *In Vitro* **14,** 715–722.

Waymouth, C. (1982), Methods for obtaining cells in suspension from animal tissues. in: *Cell Separation Methods and Selected Applications,* vol. 1, Pretlow, T. G. and Pretlow, T. P., eds., Academic Press, New York, **1,** 1–29.

Worthington Biochemical Corporation (1993), *Tissue Dissociation Guide.* Worthington Biochemical Corporation, Freehold, NJ, pp 1–78. Also available on line at http://www.tissuedissociation.com.

Zigler, J. S., Lepe-Zuniga, J. L., Vistica, B. and Gery, I. (1985), Analysis of the cytotoxic effects of light-exposed HEPES-containing culture medium. *In Vitro Cell. Devel. Biol.* **21,** 282–287.

Note: Page numbers followed by *t* and *f* refer to pages containing tables and figures respectively.




L.C. Doering (ed.), *Protocols for Neural Cell Culture*, Springer Protocols Handbooks,
DOI 10.1007/978-1-60761-292-6, © Humana Press, a part of Springer Science+Business Media, LLC 1992, 1997, 2001, 2010